NEURONS *and the* DHA PRINCIPLE

NEURONS *and the* DHA PRINCIPLE

RAYMOND C. VALENTINE
DAVID L. VALENTINE

CRC Press is an imprint of the
Taylor & Francis Group, an **informa** business

CRC Press
Taylor & Francis Group
6000 Broken Sound Parkway NW, Suite 300
Boca Raton, FL 33487-2742

First issued in paperback 2019

ISBN-13: 978-1-4398-7486-8 (hbk)
ISBN-13: 978-0-367-38074-8 (pbk)

Library of Congress Cataloging-in-Publication Data

Valentine, R. C. (Raymond Carlyle), 1936-
Neurons and the DHA principle / Raymond C. Valentine, David L. Valentine.
p. ; cm.
Includes bibliographical references and index.
ISBN 978-1-4398-7486-8 (hardcover : alk. paper)
I. Valentine, David L. II. Title.
[DNLM: 1. Docosahexaenoic Acids--metabolism. 2. Alzheimer Disease--etiology. 3. Cell Death. 4. Neurons--metabolism. QU 90]

616.8'3--dc23 2012028537

Visit the Taylor & Francis Web site at
http://www.taylorandfrancis.com

and the CRC Press Web site at
http://www.crcpress.com

Contents

SECTION II Benefits of DHA

SECTION III Risks of DHA

Preface

Although Alzheimer's disease is the most devastating form of neurodegeneration, its cause is unknown. We developed this book around a proposition: that the secret to understanding this disease is kept by deep-sea bacteria whose membrane oils are similar to neurons. Studies with bacteria and other systems suggest that the omega-3 fatty acid DHA (docosahexaenoic acid, 22:6) contributes great benefits to neurons in maximizing both speed of neural impulses and energy efficiency. But DHA is also harmful in that its oxidation damages membrane integrity. This duality leads us to propose a new model for the cause of Alzheimer's disease, in which DHA-enriched membranes of neurons become dysfunctional and energetically wasteful, triggering premature death of neurons. The challenge of this book is to understand how DHA acts as an essential building block of neurons while also conspiring for their assassination during aging.

Acknowledgments

A big hug for Cindy Anders (RCV's life partner) for the major responsibility of manuscript preparation, including checking references, permissions, typing, general editing, and great food. Also, hugs for the little bears, Isabella, Sienna, Stella, Sawyer Ray, Eva, Sarah, Eli, and Jude for being bears for a moment. Special thanks to RCV's brothers and sisters, Moses, Monie, Bobbie, Marilyn, Micky, and Paula, for lifetime support of his quest to be a molecular farmer. Thanks to Carla Valentine and Dacia Harwood for advocating that our second omega-3 book be relevant to human health.

We are greatly indebted to two interdisciplinary groups of scientists whose research provided important stepping-stones for the development of this book. First, contributions from physical chemists such as Scott Feller provide striking images of the dynamic nature of DHA, which we suggest match the extraordinary biochemical roles played by these molecules in neurons. Second, research by A. J. Hulbert and colleagues and R. Pamplona and coworkers, respectively, co-fathers of the membrane unsaturation theory of aging, stimulated us to apply similar concepts toward understanding the molecular roles of DHA in the aging brain. We are also indebted to Manabu T. Nakamura for helpful discussions on the essential roles of DHA in membrane development.

Once again we thank our sister publisher, Garland Press, and the editors of perhaps the most admired cell biology book ever written—*Molecular Biology of the Cell*—for permission to use several figures and illustrations. Editors for editions 1 through 3 are Bruce Alberts, Dennis Bray, Julian Lewis, Martin Raff, Keith Roberts, and James D. Watson; and for editions 4 and 5, Bruce Alberts, Alexander Johnson, Julian Lewis, Martin Raff, Keith Roberts, and Peter Walter. One of our favorite pictures appears on the back cover of edition 3 and shows J. D. Watson, one of the great pillars of reductionist thinking, following his fellow editors as they cross the street.

Special thanks to Hilary Rowe, our editor at Taylor & Francis/CRC Press, for advice, help, and encouragement. Thanks to Jennifer Ahringer of CRC Press for help with permissions, illustrations, and coordinating the production of this book.

The Authors

Raymond C. Valentine, Ph.D., is currently professor emeritus at the University of California Davis and visiting scholar in the Marine Science Institute at the University of California, Santa Barbara. He was also the scientific founder of Calgene Inc. (Davis, California), now a campus of Monsanto Inc. Dr. Valentine's scientific interests involve the use of reductionism to address problems of fundamental scientific and societal importance, such as agricultural productivity and aging. Some of his scientific accomplishments include the discovery of ferredoxin, the identification and naming of the nitrogen fixation (*nif*) genes, and the development of Roundup®-resistance in crops. He holds B.S. and Ph.D. degrees from the University of Illinois at Urbana–Champaign.

David L. Valentine, Ph.D., is currently professor of earth science with affiliations in ecology, evolution, and marine biology, as well as the Marine Science Institute, at the University of California, Santa Barbara. Dr. Valentine's scientific interests involve the use of a systems-based approach to investigate the interaction between microbes and the Earth, particularly in the subsurface and oceanic realms. He is best known for his research on the biogeochemistry of methane and other hydrocarbons, for his works on archaeal metabolism and ecology, and his scientific work on the *Deepwater Horizon* oil spill. Dr. Valentine holds B.S. and M.S. degrees from the University of California at San Diego and M.S. and Ph.D. degrees from the University of California at Irvine.

Introduction

BRAIN HEALTH AND DECLINE IN THE AGE OF LONGEVITY

Understanding and curing Alzheimer's disease (AD) is one of the major challenges of the 21st century. This goal is driven by the epidemic of Alzheimer's disease, which is now occurring among the aging world population (Jackson and Prince, 2009). Unless a cure is found it is estimated that by the year 2050 more than 115,000,000 patients with dementia worldwide (Figure I.1) will require care. In the United States alone, the cost of care is estimated to be trillions of dollars from now to 2050. The global cost of neurodegeneration in terms of human suffering and resources is staggering. The need for immediate action has spurred the U.S. government to embark on a "War on AD" in the year 2012. The path of the war on AD is likely to be similar to the 40-year war on cancer, which involved an initial period devoted to basic research followed by applications. Aging and neurodegeneration are tightly linked, with age being the major risk factor for dementia. Thus, both of these important fields will likely benefit from a war on AD. In this book, we explore the relationship between aging, brain health, and brain decline. It is now clear that the human brain, though sometimes lasting more than a century in rare individuals, is far more likely to decay on average starting at about ages 60 to 65. This horrible scenario of a brain that evolved for the shorter life span of our ancestors is backed by statistics that are difficult to refute (Jackson and Prince, 2009).

AGING IS RELENTLESS FOR ALZHEIMER'S DISEASE

The greatest known risk for AD is aging (Figure I.2). Most individuals diagnosed with the disease are 65 years or older. The likelihood of developing Alzheimer's doubles about every 5 years after age 65. After age 85 the risk reaches 50 percent.

However, it is important to recall that localized neuron death in AD might occur 20 years or more before diagnosis. The earliest stages of AD cannot currently be detected. Toward the end of the early stage, lasting a few years, behavioral changes may be noticed in the following areas:

- Learning and memory
- Thinking and planning

Mild-to-moderate stages of Alzheimer's generally last from 2 to 10 years. During this period, regions of the brain important in memory, thinking, and planning develop more plaque and tangles than were present in the early stages. As a result, individuals develop problems with memory or thinking, which are serious enough to interfere with work and social life. They may also become confused and have trouble

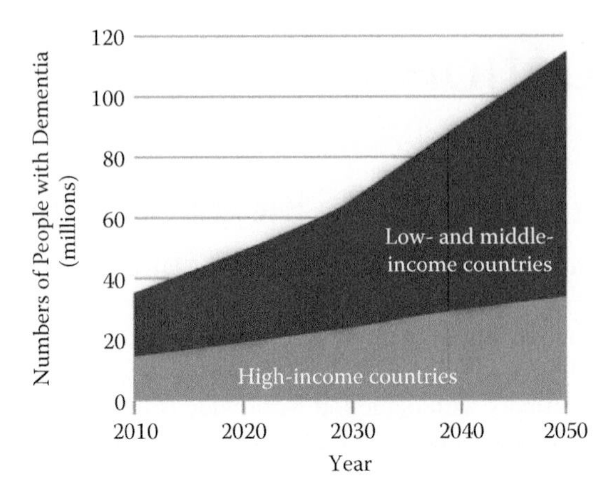

FIGURE I.1 Alzheimer's Disease International has predicted a sharp rise in dementia as the world population ages. Note that about a doubling in dementia cases is predicted every 20 years, reaching greater than 100 million cases by 2050. (Used with kind permission from Alzheimer's Disease International, *World Alzheimer Report 2010*.)

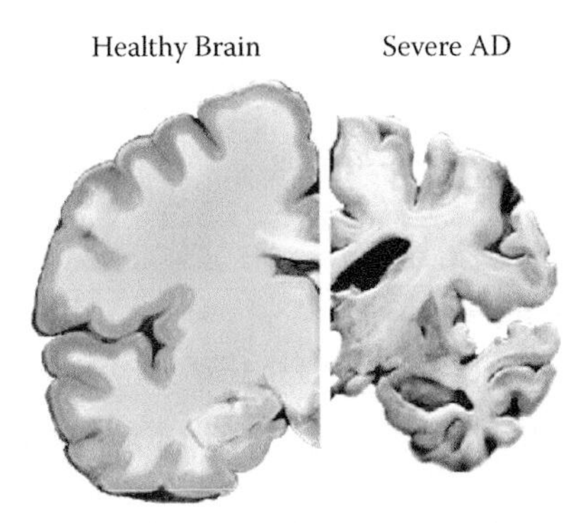

FIGURE I.2 Neuron death is extensive in the advanced stage of Alzheimer's disease. The brain shrinks appreciably. Compare the size of a normal brain on the left with the brain with AD at the right. (Image courtesy of the National Institute on Aging/National Institutes of Health.)

handling money, as well as organizing and expressing their thoughts. Many people with Alzheimer's are first diagnosed during these stages.

Plaques and tangles eventually begin to spread to other brain areas involved in the following:

- Speaking and understanding speech
- The sense of where one's body is in relation to surrounding objects

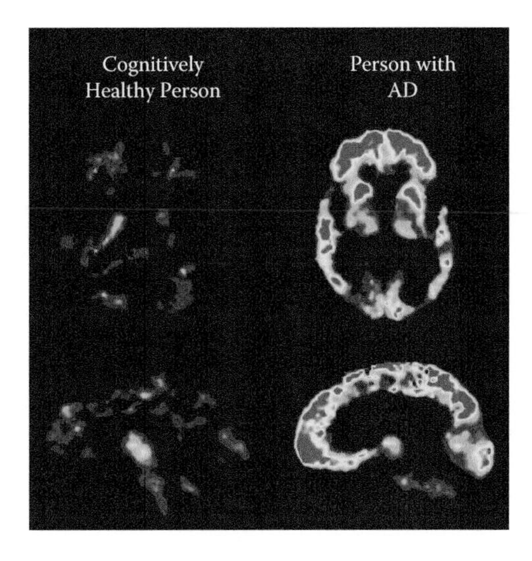

FIGURE I.3 Positron emission tomography (PET) scans of a healthy brain on the left and the brain of a patient with Alzheimer's disease (AD) on the right show high levels of amyloid plaque (red color) accumulating in advanced cases of AD. A marker dye PiB is injected to detect regions of amyloid accumulation, which are extensive in this case. (Image courtesy of the National Institute on Aging/National Institutes of Health.) **(See color insert.)**

As Alzheimer's progresses from mild to moderate stages, individuals may experience changes in personality and behavior and have trouble recognizing friends and family.

In advanced AD, many of the neurons forming the brain cortex are destroyed or seriously damaged. The brain shrinks dramatically in size due to widespread cellular death (Figure I.2), and biochemical activity is greatly reduced (Figure I.3). Individuals lose their ability to communicate and recognize family and loved ones, and require constant care.

WITH NO CURE OF AD IN SIGHT THINKING OUTSIDE THE BOX SHOULD BE ENCOURAGED

General causes of the increase in the number of patients with Alzheimer's disease are being hotly debated and include aging, modern medicine, heredity, and the environment. Unfortunately, the causes of mental diseases such as AD are still poorly understood, and there is no cure. However, brilliant, even Nobel prize–winning research (Prusiner, 1991) is being conducted in this field, leading to a certain amount of guarded optimism for the prospect of a cure. In the meantime, a great deal of basic research is still needed. Thinking outside the box toward understanding the causes of mental illness seems justified in the mental health field.

Scientists have in the past used reductionist strategies (perhaps applicable to brain health and decline) toward understanding complex biological questions. The following questions along with our responses illustrate our application of this strategy:

Q: Is AD a cellular versus, say, an organismal disease?
A: AD likely starts out as a neuron (i.e., cellular) disease and later spreads to destroy brain function, before causing death. Both our genetic makeup and our environment govern when dementia can strike.

Q: What is the fundamental cause of AD?
A: Death of neurons.

Q: Is it necessary to study neurons in order to understand neuron death?
A: Not necessarily.

Q: What are the cells or model organisms used as research tools to study neuron death?
A: Bacteria → *Caenorhabditis elegans* (nematode) →→ fruit flies →→ mice → monkeys → humans

Q: Which bacteria?
A: DHA-producing, deep-sea bacteria, and their recombinants of *Escherichia coli.*

Q: Is the use of bacteria to study complex human processes and diseases a new strategy?
A: No. For example, the foundation for understanding human heredity and hereditary diseases was largely established by using bacteria.

Q: Has sufficient forward progress been made with DHA bacteria to justify their use in understanding AD?
A: Yes. See earlier book by same authors (*Omega-3 Fatty Acids and the DHA Principle*, Taylor & Francis, 2009).

Q: What is an example of a contribution of bacterial research toward understanding the molecular pathology of AD?
A: *DHA principle* as applied to neurons predicts that neuron speed, which is essential for brain speed, comes at the expense of brain span, a term we use to define the useful lifetime of the human brain.

Q: Do bacterial models offer any clues about the biochemical basis of membranes' contributions to longevity?
A: Yes, the tripartite membrane fatty acid blending code developed with bacteria likely applies to neurons and predicts that oxidatively damaged membranes contribute to energy stress, causing aging.

In summary, the human brain is perhaps the greatest marvel of evolution, and it is not surprising that remarkable building blocks enabling speed are needed for the function of neurons. In this book we follow the fate of an essential neuron building block, an omega-3 fatty acid called DHA. The extraordinary properties of DHA are seen across a range of life forms from deep-sea bacteria to human neurons, and the take-home lesson is always the same. In enabling extreme speed, DHA offers great benefits balanced against great risks. This data led to the development of the DHA principle as applied here to brain health and decline. We explore the concept that neurons work so hard as to compromise or risk their longevity. Is there a trade-off between speed and efficiency of brain function enabled by DHA versus longevity or life span? Has modern medicine gotten ahead of itself in treating the body but not the brain? There is no doubt that we have entered the "age of longevity" and that our brain seems to be lagging behind. In this volume we seek molecular explanations for the decline in brain health during the age of longevity.

REFERENCES

Jackson, J., and M. Prince, editors. Alzheimer's Disease International, *World Alzheimer Report 2009—Executive Summary,* September 21, 2009.

Prusiner, S. B. 1991. Molecular biology of prion diseases. *Science* 252:1515–1522.

Valentine, R. C., and D. L. Valentine. 2004. Omega-3 fatty acids in cellular membranes: A unified concept. *Prog. Lipid Res.* 43:383–402.

Valentine, R. C., and D. L. Valentine. 2009. *Omega-3 Fatty Acids and the DHA Principle.* Boca Raton, FL: Taylor & Francis.

Section I

Evolution and Essential Roles of DHA in Neurons

Neurons stand out from other cells in our body based on their enormous size, permanence, extraordinary energy use, myelination, long-distance intracellular transport systems, extreme sensitivity to oxygen deprivation, and, above all, a prodigious and interconnected neural circuitry that enables learning and storing of information. Neurons link us to our environment through our senses with vision being the king of senses in humans. Somehow the omega-3 fatty acid DHA (22:6) enables neurons to perform these amazing feats. In this section we explore the evolution of neurons with an eye toward deciphering in molecular terms why DHA is essential to them. A role of DHA as an energy food in neurons is eliminated because the adult brain lacks the ability to store and burn fats for energy. A hormonal role for DHA and its sister molecules EPA (eicosapentaenoic acid, 20:5) and ARA (arachidonic acid, 20:4) is vital but not covered here. Instead, the emphasis is on DHA working as an essential membrane building block for more than 100,000 miles of neural connections present in the human brain. This circuitry physically links some 100 billion neurons forming the brain. We hypothesize that DHA allows axons to transmit electrical signals among neurons faster and more efficiently compared to other fatty acids found in nature.

The important roles played by DHA and other omega-3 fatty acids in human health and development have become a popular topic covered regularly in the popular press. However, a full accounting of the roles of DHA in neurons requires balancing the enormous benefits of these molecules against risks, the latter having been obvious to lipid chemists for decades and now being recognized by biochemists. DHA is a highly unstable molecule, and wherever oxygen is present, DHA is readily degraded to toxic forms. This is especially important in the highly oxygenated environment essential for brain function. We choose to introduce dual chemical personalities of DHA from an evolutionary perspective because all cells incorporating DHA into their membranes face this dilemma. We suggest that exploring DHA from the standpoint of benefit–risk analysis opens new windows for understanding how DHA functions in neurons.

1 Nature of Neurons and Their Specialized Membranes

The human brain is an amazing biological computer boasting one property above all that is unmatched by electronic computers—the power of lifelong learning. Of course the brain can also perform herculean tasks, such as processing visual signals that flash instantly in front of our eyes. Vision is so important for survival that a significant amount of our total brain capacity, including at least 30 separate areas, is dedicated to visual processing. Massive numbers of neurons are the computer chips behind visual processing and all other brainpower. Thus, to understand the nature of the brain and how it ages, it is necessary to explore the molecular world of neurons, especially their unique DHA membranes. We will see that neurons are extraordinary cells whose capacities can be explained in part by their extraordinary membranes. Neuron membranes have evolved to conduct electrical signals or impulses at maximum speed and efficiency as honed by Darwinian selection. It is necessary to explore the inner workings or molecular architecture of DHA-enriched neuron membranes to gain an understanding of why human neurons are so efficient and perhaps so fragile at the same time. At the membrane level we encounter DHA whose long chains are used as building blocks for neuronal membranes—especially the long connecting tubes called *axons* along with miniature vesicles called *synaptosomes*, the latter of which deliver neurotransmitters across synapses. The power of DHA resides in its conformational dynamics (see Feller, 2008), which is harnessed in our brain for great benefit.

1.1 SOME BRAIN STATISTICS

The fundamental work of neurons is to receive, conduct, and transmit electrical signals. These tasks in the human brain require an incredible number of neurons (Table 1.1) and their working parts, as described in Table 1.2. At almost every turn there is something breathtaking about these numbers. For example, note that the adult brain has about 100 billion neurons compared to infants, who have roughly double this number. Why is it necessary that 100 billion neurons present in infants must die during brain development? A typical neuron shape is shown in Figure 1.1a. In Figure 1.1b different parts of neurons are defined along with the direction of electrical flow. These pictures also provide clues about why the wholesale pruning of neurons in infants is necessary. A large adult neuron can be 100,000 times larger than a lymphocyte. As neurons grow and form hundreds and even thousands

TABLE 1.1
Hierarchy of Neuron Numbers Forming the Brains of Animals (Estimates)

Animal	Number of Neurons
Sponge	0
C. elegans (total nervous system)	302
Pond snail	111,000
Zebra fish (embryo)	101,000
Honeybee	85,000
Fruit fly	100,000
Frog	16,000,000
Rat (cerebral cortex)	15,000,000
Octopus	300,000,000
Chimpanzee	6,000,000,000
Human	100,000,000,000
Cerebral cortex	11,000,000,000
Elephant	200,000,000,000
Whale	200,000,000,000

Source: Modified from Chudler, E. H., 2009, Brain Facts and Figures, http://faculty.washington.edu/chudler/facts.html; Society for Neuroscience, 2008, Brain Facts: A Primer on the Brain and Nervous System, p. 74, http://www.sfn.org/skins/main/pdf/brainfacts/2008/brain_facts.pdf.

TABLE 1.2
Selected Brain Statistics

Property	Numbers
Weight of adult brain	~3 lbs
Number of neurons	
Infants	~200 billion
Adults	~100 billion
Neuron—Neuron connections (synapses)	60–240 trillion
Miles of myelinated neural circuitry	~100,000 miles
Total surface area of connections	~4 football fields
Number of synaptic vesicles	100 thousand trillion
Maximum neuron length	~3 feet/1 meter
Relative size of neuron compared to a human lymphocyte	~100,000 times larger
Maximum life span of a human neuron	~120 years

Source: Modified from Chudler, E. H., 2009, Brain Facts and Figures, http://faculty.washington.edu/chudler/facts.html.

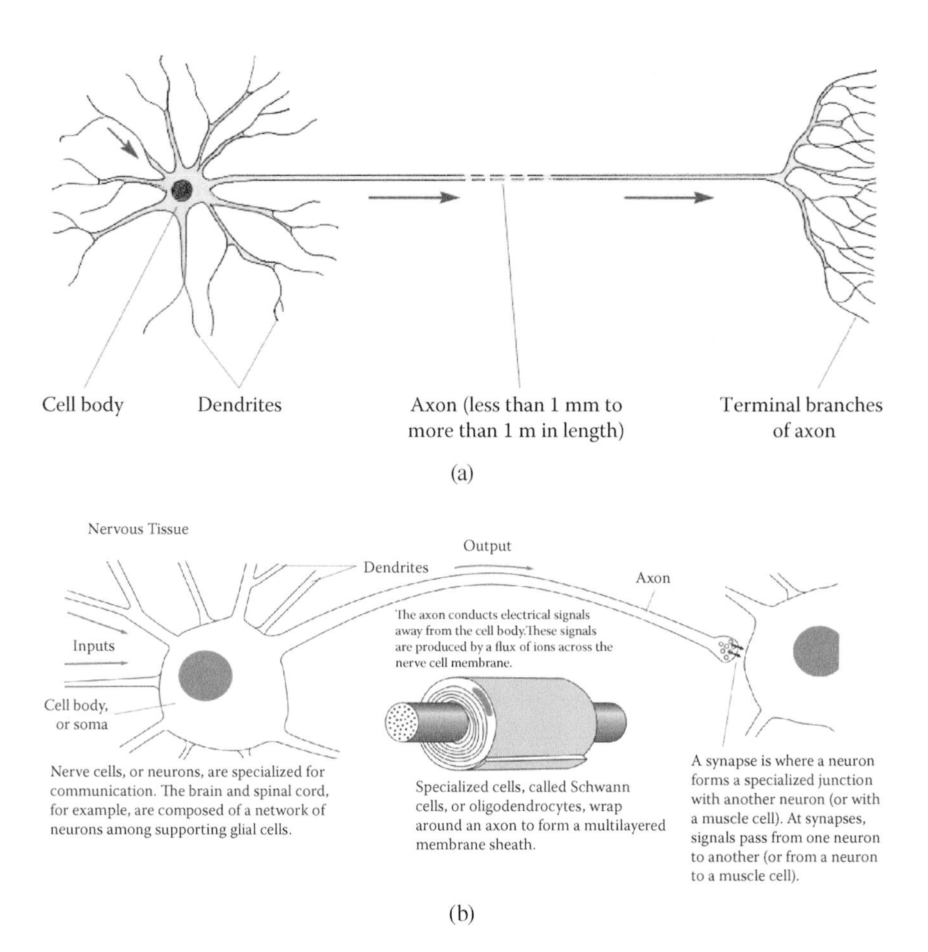

FIGURE 1.1 Neurons: (a) Neurons are giant cells sometimes up to 1 meter in length and roughly 100,000 times the volume of a human lymphocyte. The arrows indicate the direction in which neural signals are conveyed. The single axon conducts signals away from the cell body, while the multiple dendrites (and the cell body) receive signals from the axons of other neurons. The nerve terminals end on the dendrites or cell body of other neurons or on other cell types, such as muscle or gland cells. (From Alberts, B. et al., *Molecular Biology of the Cell*, Copyright 2008. Reproduced by permission of Taylor & Francis Group, LLC, a division of Informa plc.) (b) A typical neuron has a cell body containing the nucleus with dendrites, with a long central axon terminating with a synapse. Many axons are myelinated. The sophisticated biochemistry of neurons depends on numerous helper cells called *glial cells*, not shown in this drawing. (From Alberts, B. et al., *Molecular Biology of the Cell*, Copyright 1994. Reproduced by permission of Taylor & Francis Group, LLC, a division of Informa plc.)

of connections, it would not be possible to fill the 200 billion neurons present in the infant brain into our adult brain cavity. In Chapter 6, we present an energy-based argument to explain why brain size has maxed out in humans.

1.2 A CLEAN, CONSTANT, AND PROTECTIVE ENVIRONMENT ALLOWS NEURONS TO PERFORM AT MAXIMUM EFFICIENCIES

There is a remarkable similarity between the clean rooms required in manufacturing computer chips compared to the invariant biological world surrounding neurons in the brain. In one sense the term *clean* can also be applied to the brain environment where strict tolerances are set for metabolites and environmental conditions including salt, acidity, waste products such as lactic acid and ammonium ions, oxygen, energy-rich sugars, essential trace metals, and so forth. The point is that much wider deviations in these parameters are tolerated or even required of cells working in other parts of our body in contrast to neurons. One of the most dramatic differences is seen in comparing the strict dependence on oxygen for neurons compared, say, to sperm cells, the latter whose tails are enriched with DHA at levels similar to axons. Neurons begin to die after more than 2 minutes of oxygen deprivation in contrast to sperm cells, which are adapted to spend several days traveling in the oxygen-starved world of the female reproductive tract. It seems ironic that while both adult neurons and sperm are dependent on the common sugar glucose for energy, the lactic acid produced as a major by-product of sugar degradation by sperm would be lethal for neurons if allowed to accumulate inside the brain. Severe drops in glucose levels of short duration do not kill masses of neurons as oxygen starvation does. However, there is the possibility that such steep plunges in brain sugar levels, though extremely rare, might strike many years later in the form of Alzheimer's disease. Strict regulation of essential but potentially toxic metals such as copper ions, which can severely damage DHA membranes, is discussed in Chapter 10.

Thus, the brain as an organ can be considered as a small, isolated, highly integrated, and unique ecosystem housing perhaps several hundred billions of cells (i.e., neurons and glial cells) that are interdependent. Neurons are obviously dependent on the rest of the body for their energy and oxygen supplies. Even conditions found normally in circulating blood, which directly bathes most body cells, are an insult for neurons. The blood–brain barrier serves as a robust gatekeeper permitting the preferred levels of nutrients in, while turning back undesirable molecules and allowing waste products to exit. In addition to macronutrients and micronutrients, hormones and signaling molecules pass in and out of the brain. Of particular interest is the circulation and monitoring of DHA entering the brain (Rapoport, Chang, and Spector, 2001; Umhau et al., 2009). How important is the blood–brain barrier in targeting DHA and other classes of lipids into or out of the brain? Does the blood–brain barrier weaken with age and are there conditions that can overpower the strength of this barrier? How much cholesterol is allowed to enter or exit the brain? Note that cholesterol in large amounts is equally essential to DHA as a neuron building block (discussed in Chapter 6). However, cholesterol seems to be more stable in the brain environment compared to DHA.

In the end, metabolite traffic patterns entering, circulating within, and exiting the brain are amazingly complex, defining a balanced brain ecosystem. That is, cells in the brain are in constant communication, some having specialized jobs that impact others, and all activities are coordinated to perfection. Keep in mind that the harmony of ecosystems depends on each part working smoothly, defining a system in which breaking or weakening any link of the chain can lead to collapse. We suggest that understanding the brain as an ecosystem can help us to better understand the roles of DHA, its impact on energy relationships and eventually premature aging of the brain. We will have occasion to return to this topic later.

1.3 MICROGLIA DESTROY PATHOGENS AND SCAVENGE CELLULAR DEBRIS BUT HAVE A DANGEROUS SIDE

Neurons alone would be helpless without their servants, some 100 billion or more glial cells that display some remarkable properties. For example, white cells, which are the disease fighters in the rest of the body, are largely locked out of the brain by the blood–brain barrier. Some 20 billion microglial cells replace white cells as sentinels against pathogens, but in addition perform myriad other functions essential for neuron and brain health. When roaming the brain a microglial cell resembles a miniature octopus, with its many processes acting like tentacles to feel every nook and cranny, not for food but for surveying even the tiniest defect or biochemical dysfunction on or surrounding neurons. When pathogens strike the brain, microglia burst into action, growing faster, reprogramming for chemical warfare, and rushing to the site of invasion. Surrounding a pathogen, such as an HIV virus particle or a bacterial pathogen, microglia unleash their attack, which can take the form of engulfment dependent on a dramatic change in size and shape of microglia to accommodate phagocytosis. Powerful weapons of chemical warfare including potent chemicals called *reactive oxygen species* (ROS) can be generated to kill pathogens. ROS including hydrogen peroxide are too toxic to produce continuously and are synthesized by microglia in a well-known oxidative burst. Unfortunately, whereas microglia have evolved adaptations against ROS damage, neurons have not and can sustain collateral damage or even be killed by pulses of these potent chemicals. We will see later that the necessity for DHA in neurons renders the membranes surrounding neurons especially vulnerable to damage by ROS. In contrast, the membranes of microglia are more resistant.

For many years it was believed that an oxidative burst is primarily triggered by the presence of an invader from the outside, a classic response to an attack by a pathogen. Swelling and inflammation are conventional markers of the presence of pathogens being destroyed by disease-fighting cells. However, numerous examples of "sterile" inflammation have now been described throughout the body in which inflammation occurs in the absence of any foreign invader. Sterile inflammation can occur in the brain due to brain trauma or other major upheavals, including structural damage to neurons and shifts in biochemical homeostasis. One of the most amazing features of microglia is the diversity of signals besides pathogens that activate these cells. Numerous receptors on the surface of microglial cells can biochemically sense, help integrate, and govern suitable responses against changes in neurons. Sometimes, especially with aging, microglia get confused and do their jobs too well (McGeer and McGeer, 2002).

1.4 OUR BRAIN IS AN OILY COMPUTER

It has been known for many years that DHA is a major constituent of membranes of only a few highly specialized cells in the human body: neurons, rod cells of the eye, and sperm cells (Salem, Kin, and Yergey, 1986). Obviously, we cannot do without any of these cells. The numbers in Table 1.3 show that our bio-computer chips or neurons operating in the brain have a high content of oil, of which a significant amount is DHA. DHA displays extraordinary membrane properties and sits atop the hierarchy of oils or fatty acids forming membranes in nature. In short, we cannot do without DHA.

It is important to recall that a human mother's milk comes with levels of DHA and other essential oils that mirror the need for these oily building blocks during rapid brain development in infants (Cetin, Alvino, and Cardellicchio, 2009; Innis, 2007). It is proposed that the period of rapid growth in the developing brain is perhaps the most critical window demanding preformed DHA supplied in mother's milk to keep up with the demand for building new neuron circuitry (Birch et al., 2000). During early pregnancy it is estimated that about 250,000 new fetal neurons are added per minute. To avoid confusion regarding where DHA circulating through the umbilical cord ends up in the fetus or where DHA from mother's milk ends up in infants, recall that there are three major fates of DHA in our body and in animals in general. Many animals, such as king salmon, build up huge stores of DHA and other storage forms of fat later used as energy food for their long migrations. Stored oils including DHA allow king salmon to travel as far as 1200 miles upriver to their spawning grounds. Note that salmon stop eating once they enter the mouths of their home streams and must depend on stored deposits of fat. DHA in mother's milk seems far too precious to burn as fuel by infants. Instead DHA from mother's milk is transported in the bloodstream to the infant's brain and is used primarily as an essential neuron building block. Recall that the adult brain unlike other organs shuts down the use of oils as fuel, metabolizing primarily sugar. DHA is used as building blocks in the growth of new neurons and their hundreds of trillions of axons and synaptic connections

TABLE 1.3
Our Oily Brain

Oil content of dehydrated brain tissue	~50 percent
DHA content	
Percentage of total fatty acids	5
Axons	10
Synaptosomes	10

and thousands of trillions of synaptic vesicles working inside neurons (the latter are discussed in Chapter 6). Thus, while oils and fatty acids including DHA are important as energy food for many other kinds of human cells, in the brain DHA plays a structural role in building new neurons in fetuses and infants, shifting to building and repairing neural connections in adults. Sugars instead of oils are almost exclusively used to meet an enormous appetite for energy by the adult brain (Clarke and Sokoloff, 1999). Omega-3 fatty acids play a third important role in humans where small amounts of these molecules are converted to produce a powerful class of essential hormones. However, this vital hormonal role is not discussed further here.

1.5 NEURONS HAVE EVOLVED DHA-ENRICHED MEMBRANES THAT PLAY ESSENTIAL ROLES

All human cells including neurons have oily membranes, but neurons are unique in having relatively high levels of DHA. The nature of axon membranes is illustrated in Figure 1.1b. An axon is a tube-like structure surrounded by a thin oily membrane. Note that the watery inside of the axon (axoplasm) contains high K^+ (potassium) levels in contrast to the outside with high Na^+. A critical functional role of membranes surrounding axons is that they form a physical barrier preventing sodium from spontaneously rushing inside and potassium from leaking out. Nerve conductance would stop if these ions rapidly and spontaneously swapped sides (which would occur instantaneously if the membrane were compromised). Thus, one of the most critical roles of the axon membrane is to prevent this leakage of electrically active ions from happening. However, there is always some spontaneous leakage as well as conductance-triggered ion exchange, which means that ideal salt levels essential for neural impulse conductance must be constantly reset within razor-sharp limits. Reestablishing favorable salt levels across the membrane requires a molecular machine that literally pumps sodium out and potassium in (Glynn, 1993). We will see in the next section that these ion-balancing molecular machines consume an enormous amount of the total energy available to neurons.

A single-axon membrane surface is built of billions of individual repeating phospholipid units (Figure 1.2a). Note that a typical phospholipid is composed of two long tails made of fatty acids attached to a head group. Figure 1.2b shows two modeled images of a single phospholipid molecule in which the golden-colored chain is DHA. The blue tail represents another common class of saturated fatty acids shown as a more linear configuration. Phospholipids spontaneously join together in axon membranes forming a sandwich-like bilayer structure composed of top and bottom leaflets. Figure 1.2c shows a single K^+ channel embedded in a DHA-enriched neuron membrane. The K^+ gate, a membrane protein, is shown as a yellow ribbon coursing through the membrane with surrounding DHA molecules depicted as red chains. The K^+ channel viewed from above forms a molecular pore that selectively allows K^+ ions, but not Na^+, to cross the bilayer. These channels play central roles in the transmission of nerve impulses.

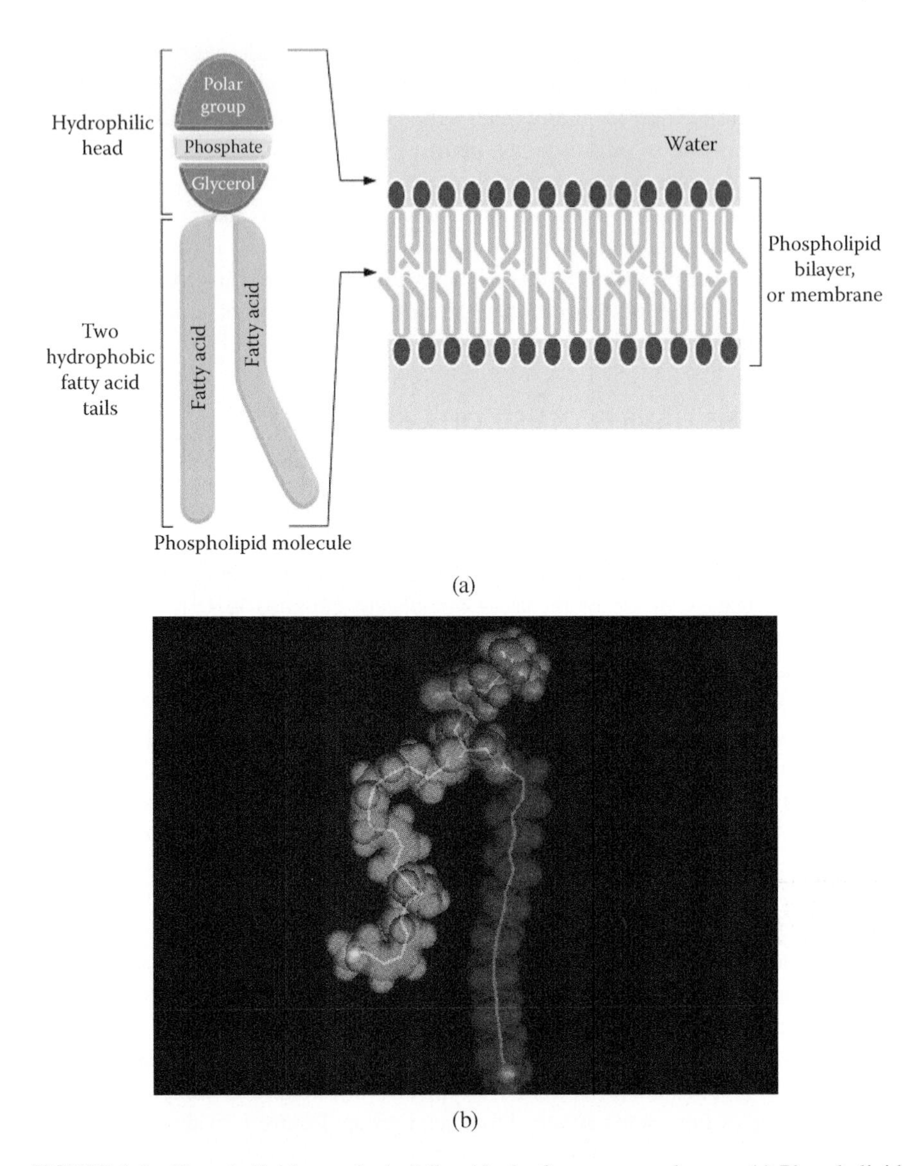

FIGURE 1.2 Phospholipids are the building blocks for axon membranes. (a) Phospholipid nomenclature. In an aqueous environment, the hydrophobic tails of phospholipids pack together to exclude water. Here they have formed a bilayer with the hydrophilic head of each phospholipid facing the water. Lipid bilayers are the basis for most cell membranes. (From Alberts, B. et al., *Molecular Biology of the Cell*, Copyright 2008. Reproduced by permission of Taylor & Francis Group, LLC, a division of Informa plc.) (b) Image of a DHA phospholipid molecule such as found in neuron membranes, as generated by chemical simulation. Gold chains are DHA; green chains are stearic acid (18:0). (Image courtesy of Scott Feller.) (c) Model of an axon-membrane-embedded K^+ channel or gate (yellow ribbon) surrounded by DHA chains (red). Potassium gates play a crucial role in electrophysiology of neurons and other cells. (Courtesy Igor Vorobyov, Scott Feller, and Toby W. Allen, unpublished communication.) **(See color insert.)**

(c)

FIGURE 1.2 *(Continued)*

As shown in Figure 1.1b, a thick myelin sheath surrounds many axons. This sheath provides a thick layer of insulation for the same reason that electrical wires running through water must be entirely coated. Axons require their insulating layer to prevent short-circuiting the flow of electricity. Myelination also offers neurons a huge speed bonus. Myelin-insulated axons allow electrical impulses to skip sharply ahead from node to node permitting a quantum leap in speed and efficiency of brain signals. In essence, this creates a much faster and more energy efficient bio-computer helping maximize brain function. The importance of insulating axons is dramatically demonstrated by the demyelinating disease multiple sclerosis in which sheathing in some regions of the central nervous system is destroyed, resulting in lack of control of muscle action.

1.6 THE BRAIN AS A MACHINE

Sugar as metabolized in the normal brain is a clean-burning fuel leaving no toxic byproducts. The key process that makes most brain energy is respiration, a process that uses oxygen. Each neuron has numerous mitochondria that carry out respiration and act as internal powerhouses (see Chapter 11). The brain has become so dependent on oxygen carried in the bloodstream for respiration that interruption of the supply of this vital molecule by heart attack, stroke, or asphyxiation is lethal beyond the duration of a few minutes.

The total amount of food and oxygen routed to the brain is staggering considering that an adult brain weighs about 3 pounds and represents about 1/40 to 1/60 the total weight of our body. The brain consumes some 20 to 25 percent (and sometimes higher) of our total oxygen and energy food budget. As already mentioned, the adult brain does not use stored fats, in spite of the fact that fats are far more energy rich than sugar.

There are many reasons why the brain is considered an energy hog with some points introduced here and others discussed in Chapter 4. First, the human brain does not or cannot sleep. Holding our thoughts and memories seems to require that the brain keep active at all times. Dreams are one product of this activity. One of the largest energy usages is for fueling the ion pumps introduced above. Humans live with a "sea within," which means that our circulating blood and blood serum is saline like seawater (i.e., actually about one-third seawater). The evolutionary process that led us to this salinity is beyond the scope of the present discussion, but the point we are exploring here concerns the high energy cost of maintaining salt balance when neurons are predominantly surrounded by Na^+ ions. Note that the kidney handles the bulk of salt balancing for the body as a whole. Neurons, like all cells bathed in blood plasma, are surrounded by high levels of Na^+. However, inside neurons K^+ is needed in high levels and sodium is maintained at low levels. In fact, sodium inside the neuron in levels similar to the outside would be highly toxic. Also, K^+ tends to leak outward while Na^+ leaks inward. The membrane is not a perfect permeability barrier, resulting in considerable spontaneous leakage of ions along these gradients. Because electrical impulses in the brain require razor-sharp control of these ions, brain signals would soon stop if the levels of these ions were allowed to become imbalanced. Note that each passing neural impulse also contributes to an ionic imbalance, but not nearly as much as spontaneous leakage (see Chapter 6). It is estimated that fueling these powerful pumps burns 50 percent or more of the energy budget of neurons (see Chapter 6). Thus, in calculating an energy budget for neurons the cost of maintaining ionic balance is one of the biggest, if not the largest energy cost, leaving many other energy-intensive processes to share the remainder. For example, about 100 thousand trillion synaptic vesicles use large amounts of energy in delivering neurotransmitters at synapses (Chapter 6).

Neurons face a tight energy budget. Energy stress can be deadly, apparently striking the hardest working neurons first. For example, the massive signaling power needed by neurons of the optic nerve for processing visual images seems to occasionally potentiate these vital neurons to premature death via energy stress. This concept helps explain a well-known hereditary disease in which an insufficient energy supply in neurons forming the optic nerve somehow creates age-related blindness (Wallace et al., 1988). This disease is called Leber's hereditary optic neuropathy (LHON). This is an energy disease attributable to a genetic defect or mutation in mitochondria, which provide most of the energy for neurons. In the case of LHON a mutation has occurred that reduces, but does not eliminate, the ability of mitochondria to manufacture cellular energy. What is remarkable is that symptoms of this disease are age related where energy stress is somehow amplified catastrophically during aging, killing enough neurons in the optic nerve to sometimes cause blindness. Studies of mitochondrial diseases such as LHON provide important clues about energy as a pacemaker for dementia, as discussed in Section IV.

Brain imaging science has come of age and has already contributed many important advances in understanding bioenergetics of neurons in health and disease (Li, Gong, and Xu, 2011; Shulman, Hyder, and Rothman, 2009; Shulman, Rothman, and Hyder, 1999; Shulman et al., 2004). Some selected milestones include:

- The high energy cost of brain function is confirmed.
- Metabolic pathways of glucose utilization in the brain are elucidated.
- Energy use by a sleeping brain is similar to that of an awake brain.
- Brain energy use is halved during anesthesia.
- Localized differences in energy utilization are seen in different regions of the brain.

The last point is particularly important from the standpoint of neurodegeneration because these data suggest that certain neurons and their mitochondria work harder than others. This point is discussed in detail in Sections IV and V.

1.7 PROTECTING DHA MEMBRANES OF NEURONS IS ONE OF THE SECRETS TO THE HEALTH AND LONGEVITY OF THE BRAIN

The basis of neuron permanence in the adult brain lies in their ability to continuously repair or rebuild any worn out parts. This ability to continuously fix themselves without shutting down the whole machine is one of the cardinal distinctions between the brain and electronic computers. A partial list of components that frequently wear out in neurons is as follows:

- Mitochondria, the main energy generators for neurons, constantly divide and sometimes become dysfunctional with severely damaged mitochondria being digested. (Note that mitochondria are repaired, usually every few weeks, during division.)
- Thousands of different enzymes and proteins working in neurons wear out and require continuous resynthesis.
- Components of protein synthesis machinery such as ribosomes also require renewal.
- Massive areas of membranes are monitored for damage and are repaired when necessary.
- Dysfunctional axons, synaptic vesicles, and synapses die and new ones grow.

We focus next on why it is necessary to constantly resurface or rebuild axonal and synaptic membranes (see Section III), a process that also applies to tiny synaptic vesicles found inside neurons. As background, a chemist seeing the high composition of DHA as a structural element in membranes of neurons might be surprised to learn that neurons can live as long as 100 or more years. How is this possible when

DHA is widely recognized as being the most chemically unstable of all membrane fatty acid building blocks in nature? That is, oil chemists have known for many years that whenever DHA comes into contact with oxygen the process of peroxidation occurs. When foods with high DHA content are fried or stored at room temperature, their oils begin to turn rancid, giving a pungent odor characteristic of 3-day-old fish (Gunstone, 1996). In Chapters 8 through 10, we describe the improbability of DHA lasting in membranes for days, let alone a century. The specialized rhodopsin disk membranes located in the outer segment of a rod cell and behaving as critical light-sensing antennae for vision have the shortest functional life span of any membrane found in human cells. Each rod cell contains stacks of about 1000 membrane disks. Each day about 100 of these aged disks are removed from the top of the pile with an equal number of newly minted disks added to the bottom. Thus, the lifetime of an individual disk is just 10 days, a process that can be catastrophically sped up if you look for too long directly into the sun. Oxidative damage to DHA is discussed in more detail in Section III.

DHA-enriched membranes surrounding axons though continuously in contact with oxygen appear to last as much as 180 times longer compared to rod cells (Chapter 10). One explanation of their longer duration is that DHA membranes of neurons are obviously never exposed to light. The absence of light radiation is clearly a great advantage to neurons because light dramatically amplifies the oxidative destruction of DHA and can quickly lead to a chemical chain reaction, which must be avoided in the brain at all costs.

In summary, neurons belong to a rare class of human cells including rod cells of the eye and ciliated auditory cells in the ear characterized or classified as being stuck in a nondividing or permanent mode. Thus, a centenarian can boast of 100-year-old neurons. However, a closer look shows that whereas some parts or molecules of an old neuron might have lasted a century, much of the inner workings of the cell have undergone numerous cycles of rebuilding. DHA-enriched membranes are also continuously resurfaced, but there are clues that neurons manage protection of their membranes far better than rod cells even after considering the absence of light. We suggest that conventional means of protecting membranes, as practiced by other cells in our body, are inadequate for this task in neurons because of a higher risk associated with the presence of DHA in specialized neural membranes. That is, we propose that a variety of undiscovered protective mechanisms preventing or avoiding DHA oxidation likely exist in the brain and nervous system (see Chapter 10). We suggest that this putative novel battery of protective defenses is needed to sustain the long-term viability of neurons, essentially to protect these cells from their own potentially toxic, DHA-enriched membranes, whose oxidation might fast-forward neurodegeneration. Some possible new defenses that come to mind include maintenance by the whole brain of reduced free oxygen levels due to the powerful oxygen scavenging powers of so many concentrated brain cells. Axons tightly wrapped with sheaths of myelin might also block exposure of a large proportion of DHA-enriched membrane surfaces to attack by oxygen as discussed in detail in Chapter 10. Note that myelin is white when observed in brain tissues stored in preserving solution, and the region of the brain containing large numbers of long myelinated axons is called *white matter*. In contrast, axons of gray matter are not myelinated. We suspect that a highly sophisticated, well-placed and multilayered set of defenses is in place to preserve the integrity of neuronal DHA membranes as long as possible against damage caused by peroxidation (see Chapters 8 through 10). We also suspect that should any of the processes geared to extending the effective working life of neural membranes become enfeebled, DHA might become unmasked as a harmful pro-aging molecule perhaps leading to neurodegeneration as discussed in Chapters 11 through 17.

Finally, neurons in the brain sometimes flawlessly perform a series of amazing biochemical tasks for a century. The term *extraordinary* somehow seems inadequate when considering the giant size of neurons, their enormous appetite for energy, the evolution of sophisticated excitatory biochemistry, their self-healing powers, and above all their ability to learn, remember the past, and envision the future. Something simply has to go wrong in this "perfect world" of neurons, an idea developed in this book. However, before moving to the dark side, we explore the great benefits of DHA in specialized membranes of neurons, membranes that play essential roles in human health and development.

REFERENCES

Birch, E. E., S. Garfield, D. R. Hoffman et al. 2000. A randomized controlled trial of early dietary supply of long-chain polyunsaturated fatty acids and mental development in term infants. *Dev. Med. Child Neurol.* 42:174–181.

Cetin, I., G. Alvino, and M. Cardellicchio. 2009. Long chain fatty acids and dietary fats in fetal nutrition. *J. Physiol.* 587:3441–3451. Epub June 15, 2009. Review.

Chudler, E. H. 2009. Brain Facts and Figures. http://faculty.washington.edu/chudler/facts.html.

Clarke, D. D., and L. Sokoloff. 1999. Circulation and energy metabolism of the brain. In *Basic Neurochemistry: Molecular, Cellular, and Medical Aspects*, 6th edition. Siegel, G. J. et al. Philadelphia, PA: Lippincott Williams & Wilkins.

Feller, S. E. 2008. Lipid animation: 500 ps dynamics of a stearoyl docosohexaenoyl PC lipid molecule (14MB), http://lipid.wabash.edu/, Wabash College Chemistry Department, Crawfordville, IN.

Glynn, I. M. 1993. Annual review prize lecture. All hands on the sodium pump. *J. Physiol.* 462:1–30.

Gunstone, F. D. 1996. *Fatty Acid and Lipid Chemistry*. London: Blackie Academic and Professional; 196. p. 252.

Innis, S. M. 2007. Dietary (n-3) fatty acids and brain development. *J. Nutr.* 137:855–859. Review.

Li, A., L. Gong, and F. Xu. 2011. From the cover: Brain-state-independent neural representation of peripheral stimulation in rat olfactory bulb. *Proc. Natl. Acad. Sci. USA* 108:5087–5092.

McGeer, P. L., and E. G. McGeer. 2002. Local neuroinflammation and the progression of Alzheimer's disease. *J. Neurovirol.* 8:529–538.

Rapoport, S. I., M. C. Chang, and A. A. Spector. 2001. Delivery and turnover of plasma-derived essential PUFAs in mammalian brain. *J. Lipid Res.* 42:678–685.

Salem, N., H.-Y. Kim, and J. A. Yergey. 1986. Docosahexaenoic acid: Membrane function and metabolism. In: Martin, R., editor. *Health Effects of Polyunsaturated Fatty Acids in Seafood*. New York: Academic Press, pp. 263–317.

Shulman, R. G., F. Hyder, and D. L. Rothman. 2009. Baseline brain energy supports the state of consciousness. *Proc. Natl. Acad. Sci. USA* 106:11096–11101.

Shulman, R. G., D. L. Rothman, K. L. Behar et al. 2004. Energetic basis of brain activity: Implications for neuroimaging. *Trends Neurosci.* 27:489–495.

Shulman, R. G., D. L. Rothman, and F. Hyder. 1999. Stimulated changes in localized cerebral energy consumption under anesthesia. *Proc. Natl. Acad. Sci. USA* 96:3245–3250.

Society for Neuroscience. 2008. Brain Facts: A Primer on the Brain and Nervous System. 74 pages. http://www.sfn.org/skins/main/pdf/brainfacts/2008/brain_facts.pdf.

Umhau, J. C., W. Zhou, R. E. Carson et al. 2009. Imaging incorporation of circulating docosahexaenoic acid into the human brain using positron emission tomography. *J. Lipid Res.* 50:1259–1268.

Wallace, D. C., G. Singh, M. T. Lott et al. 1988. Mitochondrial DNA mutation associated with Leber's hereditary optic neuropathy. *Science* 242:1427–1430.

2 DHA Contributes Both Benefits and Risks to Mammals

Much of the research on omega-3 fatty acids focuses on the benefits of these unique compounds toward human health and development. This line of research began about eight decades ago and led to the definition of polyunsaturated fatty acids as being "essential oils" for animal and human health (Burr and Burr, 1929; also see comprehensive review by Bourre, 2006, a pioneer in this field). Research on DHA in human nutrition continues today with some selected recent advances highlighted in this chapter. The dietary approach to DHA is on solid chemical ground because of the fact that rods and cones of the eye, sperm, neurons, and certain other cells such as colon epithelial cells enrich their membranes with DHA contributed in part by diet. Helping define the molecular roles of DHA is a long-range goal of human nutritionists and involves animal models including transgenic and mutant mice. Using these modern tools nutritionists have now established that DHA offers both benefits as well as risks to mammals, with studies described below. In addition, nutritionists are contributing toward understanding the evolution of the human brain size, efficiency, and durability—studies that link us to the sea where much of the global supply of DHA originates (see Chapter 4).

In one case history discussed below, nutritionists have found that seafood rich in DHA helps prevent colon cancer. This is an interesting example in which the harmful chemical properties of DHA are harnessed as an assassin against senescing colon cells, in essence preventing these cells from gaining a foothold as precancerous cells. In this case a risk to the cell is a benefit to the organism. However, there is increasing information that shows that too much DHA incorporated into certain classes of membranes represents a pathological state.

2.1 FERTILITY

We consider DHA-enriched tail membranes of sperm to be surrogates for axons and expect that understanding of the essential roles of DHA in sperm membrane biochemistry and development has major implications for deciphering the molecular biology of DHA in neurons. DHA knockout mutants of mice have been found to be infertile due to blocked sperm production (Stroud et al., 2009), a complex developmental process as illustrated in Figure 2.1. Null or knockout mutants of delta-6 desaturase (D6D) of mice were used in this research. This enzyme, D6D, catalyzes the first step in the synthesis of highly unsaturated fatty acids, including DHA as

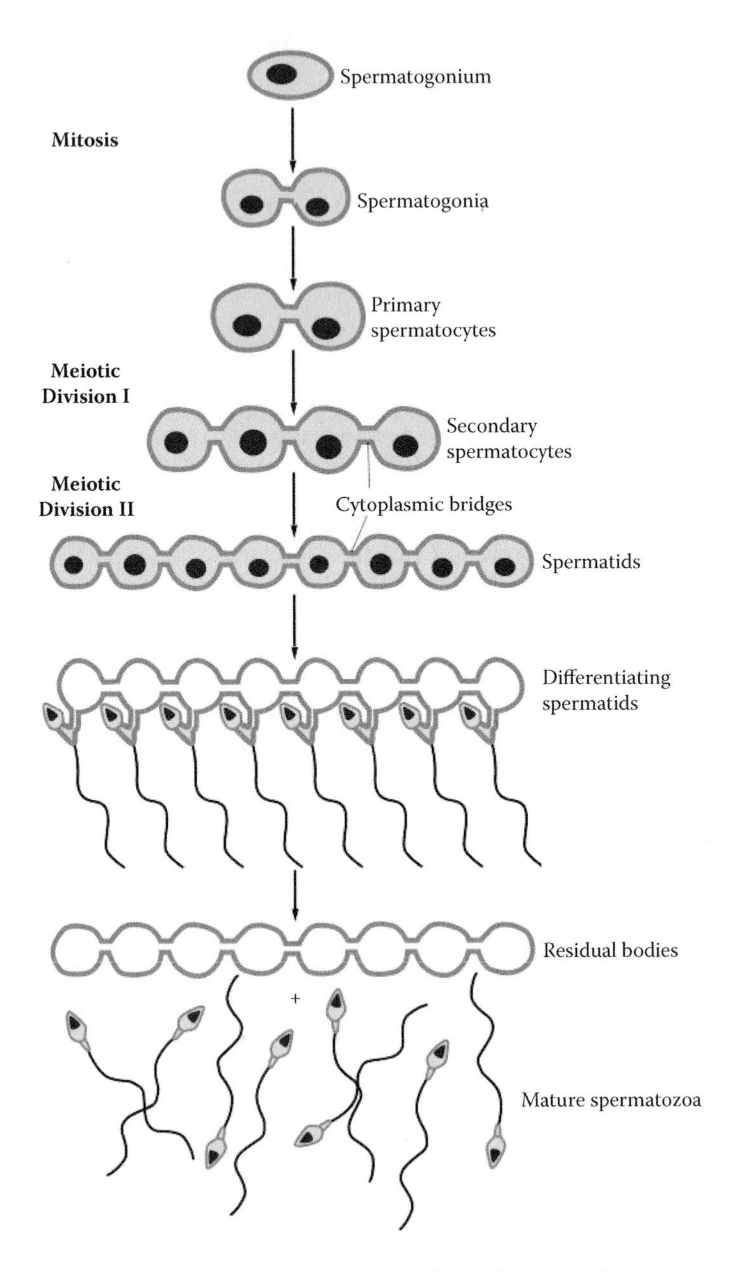

FIGURE 2.1 Sperm differentiation is an example of a sophisticated developmental process requiring DHA. Data from studies of DHA knockout mutants of mice show that spermatogenesis is blocked in the absence of DHA. See text for details and also Chapter 4 for implications for axon development. (Alberts, B. et al., *Molecular Biology of the Cell*, Copyright 1994. Reproduced by permission of Taylor & Francis Group, LLC, a division of Informa plc.)

well as the last desaturation step for DHA. Biochemical studies show that D6D enzymatic activity in mutants is blocked. Male mutant mice with a severe impairment of spermatogenesis produced few and abnormal spermatozoa. DHA levels in the liver are 20 times lower in DHA minus mice compared to control animals. DHA levels are also decreased in the brain and testes. Interestingly, a human case of D6D deficiency has been identified (Williard et al., 2001). The patient exhibited several abnormalities including corneal ulceration, symptoms that were largely reversed by dietary DHA in combination with arachidonic acid (20:4).

In a follow-up study of the report by Stroud et al. (2009), Nakamura and colleagues (Roqueta-Rivera et al., 2010) showed that dietary DHA restored the breeding success rate in DHA knockout mice to normal levels. These data show that a dramatic reversal occurs with normal sperm counts and robust spermatogenesis. Feeding arachidonic acid led to normal levels of 20:4 in tissue with some further elongation to 22:4n6. However, DHA was far more effective in restoring male fertility. Thus, these studies establish the essential nature of DHA over 20:4 and 22:4 in spermatogenesis. These data also substantiate the biochemical routes of DHA synthesis in mammals.

Recent data published by Nakamura and colleagues (Roqueta-Rivera et al., 2011) show that spermatogenesis in DHA null mutants of mice is arrested at the level of acrosome formation, a defect corrected by dietary DHA. A defect in the fusion of acrosin-containing vesicles or failed transport and release of syntaxin 2 vesicles from Golgi is suggested. Acrosin and syntaxin 2 are proteins targeted to transport vesicles and are essential for acrosome structure and function in sperm cells.

The data of Nakamura and colleagues are consistent with our recent mechanical stress model of the unique molecular roles played by DHA in bioenergetics of sperm tails (Valentine and Valentine, 2009). Their important research on the essential role of DHA in sperm development not only sets the stage for further studies of the molecular roles of DHA in specific steps of sperm differentiation but also has implications for understanding how neurons develop. A spin-off of this research involves the mechanism of aging. The DHA theory of aging discussed in Chapter 12 predicts that a mouse with decreased levels of DHA in its membranes would live longer, and vice versa. Thus, nutritionists studying the essential roles of DHA in humans might be on the threshold of yet another major milestone concerning the roles of DHA in aging. The impact of a diet rich in seafood on the evolution of human brain size is discussed in Chapter 4.

2.2 VISION

Sperm, rod cells, and neurons share the common property of high levels of DHA in their membranes. Connor and colleagues (Neuringer et al., 1984) have shown that dietary DHA deficiency in rhesus monkeys results in sharply lowered plasma levels of DHA phospholipids accompanied by visual loss. Studies with monkey sperm show that DHA is targeted primarily to the tail membrane (Connor et al., 1998). Thus, deprivation of DHA is predicted to result in abnormal or defective sperm. To test this hypothesis in humans, Connor and colleagues took advantage of the lowered plasma levels of DHA seen in patients with retinitis pigmentosa. Retinitis pigmentosa is a type of progressive retinal dystrophy or a group of inherited disorders in which abnormalities of rods and cones of the retina lead to gradual visual loss. Affected individuals carrying this class of mutations first experience night blindness, followed by tunnel vision and eventually sometimes blindness. Sperm of patients with retinitis pigmentosa had a much lower DHA concentration and displayed reduced motility and abnormal structure, and patients had lower sperm counts (Connor et al., 1997). However, dietary DHA has not been found to restore to normal symptoms of retinitis pigmentosa in patients. This result is not surprising given the complex molecular pathology of this disease.

2.3 HEALTHY HEART

For a month or so after a heart attack patients are at an elevated risk of a second fatal heart failure. Great international interest has led to large-scale studies with humans showing that dietary fish oil can greatly reduce the risk of a fatal second heart attack (Albert, Campos, and Stampferetal, 2002; Daviglus, Stamler, and Orenciaetal, 1997). Data from these large human trials are backed by research on animal models showing the benefits of dietary fish oils against heart failure (Billman, Nishijima, and Belevychetal, 2010; Duda, O'Shea, and Tintinuetal, 2009; O'Shea et al., 2010). Data from these studies not only show the benefits of fish oil against various cardiac pathologies but also the steady increase in DHA levels in heart cell membranes as a function of fish oil supplementation. For example, DHA levels in the cardiac tissue of mice rose from 30 to 50 percent of total fatty acids. These extraordinarily high levels of DHA in mice might be explained by the fact that mice seldom encounter DHA in their diet, which is rich in plant matter devoid of DHA. Thus, mice may not have evolved systems for modulating levels of DHA intake or for mitigating any adverse effects of excess DHA. Increases of DHA in rats and dogs fed fish oil, where basal levels are much lower than mice, showed roughly a doubling of DHA in phospholipids, though total levels of DHA are much lower compared to mice. In humans, where data on cardiac DHA levels are still scarce, a 4.3 percent basal level of DHA with an increase with supplementation to 8.5 percent was reported (Metcalf, Cleland, and Gibsonetal, 2010; Metcalf, James, and Gibsonetal, 2007). It is generally agreed that fish oil supplementation causes significant increases in cardiac DHA content as well as other fatty acids in membrane phospholipids of human heart cells, changes that might mean survival for patients following a first heart attack.

2.4 CANCER-FREE COLON

We now switch gears from benefits to risks using the case history of DHA acting as a pro-apoptotic molecule, assisting in a beneficial way to the death of potentially dangerous senescing colon cells. Ironically, the death of colon cells in this case is beneficial to the organism in that aged colonocytes represent a pool of precancerous cells whose assassination is preventative. These data illustrate in dramatic fashion the concept of the dual biochemical modes of DHA working in human cells. In this case, the whole organism benefits from the death of rogue colonic cells. Senescing colon cells sometimes escape the natural cascade of programmed cellular death, which serves as a first line of defense preventing these cells from surviving and then causing polyps and later colon cancer. Colon cells are characterized by their rapid growth, rapid aging, and eventual death and replacement. Programmed cellular death or apoptotic cascades are honed to prevent the survival of any colon epithelial cells and initiation of the tumor process. DHA comes into the picture as a dietary-derived insurance policy against colon cells escaping their fate.

Ironically, many molecules central to life display diametrically opposed chemical modes, being absolutely essential on the one hand and turning deadly on the other. Even water (H_2O) becomes a deadly energy uncoupler when incorporated into our membranes (see Chapter 8). The oxygen that we breathe also falls under this category, turning deadly in the form of toxic radicals discussed widely in this book. At the same time reactive oxygen species are utilized by disease-fighting cells in a beneficial way to rid the body of pathogens.

Colorectal cancer is the third most common form of cancer and the second leading cause of cancer-related death in the Western world. Based on current rates, 5.2 percent of men and women born today will be diagnosed with cancer of the colon and rectum at some point during their lifetime. In 2004, in the United States there were greater than 1,000,000 men and women who had a history of colon cancer with greater than 100,000 new cases added per year. Many cases of colon cancer begin as small, benign clumps of cells or polyps that originate from cells lining the walls of the colon (Figure 2.2). Recent data also show that "pancake-like" clumps of cells that are much more difficult to detect than polyps can also become cancerous, opening a second cellular-based pathway toward colon cancer. Like other cancers, colon cancer requires a series of genetic changes (see Chapter 13).

Billions of colonic epithelial cells (colonic cells or colonocytes) join together to form the surface or lining of the colon, an organ that is essential for our survival. As in the case of sperm cells functioning in the female reproductive tract, colonic cells have evolved to inhabit a location contacting the external environment. This location allows them to obtain nutrients directly from the gut. Colonic cells are also exposed to a rapidly changing and potentially lethal environment determined by conditions and contents of the colon, a largely anaerobic world. Colonocytes have evolved to benefit from DHA present in the diet and transported via the circulatory system, or alternatively, transported via the gastrointestinal (GI) tract and incorporated into membranes of these cells. Uptake of DHA is believed to prevent colon cancer. It is estimated that 70 percent of colorectal cancers are preventable by moderate changes in diet. Epidemiological studies indicate that populations ingesting higher amounts of fish are at a lower risk for colon cancer and a lower mortality rate of colorectal cancer, compared to those with diets high in saturated fat. Experts in this field suggest that the combination of dietary fish oil and fermentable fiber may provide the basis for a food-based strategy for maintaining a healthy colon (Chapkin et al., 2008).

Colon cancer is a major killer triggered by mutations that upset the normal life and death cycle of colonic cells. It is not surprising that colon cancer cells have mutations that allow them to escape from undergoing programmed cellular death, a natural form of cellular suicide. Mutated genes are responsible for the first stage of colon cancer involving the formation of an uncontrolled growth of polyps (see Figure 2.2a and Figure 2.2b). A polyp can grow very large, meaning that many precancerous cells that have escaped suicide are being generated in large numbers. Polyps are

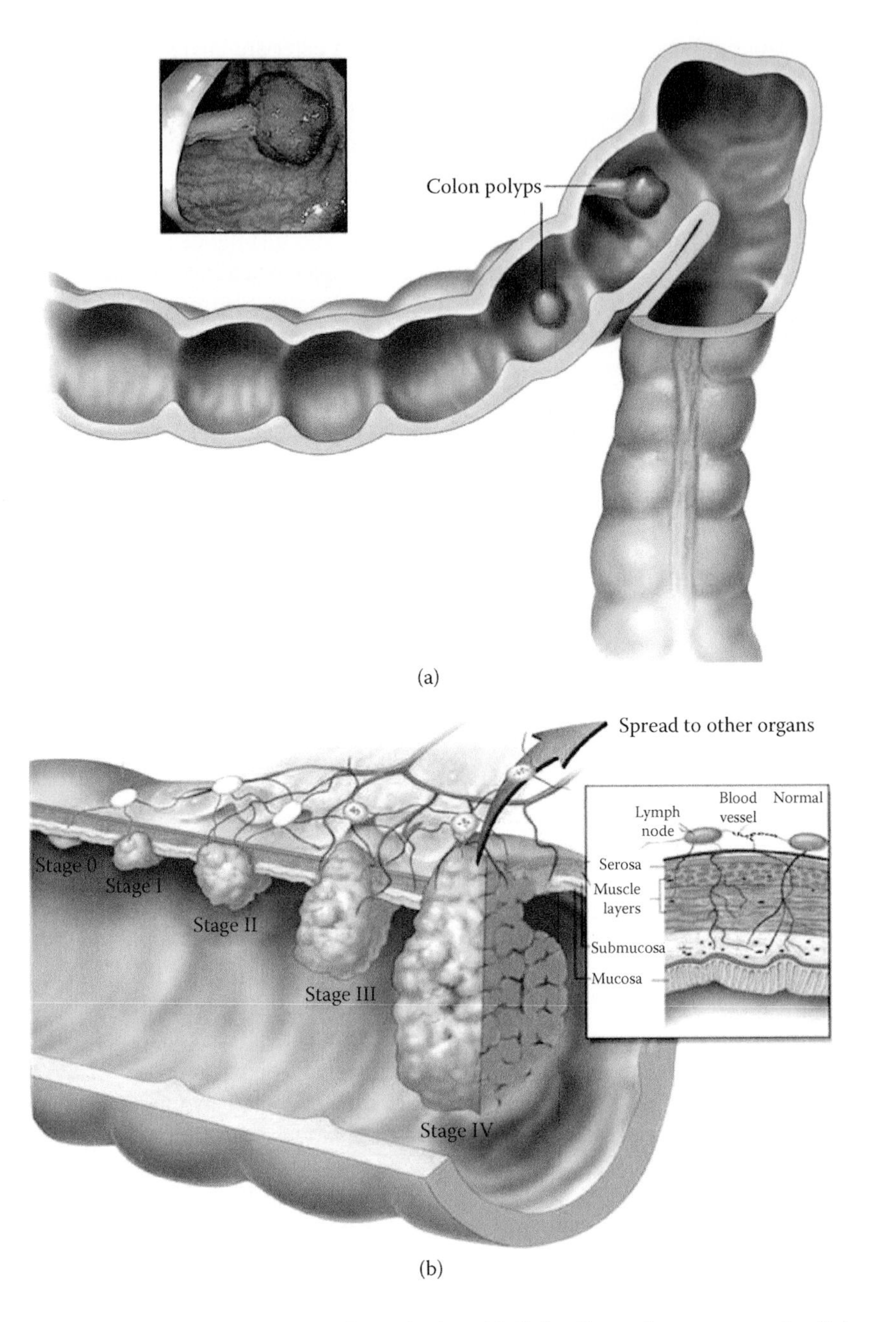

FIGURE 2.2 Rogue lines of senescing colonic epithelial cells evade programmed cellular death (apoptosis) and form polyps that can become cancerous. (a) and (b) It is estimated that moderate changes in diet including eating adequate fiber and marine oils can potentiate apoptosis of precancerous cell lines and help prevent a majority of colon cancer. This is a dramatic example of a crucial link between DHA in the diet and prevention of colon cancer. (Courtesy of the National Cancer Institute.)

not deadly by themselves but can become deadly as new mutations arise. Thus, new rounds of mutagenesis can occur in already precancerous cells, forming polyps. Even a second round of mutations may not be deadly, but a third might convert cells held fast in the polyp into a mobile form carried in the bloodstream to cause cancer in other vital organs such as the liver.

Colon cells forming the lining of the colon not only turn over rapidly but do so in a layered fashion with the outer layer of cells exposed to the colon environment being the oldest. As the youngest cells move with time from the innermost layer toward the outermost or oldest layer their fate is predestined to end in suicide. Because billions of cells are involved in this continuous cycle of birth and death, mutations blocking suicide can arise, leading to the formation of polyps.

Researchers studying familial adenomatous polyposis (FAP) have identified a possible candidate gene called APC proposed to increase the risk of colon cancer (Laken et al., 1999). People with FAP have a rare inherited disorder leading to thousands of polyps lining their colon. Because this condition frequently progresses to cancer, some patients diagnosed with FAP opt to have their colon removed to prevent colon cancer. People with FAP inherit a mutation in one of their two APC genes. In this case a single defective APC gene does not trigger FAP and almost certain colon cancer. Instead, when mutations strike the second still functioning APC gene FAP appears. In one sense these unfortunate FAP carriers represent a living colon laboratory that shows how readily mutations leading to cancer occur in this unusual colonic ecosystem.

How DHA helps assassinate senescing colonic cells has been studied in some detail. Immortalized human colonic, adenocarcinoma (HT-29) cells, as well as "normal colon cells," fed exogenous supplies of DHA have been found to target DHA to cardiolipin (CL) of their mitochondria (Kato et al., 2002; Ng et al., 2005; Watkins, Carter, and German, 1998; Watkins et al., 2001). The role of DHA-cardiolipin in aging and neurodegeneration is discussed in Chapter 11, and the chemical structure of cardiolipin is shown in Chapter 4. See Schlame (2008) for a comprehensive review of the structures, biosynthesis, and function of cardiolipin. DHA in cultured cells reaches levels as high as 48 percent of the total acyl chains in cardiolipin. Cardiolipin is present primarily in the inner mitochondrial membrane, and in tissues with high respiration rates, such as heart cells, CL can account for 25 percent of the phospholipids. The HT-29 cell line is nonspecific regarding the incorporation of unsaturated fatty acids into phospholipids in general, with values of 38, 18, 25, 13, or 11 mole percent for 18:1, 18:2, 20:4, 20:5, or 22:6 in cells fed with these fatty acids, respectively. Cardiolipin from fatty acid–enriched HT-29 cells contained 26, 41, 6, 9, or 48 mole percent for 18:3, 18:2, 20:4, 20:5, and 22:6, respectively. Thus, if HT-29 cells in culture are representative of colonic endothelial cells lining the colon, then one would expect to find selective incorporation of DHA along with linoleic acid from dietary sources into lipids of these cells.

Cardiolipin in heart cells is often highly enriched with linoleic acid (18:2) with CL species composed of four 18:2 chains being common. The presence of two DHA chains replacing two of the four 18:2 chains, as occurs in colonic cell cultures, generates one of the most unsaturated and chemically unstable phospholipid species found in humans—a total of 16 double bonds per CL molecule. Synthesis of such highly unsaturated CL structures is likely carried out by a combination of conventional fatty acid pathways in concert with active remodeling steps in which DHA is substituted for more saturated chains. It seems unlikely that dietary sources of DHA alone will drive such high levels of 22:6 incorporation into CL in the intact colon. However, feeding studies with whole animals confirm that DHA is targeted to CL, in agreement with studies of cells in culture.

Animals have evolved three different apoptotic mechanisms to kill unwanted cells, such as precancerous cells forming in the colon. Two of the three are mediated by oxidative stress, and this is likely where DHA enters the picture (Chicco and Sparagna, 2007; Gonzalvez and Gottlieb, 2007; McMillin and Dowhan, 2002; Nakagawa, 2004; Orrenius and Zhivotovsky, 2005). As summarized in Figure 2.3, DHA-cardiolipin is highlighted as a target of lipoxidation resulting in energy-oxidative stress (see Chapter 11). According to this model, incorporation of DHA into cardiolipin increases levels of oxidative stress as well as energy stress and induces apoptosis.

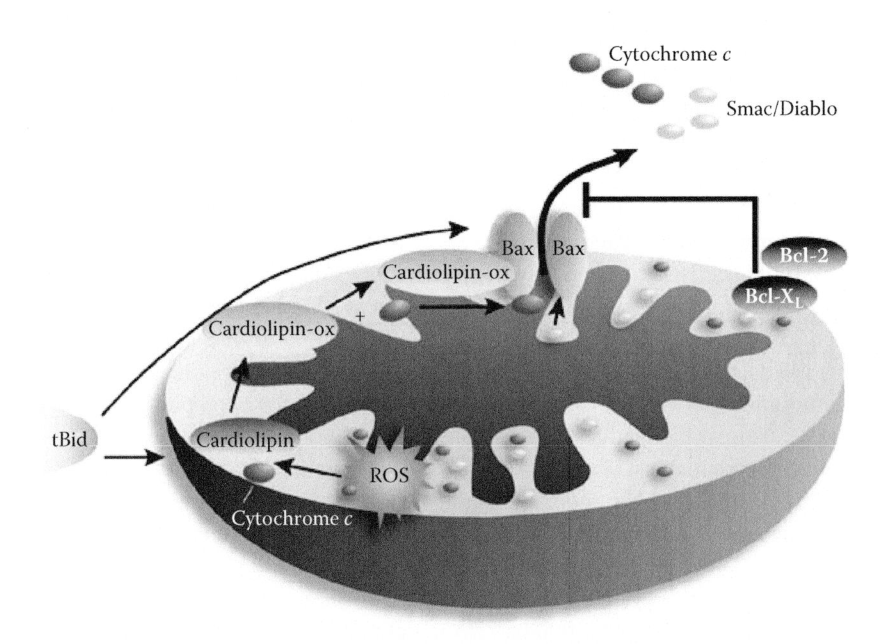

FIGURE 2.3 DHA may exert its anticancer effects following incorporation into mitochondrial cardiolipin of senescing colonic epithelial cells. DHA incorporated into cardiolipin of mitochondrial membranes is proposed to intensify apoptosis, effectively ridding the colon of potentially precancerous cells. (Cardiolipin ox, oxidized cardiolipin; tBid, 16 kDa truncated peptide of Bid, which is a pro-apoptotic effector protein. Bax is a pore-forming protein in the outer mitochondrial membrane. Bcl or Bclx are proteins that inhibit pore formation. Smac and Diablo, like cytochrome C, are pro-apoptotic proteins that are released into the cytoplasm through Bax.) (Reprinted by permission from Macmillan Publishers Ltd: *Nature Chem. Biol.*, Orrenius, S. and B. Zhivotovsky, Cardiolipin Oxidation Sets Cytochrome c free, 1:188–189, © 2005.) **(See color insert.)**

Thus, there is growing evidence that DHA incorporated directly from dietary sources alters mitochondrial membrane composition and function in colonocytes, thereby creating a permissive environment for apoptosis. This is proposed to widen the spectrum of potential precancerous colon epithelial cells targeted for apoptosis, and serves as a defense against colon cancer. Finally, it is ironic to note that one of the riskiest properties of DHA is proposed to prevent colon cancer in humans by selectively increasing the risk to senescing colon cells. The molecular basis of DHA, fiber, and other dietary constituents as anticancer molecules is being actively researched in several laboratories. These studies continue to shed light on the fundamental nature of cancer and highlight how dietary preferences can dramatically lower risks of colon cancer. It is also important to note that aging is a major risk factor in colon cancer. This fact helps unite the fields of aging, cancer, and dementia, a point which will be discussed again in Chapter 13.

2.5 RISKS OF DHA INCLUDING VISION IMPAIRMENT

One of the clearest examples yet concerning the potential harmful effect of DHA comes from data using transgenic mice, which overexpress DHA. It is well known that looking too long directly into the sun can cause photo-oxidative damage to the eye. It has been established that such strong light causes photochemical damage to photoreceptor cells (i.e., rods and cones) at the level of DHA membranes. Molecular events are detailed in Chapters 8 and 10. Recently, a critical test of the DHA principle was performed using mice. Transgenic mice were constructed by transfer and expression of an omega-3 gene from a tiny nematode.

The *fat*-1 gene cloned from *Caenorhabditis elegans* encodes an N-3 fatty acid desaturase that elevates DHA levels when expressed in transgenic mice (Tanito et al., 2009). Fatty acid profiles of various tissues including rod cell outer segments, cerebellum, plasma, and liver demonstrated higher levels of DHA in the transgenic mice compared to controls. When exposed to light stress the predicted higher levels of DHA-mediated lipoxidation in the retina were observed in mice expressing the *fat*-1 gene compared to controls. For example, levels of 4-hydroxyhexenol, a by-product of DHA oxidation, were increased. Also, the numbers of apoptotic photoreceptor cells were greater. Results of this important experiment show a positive correlation between the levels of DHA in the retina of a living mouse, the degree of DHA oxidation, and the vulnerability of the retina to photo-oxidative damage. These results highlight the dual roles played by DHA as an essential molecule in vision on the one hand, becoming harmful when levels are too high on the other. The laboratory of R. E. Anderson has pioneered the concept of the importance and rapid turnover of DHA membranes in rod cells, and this is another important contribution from his laboratory (also see Anderson and Penn, 2004).

In humans, DHA is essential for biochemical functions in membranes of neurons, rod cells, and sperm, where levels of this highly unsaturated chain are high. In contrast, many human cells contain low to modest levels (e.g., a few percent of total membrane phospholipids). For example, human red cell membranes contain 1.9 percent DHA in contrast to white blood cells with 0.6 percent DHA (Witte, Salazar, and Ballesteretal, 2010). Dietary supplementation with omega-3 resulted in a linear increase in red cells to about 10 percent DHA in contrast to white cells where an initial increase to about 3 percent DHA is observed with little further enrichment occurring even with a tripling in dietary levels of DHA. Furthermore, Witte and colleagues (2010) show that in human white cell membranes 18:0 (42 percent) and 18:1 (24 percent) are the major fatty acids; both fatty acids are resistant to degradation by reactive oxygen species (lipid peroxidation). The dominant fatty acid of cardiolipin of human lymphoblasts is 18:1 (greater than 50 percent) (Schlame, 2008). Because white cells as disease fighters routinely produce and handle reactive oxygen species (ROS) in their chemical warfare against pathogens, it makes sense that their membranes contain low levels of DHA, replaced by oxidatively stable phospholipids. Microglial cells in the brain, like white cells, generate ROS via an "oxidative burst," and there are data that show that membranes of these macrophages are resistant to peroxidation (De Smedt-Peyrusse, Sargueil, and Moranisetal, 2008; Omodeo-Salè, Basilico, and Folinietal, 1998). One interpretation of this data is that macrophages, due to their routine handling of ROS, have evolved membranes less likely to be chemically damaged by these harsh chemicals. According to the DHA principle, enrichment of membranes of microglia in the human brain with levels of DHA comparable to those found in axonal membranes and synaptosomes would be pathological. Thus, the absence of significant amounts of DHA in many classes of human cells seems to be beneficial.

In summary, the dual chemical modes of DHA have now been observed in mice and men. Studies of DHA-null mutants in mice highlight the beneficial roles of DHA, while studies of DHA transgenic mice overproducing DHA show the potential risks of these molecules. These data show that DHA offers both benefits and risks to a living mammal. In a brain-twister, DHA present in the diet acts as a cellular assassin of senescing colon cells whose continued presence might seed colon cancer in humans. In this way the risk of DHA to colon cells is a benefit to the organism. Finally, over 80 years ago Burr and Burr (1929) first defined the importance of "essential fatty acids" in the mammalian diet. Today human nutritionists armed with modern research tools are closing in on the molecular roles played by an important member of the omega-3 family of essential oils—DHA. The benefits and risks of DHA in neurons are highlighted in this book.

REFERENCES

Albert, C. M., H. Campos, and M. J. Stampferetal. 2002. Blood levels of long-chain n-3 fatty acids and the risk of sudden death. *N. Engl. J. Med.* 346:1113–1118.

Anderson, R. E., and J. S. Penn. 2004. Environmental light and heredity are associated with adaptive changes in retinal DHA levels that affect retinal function. *Lipids* 39:1121–1124.

Billman, G. E., Y. Nishijima, and A. E. Belevychetal. 2010. Effects of dietary omega-3 fatty acids on ventricular function in dogs with healed myocardial infarctions: *In vivo* and *in vitro* studies. *Am. J. Physiol. Heart Circ. Physiol.* 298:H1219–H1228.

Bourre, J. M. 2006. Effects of nutrients (in food) on the structure and function of the nervous system: Update on dietary requirements for brain. Part 2: Macronutrients. *J. Nutr. Health Aging*10:386–399.

Burr, G. O., and M. M. Burr. 1929. A new deficiency disease produced by the rigid exclusion of fat from the diet. *J. Biol. Chem.* 82:345–367.

Chapkin, R. S., J. Seo, D. N. McMurray et al. 2008. Mechanisms by which docosahexaenoic acid and related fatty acids reduce colon cancer risk and inflammatory disorders of the intestine. *Chem. Phys. Lipids* 153:14–23.

Chicco, A. J., and G. C. Sparagna. 2007. Role of cardiolipin alterations in mitochondrial dysfunction and disease. *Am. J. Physiol., Cell Physiol.* 292:33–44.

Connor, W. E., D. S. Lin, D. P. Wolfe et al. 1998. Uneven distribution of desmosterol and docosahexaenoic acid in the heads and tails of monkey sperm. *J. Lipid Res.* 39:1404–1411.

Connor, W. E., R. G. Weleber, C. DeFrancesco et al. 1997. Sperm abnormalities in retinitis pigmentosa. *Invest. Ophthalmol. Vis. Sci.* 38:2619–2628.

Daviglus, M. L., J. Stamler, and A. J. Orenciaetal. 1997. Fish consumption and the 30-year risk of fatal myocardial infarction. *N. Engl. J. Med.* 336:1046–1053.

De Smedt-Peyrusse, V., F. Sargueil, and A. Moranisetal. 2008. Docosahexaenoic acid prevents lipopolysaccharide-induced cytokine production in microglial cells by inhibiting lipopolysaccharide receptor presentation but not its membrane subdomain localization. *J. Neurochem.* 105:296–307.

Duda, M. K., K. M. O'Shea, and A. Tintinuetal. 2009. Fish oil, but not flaxseed oil, decreases inflammation and prevents pressure overload-induced cardiac dysfunction. *Cardiovasc. Res.* 81:319–327.

Gonzalvez, F., and E. Gottlieb. 2007. Cardiolipin: Setting the beat of apoptosis. *Apoptosis* 12:877–885.

Kato, T., R. L. Hancock, H. Mohammadpour et al. 2002. Influence of omega-3 fatty acids on the growth of human colon carcinoma in nude mice. *Cancer Lett.* 187:169–177.

Laken, S. J., N. Papadopoulos, G. M. Petersen et al. 1999. Analysis of masked mutations in familial adenomatous polyposis. *Proc. Natl. Acad. Sci. USA* 96:2322–2326.

McMillin, J. B., and W. Dowhan. 2002. Cardiolipin and apoptosis. *Biochim. Biophys. Acta.* 1585:97–107.

Metcalf, R. G., L. G. Cleland, and R. A. Gibsonetal. 2010. Relation between blood and atrial fatty acids in patients undergoing cardiac bypass surgery. *Am. J. Clin. Nutr.* 91:528–534.

Metcalf, R. G., M. J. James, and R. A. Gibsonetal. 2007. Effects of fish-oil supplementation on myocardial fatty acids in humans. *Am. J. Clin. Nutr.* 85:1222–1228.

Nakagawa, Y. 2004. Initiation of apoptotic signal by the peroxidation of cardiolipin of mitochondria. *Ann. NY Acad. Sci.* 1011:177–184.

Neuringer, M., W. E. Connor, C. VanPetten et al. 1984. Dietary omega-3 fatty acid deficiency and visual loss in infant rhesus monkeys. *J. Clin. Invest.* 73:272–276.

Ng, Y., R. Barhoumi, R. B. Tjalkens et al. 2005. The role of docosahexaenoic acid in mediating mitochondrial membrane lipid oxidation and apoptosis in colonocytes. *Carcinogenesis* 26:1914–1921.

Omodeo-Salè, F., N. Basilico, and M. Folinietal. 1998. Macrophage populations of different origins have distinct susceptibilities to lipid peroxidation induced by beta-haematin (malaria pigment). *FEBS Lett.* 433:215–218.

Orrenius, S., and B. Zhivotovsky. 2005. Cardiolipin oxidation sets cytochrome c free. *Nat. Chem. Biol.* 1:188–189.

O'Shea, K. M., D. J. Chess, R. J. Khairallah et al. 2010. ω-3 Polyunsaturated fatty acids prevent pressure overload-induced ventricular dilation and decrease in mitochondrial enzymes despite no change in adiponectin. *Lipids Health. Dis.* 9:95.

Roqueta-Rivera, M., T. L. Abbott, M. Sivaguru et al. 2011. Deficiency in the omega-3 fatty acid pathway results in failure of acrosome biogenesis in mice. *Biol. Reprod.* (doi: 10.1095/biolreprod. 110.089524).

Roqueta-Rivera, M., C. K. Stroud, W. M. Haschek et al. 2010. Docosahexaenoic acid supplementation fully restores fertility and spermatogenesis in male delta-6 desaturase-null mice. *J. Lipid Res.* 51:360–367.

Schlame, M. 2008. Cardiolipin synthesis for the assembly of bacterial and mitochondrial membranes. *J. Lipid Res.* 49:1607–1620.

Stroud, C. K., T. Y. Nara, M. Roqueta-Rivera et al. 2009. Disruption of FADS2 gene in mice impairs male reproduction and causes dermal and intestinal ulceration. *J. Lipid Res.* 50:1870–1880.

Tanito, M., R. S. Brush, M. H. Elliott et al. 2009. High levels of retinal membrane docosahexaenoic acid increase susceptibility to stress-induced degeneration. *J. Lipid Res.* 50:807–819.

Valentine, R. C., and D. L. Valentine. 2009. *Omega-3 Fatty Acids and the DHA Principle.* Boca Raton, FL: Taylor & Francis.

Watkins, S. M., L. C. Carter, and J. B. German. 1998. Docosahexaenoic acid accumulates in cardiolipin and enhances HT-29 and cell oxidant production. *J. Lipid Res.* 39:1583–1588.

Watkins, S. M., T. Y. Lin, R. M. Davis et al. 2001. Unique phospholipid metabolism in mouse heart in response to dietary docosahexaenoic or alpha–linolenic acids. *Lipids* 36:247–254.

Williard, D. E., J. O. Nwankwo, T. L. Kaduce et al. 2001. Identification of a fatty acid delta-6-desaturase deficiency in human skin fibroblasts. *J. Lipid Res.* 42:501–508.

Witte, T. R., A. J. Salazar, and O. F. Ballesteretal. 2010. RBC and WBC fatty acid composition following consumption of an omega 3 supplement: Lessons for future clinical trials. *Lipids Health Dis.* 9:31.

3 Use It or Lose It Concept of Brain Health Is Linked to DHA

The "use it or lose it" rule for maintaining a healthy brain is perhaps the most highly popularized and widely followed practical application to come from the field of neurobiology. Virtually everyone intentionally or unintentionally practices some form of brain exercise, which is believed to help maintain a healthy brain—which is especially important with aging. Brain exercise, which is as simple as handwriting, is believed to stimulate the development of brain circuitry, though there are still deep mysteries concerning the molecular mechanisms involved. DHA plays an important role as a building block and in the dynamics of axon growth and development in both infants and adults. Thus, DHA seems to be fundamentally linked to brain health, plasticity, and longevity.

In one of the most interesting paradoxes in human development, the brain begins with a great excess of neurons that must be pruned to make way for growth of the adult brain (Buss, Sun, and Oppenheim, 2006; Williams and Herrup, 1988). Mature neurons can make thousands of connections to other neurons, and the brain cavity ultimately must provide space for more than 100,000 miles of myelinated transmission lines. Failure of this natural pruning process called *programmed neuron death* would create a monster organ that would not fit into the brain cavity. Even if a doubling of head size were possible, feeding such a mega-size brain would be expected to consume an insupportable amount of the energy food and oxygen carried in the bloodstream. It is estimated that early in life up to 100 billion excess neurons must be pruned away. At about age 10 to 11 years the excess in abundance of connections is pruned away, and this process lasts into the mid-twenties. New axon circuitry is laid down throughout life.

3.1 USE IT OR LOSE IT MECHANISM OF BRAIN DEVELOPMENT IS A MARVEL OF NATURE

In the 1960s, David Hubel and Torsten Wiesel discovered that visual experience actually directs the wiring of the visual brain (Wiesel, 1982). These neuroscientists won the Nobel Prize for their research. By suturing one eye of a kitten shut at birth and thus depriving this eye of any visual stimulation, these researchers showed that neural circuitry became robust and dominant in the open eye while remaining immature in the closed eye. This showed the powerful role played by the environment on visual experience in the wiring of essential regions of the brain. In essence, this experiment stands out as being the science behind the now widely accepted "use it or lose it" concept of infant brain development and adult brain health. Interestingly, a kitten exposed to a period of visual experience in both eyes after birth and then having one sutured shut did not show the dramatic effects on brain circuitry compared to a kitten whose eye was sewn shut at birth. Thus, timing is an important variable in the development of visual circuitry in cats. The lack of development of visual circuitry on the left side of the brain (note that neurons supporting vision of the right eye are located in the left side of the brain) was dramatic, surprising the future Nobel Prize winners who conducted this experiment.

Neuroscientists studying visual development during the past four decades have expanded on this Nobel Prize–winning research. For example, following visual deprivation in kittens, rewiring of circuitry in the upper layers of the visual cortex occurs rapidly within two or fewer days (Trachtenberg and Stryker, 2001). The albino tadpole is used as an important research tool allowing neurobiologists to visualize synaptic plasticity of individual optic tectal interneurons (Figure 3.1) (Wu and Cline,

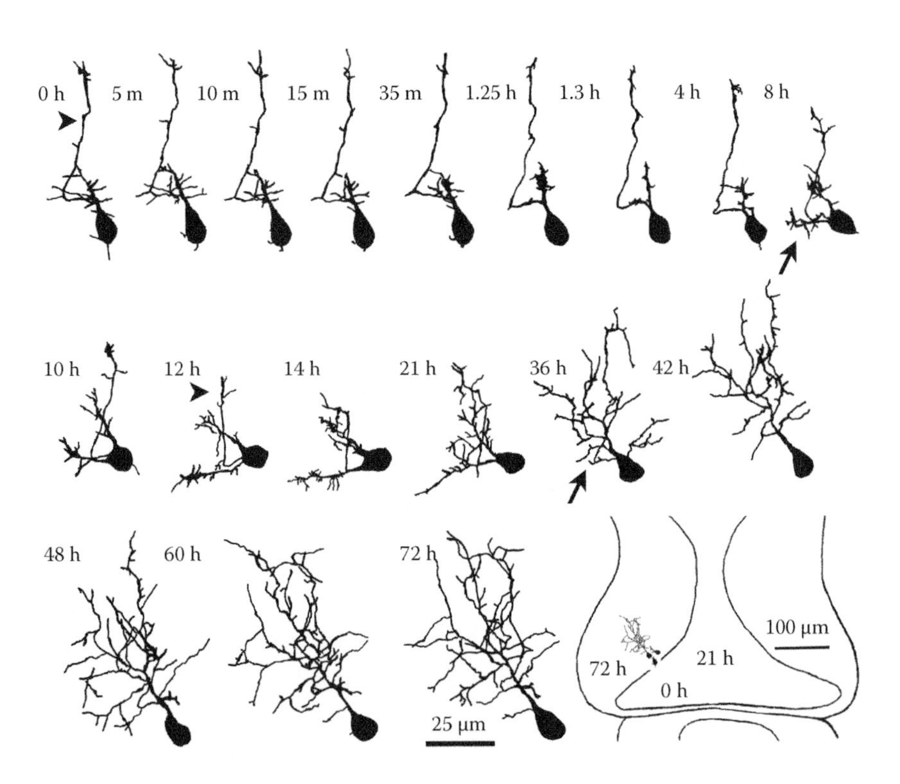

FIGURE 3.1 Time-lapse images of synaptic plasticity in tadpoles (*Xenopus laevis*). A dendritic process (arrow) first seen after 8 hours retracts within 36 hours. Also note the explosive growth and development of new neuron circuitry, which consumes DHA as membrane building blocks. (Reprinted with permission from Wu and Cline, Time-Lapse *in vivo* Imaging of the Morphological Development of Xenopus Optic Tectal Interneurons, *J. Comp. Neurol.* Copyright 2003. John Wiley & Sons.)

2003). The location of the optic neuron being analyzed is shown at the lower right of Figure 3.1. To generate this series of drawings an *in vivo* time-lapse imaging technique was used following intact individual neurons in anesthetized tadpoles. The drawings, which show the development of a dendrite arbor structure over a 72-hour time period, were derived from these images. Short-interval time-lapse images reveal that tectal interneuron arbors have rapid rates of branch additions and retractions. Many interneurons extend transient processes so that neuronal structure is dramatically different from one day to the next. These studies show that optic tectal circuits are extremely plastic during early development with dendrites forming the majority of synaptic contacts. Further refinement of these techniques has led to a deeper understanding of molecular events occurring during synaptic plasticity (Shen et al., 2009).

The pattern of pruning of neural circuits seen in monkeys, cats, mice, and frogs is consistent with the popular rule of thumb of "use it or lose it" as applied to the developing human brain. Thus, utilizing certain neural connections in the adolescent brain through everyday experiences strengthens some connections while eliminating others. In the decades since Hubel and Wiesel carried out their experiment, the field has moved to the molecular level where numerous enabling molecules and biochemical processes have been implicated in the cascades responsible for visual rewiring (Figure 3.2) (Tropea, VanWart, and Sur, 2009). This sculpting, governed in part by environmental signals, is believed to size and create a more efficient and adaptable brain. This means that experience and interaction with the world around us can shape our personalities. In essence, this adaptability is at the heart of lifelong learning and includes dynamic processes involving axon/synapse growth and repair, which require DHA. Pruning of excess axons recycles DHA. The nature of

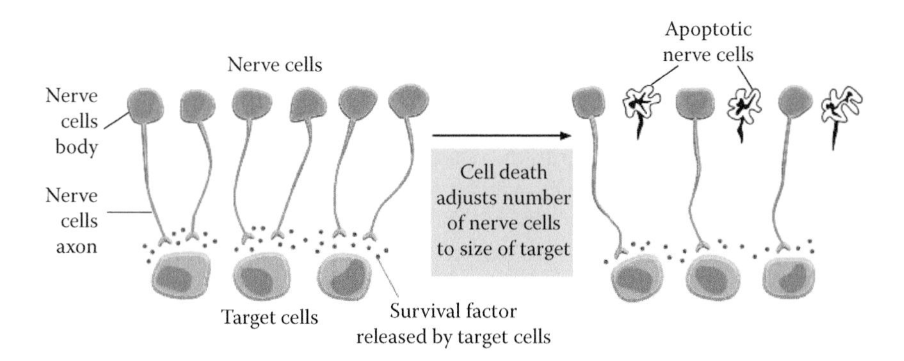

FIGURE 3.2 Billions of excess neurons are pruned away during brain development in children by the process of apoptosis (programmed cellular death). Target cells including neurons are shown releasing limiting amounts of survival factors that act as inhibitors of apoptosis of neurons. In the brain this means that there are not enough survival factors to go around causing many unused neurons to be destroyed. This culling process ensures that all target cells are contacted by nerve cells. (Alberts, B. et al., *Molecular Biology of the Cell*, Copyright 2008. Reproduced by permission of Taylor & Francis Group, LLC, a division of Informa plc.)

the pruning process active during childhood is of great interest especially in view of new data that show neuron pruning may be both beneficial and harmful.

The importance of neuron pruning during childhood is reinforced by recent data showing a linkage between defective pruning and epilepsy in a mouse model (Chu et al., 2010). Genes that govern the pruning mechanism have been identified in mice. A knockout of one critical gene (C1q) blocked synaptic pruning causing epilepsy. C1q encodes a protein ordinarily associated with immune and inflammatory processes. C1q is a gene involved in the classic complement cascade and is known to be involved in nonimmune roles in synaptic elimination in the retinogeniculate pathway during development (Stevens et al., 2007). Activation of C1q protein by astrocytes leads to "tagging" of synapses that are likely eliminated by microglia. Lateral geniculate relay neurons in C1q knockout mice deficient in these proteins retain multiple synaptic contacts from retinal ganglion cells that are normally eliminated during development (Stevens et al., 2007). Although C1q is developmentally downregulated, it is reactivated, along with other genes in the complement cascade after brain trauma, where it takes part in a sustained inflammatory response to injury. C1q has also been linked to Alzheimer's disease (D'Andrea, 2005) and glaucoma (Stevens et al., 2007). Chu et al. (2010) conclude that activation of C1q might be beneficial in balancing excessive synaptic sprouting for retarding epileptogenesis, or harmful in eliminating newly formed synapses, which can adversely affect recovery from injury. This mechanism might be important in brain trauma–induced dementia discussed in Chapter 16.

3.2 HOW MANY NEURONS TO MAKE A BIRDSONG?

Humans are not the only animals to employ a use it or lose it strategy of neuron development. When male white-fronted sparrows migrate to Alaska in the spring to breed, thousands of new neurons develop in a region of their brain devoted to songs that are important in attracting a mate. This same cluster of neurons withers and dies following the breeding season and the fall migration to California. Thus, up to 20,000 neurons that control learned song behavior in males expand and shrink seasonally. North-migrating birds encounter extensive daylight, characteristic of Alaskan spring and early summer, which prompts the release of hormones controlling the growth of their song neurons. Then a built-in neural suicide program is activated at the end of the mating season. There are numerous ways to rationalize the benefits of this neuron death and rebirth cycle. Recently, researchers have been able to duplicate programmed cellular death and the birth of song-related neurons in the laboratory using hormone injections (Thompson and Brenowitz, 2009). Striking results are obtained by these methods such as up- and down-modulation over the complex processes governing programmed cellular death of neurons and, more importantly for this discussion, the prevention or prolongation of neuron death. These researchers (Thompson and Brenowitz, 2009) consider this songbird as a suitable model that may mimic an age-related drop in hormone levels accompanied by the death of neurons in human neurodegenerative diseases such as Alzheimer's.

3.3 LIFELONG LEARNING REQUIRES SYNAPTIC PLASTICITY AND SELECTIVE AXON GROWTH

Recall that neurons belong to a rare class of cells in our body that have evolved to last a lifetime—human neurons seldom divide. Whereas exciting studies in recent years show that a small percentage of human neurons do divide, the general rule that neurons behave as permanent cells still holds true. However, there is a major caveat to this rule that becomes central to this discussion. Yes, the neuron head containing the nucleus does not divide, but this statement does not apply to axons and synapses. By their very nature axons must be continuously repaired (see Chapters 8 through 10), and when they die or become weakened, they can be replenished.

The concept of synaptic plasticity was first proposed as a mechanism for lifelong learning and memory on the basis of theoretical research by Hebb (1949). The plasticity or Hebbian rule postulates that each time a nerve impulse passes across a synapse the connection between the two neurons is potentiated or reinforced. Experimentalists who study the biochemical basis of learning use the name *long-term potentiation* to describe this process. Thus, according to the use it or lose it rule, neuron pruning in infancy is triggered by the lack of long-term potentiation. By the same token, often-used connections are reinforced and can eventually become immortalized as very long-term memories.

The removal of unnecessary neurons has already been discussed in the context of sculpting the brain for developmental purposes. Synaptic or axonal pruning is believed to be a lifelong process that revolves around several benefits as follows:

- Removal of weak connections
- Removal of damaged connections
- Removal of unnecessary connections

The removal of damaged connections is likely important for neuron health and plasticity. The same can be said for the removal of weak or unnecessary connections.

Much of what is known about axon pruning comes from studies of vertebrate muscular junctions, a convenient research tool allowing direct observation (Lichtman and Colman, 2000). During development motor neurons first extend axons to a far greater number of their target muscle fibers than seen in adults. Multiple axons must compete for a foothold. During the early postnatal period all but one of the input connections are eliminated. During this process the neuron head survives in contrast to many of its axons, which are pruned away. Thus, one connection is strengthened while many others are restricted.

Many of the molecular steps in axon pruning remain unclear, but recent studies indicate that retracting axon branches shed fragments rich in normal synaptic organelles whose membranes we will see in Section II are enriched in DHA (Bishop, Misgeld, and Walshetal, 2004). These authors propose that spent fragments of retracting axons, *axosomes*, are engulfed by adjacent Schwann cells and may be assimilated into the glial cytoplasm. Shedding of axosomes and glial engulfment may represent a widespread mechanism of synapse elimination and is likely an important process for DHA cycling in the brain.

Whereas these studies address axon retraction in motor neurons outside of the brain, they provide an important foundation for understanding events in the dynamics of brain rewiring. Signaling between senescing axons and waiting phagocytic cells in the brain is discussed in Chapters 8 and 17. The role of DHA as a possible risk factor for brain cancer in children is discussed in Chapter 13. For a recent review of axon retraction and degeneration see Luo and O'Leary (2005).

3.4 MEMBRANE EXOCYTOSIS/ENDOCYTOSIS AS A MEANS TO MODULATE SYNAPTIC STRENGTH

The intensity of synaptic transmissions can be regulated by the addition or removal of more or fewer receptors for neurotransmitters on the postsynaptic side of the synapse. A process called *exocytosis* lowers the number of receptors, thinning their density, while *endocytosis* delivers more receptors. Having more receptors strengthens the synapse, and vice versa. Delivery or removal of receptors involves lipid vesicles that contain unsaturated fatty acids including DHA, which are recognized as modulators of endocytosis/exocytosis. The role of DHA in determining rates of cycling of synaptic vesicles on the presynaptic membrane surface is discussed in Chapters 4 and 5. These data provide a molecular mechanism to explain how DHA modulates rates of exocytosis and endocytosis and likely has implications for long-term potentiation.

3.5 DHA AS A MOMENTARY MEMORY MOLECULE?

In the early days of the development of the theory of quantum mechanics in physics, a controversy emerged regarding whether an electron should be considered a particle or a wave. This led to some absurd notions such as treating a battleship as a wave. Experimental physicists of this period were often stymied because any technique they used to measure the properties of electrons seemed to end up altering the properties of the electron that they were trying to measure. We seldom think about phenomena of this sort in biology, but it now seems that it might be fruitful to treat DHA in this manner. That is, the ordinary environmental signals surrounding a DHA molecule in say a rod cell or neuron membrane—light energy, thermal energy, hydrostatic pressure, salinity, ionic milieu, pH, neighboring chains—seem to change the conformations or shapes of individual DHA chains (see Chapter 5). It is as though DHA was selected through evolution to practice this sort of "molecular contortion" mode of action in the specialized human membranes where it is present. Physical chemists have proposed, based on its sensitivity to change in the world around it, that DHA chains might play a role as a sort of instantaneously responsive memory molecule working on a picosecond to nanosecond time scale and changing its shape in response to various environmental signals. Of course the major "signal" we have in mind here is the passage of an electrical impulse. An electric impulse acts directly to modulate activity of ionic gating proteins and also likely generates changes in the shapes of DHA chains as the impulse passes. Thus, a fast-acting cascade can be imagined triggered by electrically induced changes in DHA phospholipids. Conformational changes in DHA take place so fast that they might occur at the very leading edge of a neural impulse and decay as the impulse passes, perhaps chemically communicating that an electrical impulse has passed to the next member of some sort of a fast response cascade. There is no doubt that electrically activated or stretch-gated membrane proteins quickly respond to the passage of an impulse, and there is good reason to believe that the immediate lipid structure is also altered during passage of an electrical signal. Thus, DHA seems to have the biochemical conformational power to store information momentarily before passing it on to trigger some sort of fast-paced cascade. The biological significance of such rapid lipid conformational adaptations on synaptic plasticity or any other biologically relevant process remains unclear but is worth considering. Momentary electrochemical perturbation of membrane asymmetry (Chapters 5 and 17) is one possible mechanism.

In summary, learning begins in the womb and continues to death. The time scale for mechanisms of learning might span nanoseconds to a century. It is likely that the act of writing this sentence sends a signal to strengthen some neuron circuits. Studies on the molecular roles played by DHA in learning are in their infancy, and the main goal of this chapter is to raise awareness that existing data on the conformational dynamics and biochemistry of DHA suggest potential roles in processes central to axon/synaptic dynamics and learning.

REFERENCES

Bishop, D. L., T. Misgeld, and M. K. Walshetal. 2004. Axon branch removal at developing synapses by axosome shedding. *Neuron* 44:651–661.

Buss, R. R., W. Sun, and R. W. Oppenheim. 2006. Adaptive roles of programmed cell death during nervous system development. *Annu. Rev. Neurosci.* 29:1–35.

Chu, Y., X. Jin, I. Parada et al. 2010. Enhanced synaptic connectivity and epilepsy in C1q knockout mice. *Proc. Natl. Acad. Sci. USA* 107:7975–7980.

D'Andrea, M. R. 2005. Evidence that immunoglobulin-positive neurons in Alzheimer's disease are dying via the classical antibody-dependent complement pathway. *Am. J. Alzheimers Dis. Other Demen.* 20:144–150.

Hebb, D. O. 1949. *The Organization of Behavior: A Neuropsychological Theory*. New York: Wiley.

Lichtman, J. W., and H. Colman. 2000. Synapse elimination and indelible memory. *Neuron* 25:269–278.

Luo, L., and D. D. O'Leary. 2005. Axon retraction and degeneration in development and disease. *Annu. Rev. Neurosci.* 28:127–156.

Shen, W., J. S. DaSilva, H. He et al. 2009. Type A GABA-receptor-dependent synaptic transmission sculpts dendritic arbor structure in Xenopus tadpoles *in vivo*. *J. Neurosci.* 29:5032–5043.

Stevens, B., N. J. Allen, L. E. Vazquez et al. 2007. The classical complement cascade mediates CNS synapse elimination. *Cell* 131:1164–1178.

Thompson, C. K., and E. A. Brenowitz. 2009. Neurogenesis in an adult avian song nucleus is reduced by decreasing caspase-mediated apoptosis. *J. Neurosci.* 29:4586–4591.

Trachtenberg, J. T., and M. P. Stryker. 2001. Rapid anatomical plasticity of horizontal connections in the developing visual cortex. *J. Neurosci.* 21:3476–3482.

Tropea, D., A. VanWart, and M. Sur. 2009. Molecular mechanisms of experience-dependent plasticity in visual cortex. *Philos. Trans. R. Soc. Lond. B Biol. Sci.* 364:341–355.

Wiesel, T. N. 1982. Postnatal development of the visual cortex and the influence of environment. *Nature* 299:583–591.
Williams, R. W., and Herrup, K. 1988. The control of neuron number. *Annu. Rev. Neurosci.* 11:423–453.
Wu, G. Y., and H. T. Cline. 2003. Time-lapse *in vivo* imaging of the morphological development of Xenopus optic tectal interneurons. *J. Comp. Neurol.* 459:392–406.

4 Evolution and DHA
Redefining the DHA Principle

Brain signals are electrical signals or impulses moving from neuron to neuron throughout our elaborate web of brain circuitry. However, the passage of electrical impulses along neuronal wires in the brain is very different from electricity passing through a copper wire such as used in our houses. Perhaps the greatest difference is that "wires" in the brain and nervous system are surrounded inside and out by a watery milieu, a condition that would instantly short-circuit the passage of electricity along a copper wire. Thus, a radically different electrical design is needed by neurons. Evolution has honed a remarkable oil-based "wire," the axon, enabling electrical flow from neuron to neuron. However, electrical transmission as occurs in our brain is faced with a huge handicap. Speed of electrical flow through neural connections is dramatically slower compared to electrical transmission lines. This slow pace of bioelectrical transmission has never been fully overcome (Fox, 2011), although numerous evolutionary adaptations have occurred toward maximizing brain speed. In other words, brain speed seems to have maxed out in humans. Recall from Chapter 1 that oils or lipids are the predominant structural molecules in our neural wires. It is now clear that Mother Nature tinkered with every conceivable class of biological raw material, especially fatty acids, as building blocks toward maximizing brain speed along axon/synaptic circuits (Crawford et al., 1999; Cunnane and Crawford, 2003; Cunnane and Stewart, 2010). It is also clear that our survival as a species is tied in no small way to Mother Nature's selection of DHA for optimizing speed of our neurons. But speed is only one of several key biochemical parameters needed in neuron membranes. Energy is often in short supply because of its prodigious consumption in neurons. Consequently, neurons are under great selective pressure not only to produce as much energy as possible but also to build transmission lines that conserve energy.

Scientists have estimated that the speed of neural signals or impulses in the human brain travel about 2,000,000 times slower compared to electronic computers. Also, electronic computers are believed to be some 10,000,000 times faster in terms of signals per second. However, the human brain overcomes these enormous handicaps in numbers of neurons (i.e., 100 billion) and synaptic interconnections (i.e., 60 to 240 trillion). Nevertheless, we believe that Darwinian selection for maximizing the speed of neurons was one of the great evolutionary hurdles shaping the ultimate fate of the human brain. That is, without DHA our amazing biological computer would likely be much slower and unable to compute at rates fast enough to keep up with vision and other neurosensory processes essential for survival of humans. For a molecular description of how DHA maximizes speed and efficiency of neurons see Chapters 5 through 7.

4.1 DHA MAXIMIZES ENERGY EFFICIENCY IN HUMAN NEURONS

The importance of energy efficiency in neurons is reviewed by Niven and Laughlin (2008). We consider their assessment as the standard in describing how a series of Darwinian adaptations eventually led to the evolution of highly energy efficient human neurons. These authors clearly define brain energy efficiency and its impact on human evolution. The following key points from their review set the stage for this discussion:

- Darwinian selection for brain energy efficiency was a cardinal force in the evolution of humans.
- The enormous cost of brainpower is well spent based on a cost-benefit analysis for human survival.
- A reductionist approach led Niven and Laughlin (2008) to analyze individual sensory systems such as vision (Niven, Anderson, and Laughlin, 2007) that confirm the high-energy cost of operating our brain.
- Having convincingly shown the high-energy cost of brain power the authors take us step by step through a series of evolutionary adaptations that improve brain speed and energy efficiency.
- Following this trail of adaptations the authors focus on neurons as being the focal point for brain energy efficiency.
- The authors end by discussing how through evolution adaptations of neuron structure, including shorter axons and axon myelination, act as mechanisms to gain speed and energy efficiency.

In a *Scientific American* article entitled "The Limits of Intelligence," Fox (2011) concludes that the human brain is likely the most advanced among mammals. This author further proposes that due to biophysical limitations on axons, further gains in speed and energy efficiency of human neurons face serious negative trade-offs. For example, thinning the diameter of axons is a mechanism to increase packing density and increase energy efficiency but comes at the expense of futile or unintended neuron signals. Thus, axons with diameters in the range of 150 to 200 nanometers become impossibly noisy (Faisal, White, and Laughlin, 2005; Niven, Anderson, and Laughlin, 2007). Physically thin axon membranes are embedded with far fewer ion gates compared to thicker axons, and the paucity of gates increases the accidental opening of a channel, which can trigger a chain reaction of neighboring gates firing the neuron unintentionally. Thicker axons avoid such spurious signals but consume more energy than thin axons. The speed of impulse transmission is faster in thick axons but is overshadowed by the much higher energy cost. This physical analysis of the limitations of neuron function mirrors conclusions drawn from biochemical analysis as presented in this book. The speed of neuron signals versus energy efficiency in humans seems to have become maximized, and the result is a cauliflower-sized organ (i.e., about 3 pounds) that consumes about 20 to 25 percent of our food intake. Brain energy consumption may be as high as 65 percent total food in infants. Clearly, significantly increasing the brain size of humans would hijack energy supply. Thus, energy can be considered the mother of limitations in neuron electrophysiology (Fox, 2011).

4.2 DHA LIKELY CAME FROM THE SEA

The evolutionary history of oil composition of neurons is traced here with emphasis on how, when, and why DHA came to be essential for efficient human brain function. Some important milestones of evolution of membrane lipids terminating with DHA are summarized in Figure 4.1. This drawing lays out a rough time scale for when DHA might have first appeared on Earth and how long it took to spread to neurons. It is clear that neurons belong exclusively to the world of animals; but DHA seems a better evolutionary fit with membranes of tiny marine plants called *phytoplankton*. Today most of the global primary production of DHA occurs in marine, eukaryotic, photosynthetic plants. We hypothesize that DHA first evolved perhaps one billion years ago for the purpose of maximizing light harvesting by marine plants living in a cold oceanic world (Valentine and Valentine, 2009). These plants provided enough DHA for all creatures in the marine food chain, making DHA available as building blocks for neurons in the first multicellular marine animals and eventually vertebrates. Note that even the tiny worm *Caenorhabditis elegans* needs eicosapentaenoic acid (EPA), a close relative of DHA, for neuron efficiency (see Chapter 5).

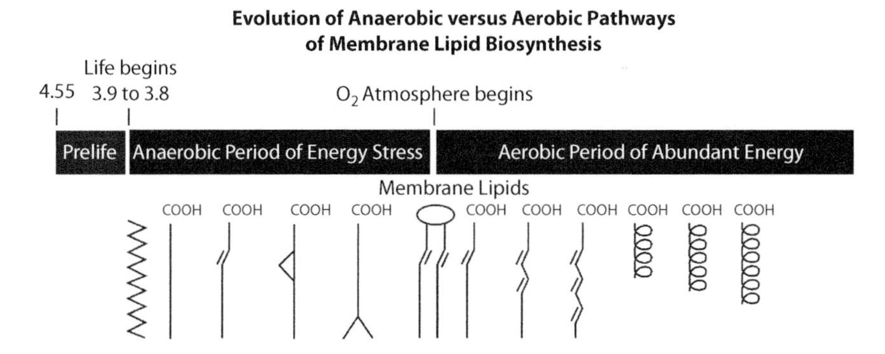

FIGURE 4.1 Evolution of DHA as a building block for human neurons: geologic timeline. The rise of oxygen (O_2) in the biosphere is suggested to have led to an explosion of polyunsaturated fatty acid conformations especially important for enhancing energy transduction in plants and animals. DHA, expected to have the wildest conformational dynamics among fatty acids in nature, is depicted at the far right of the diagram. Initially the favorable conformational dynamics of DHA were likely harnessed for maximizing photosynthesis in marine phytoplankton living in cold ocean water. Today massive amounts of DHA are produced by phytoplankton, which transfer this essential fatty acid throughout the food web, including to humans for brain development. (From Valentine, R., and Valentine, D., *Omega-3 Fatty Acids and the DHA Principle*. Copyright 2009. Reproduced by permission of Taylor & Francis Group, LLC, a division of Informa plc.)

DHA was likely plentiful in the marine world when mammals began to inhabit the earth. This is an argument in favor of the oceanic world as a proving ground for building blocks needed to construct the most efficient neurons. Several ocean-land cycles of animals have occurred during the age of neurons. The terrestrial environment especially in the interior regions of continents does not contain abundant DHA, because DHA in the presence of warm temperatures, direct sunlight, and ambient oxygen seems a lethal combination and is strictly avoided by land plants (Valentine and Valentine, 2009). We propose that animals leaving the sea and its rich source of DHA were faced with adapting their neurons with oils at hand (predominantly plant oils with far less unsaturation than found in DHA) that were abundant in the terrestrial world. Taking an evolutionary perspective, we note several adaptations made by land plants that allowed them to move away from a DHA-based photosynthesis typical of marine plants. Land plants avoid DHA/EPA altogether, replacing these highly unsaturated chains in their light-harvesting membranes with polyunsaturated chains featuring a maximum of two to three double bonds. The main point is to recognize that animals leaving the sea and its rich source of DHA could still build relatively fast and efficient neurons using more stable, unsaturated fatty acids of land plants by making a few minor biochemical adjustments to their phospholipid structure. Perhaps the most important adaptation used today by many animals including terrestrial insects involves a biochemical trick of producing membrane building blocks (i.e., phospholipids) with two polyunsaturated chains rather than one typically found in animal membranes. For generations biochemistry texts have propagated the myth that a typical phospholipid in nature has one unsaturated chain paired with a saturated chain. This definition generally fits animal membranes including those of most human cells but fails to address phospholipid structures essential for neurosensory perception. Also, from a global ecological perspective, nothing could be farther from reality, and we now know that a "typical" phospholipid in nature has two unsaturated chains per molecule, not just one. Because of the massive levels of di-unsaturated molecular species operating in plants, animals, and even specialized membranes of humans, the total quantity in the biosphere of these di-unsaturated molecular species of phospholipids far outweighs structures with only one unsaturated chain. In Chapter 5 we explain from the standpoint of enabling extreme membrane motion (i.e., rapid rotational and lateral movement of membrane components) why two unsaturated tails are better than one. Note that fatty acids with fewer double bonds compared to DHA offer their own advantages because they are more stable against photo-oxidation. This greater stability is likely of critical importance for photosynthesis in land plants.

The following events help explain how DHA evolved as the signature omega-3 fatty acid in human neurons: DHA first became abundant in marine plants to turbocharge their light harvesting membranes in cold seawater → marine zooplankton including krill graze on DHA-enriched algae to build efficient, cold-adapted neurons → sea to land migration of animals → more saturated, fatty acids available from land plants result in less efficient neurons in land animals → evolution of a biochemical pathway for synthesis of doubly polyunsaturated molecular species of neuron phospholipids (e.g., di-18:2) partially overcomes the lack of DHA in land plants used as food by animals → → → our human ancestors harvest DHA from seafood, evolve a faster, more efficient, large brain → humans maximize benefits of DHA by selective targeting to specialized membranes of neurons (also sperm tails and rods/cones of the eye) → mother's milk contains sufficient DHA to jump start rapid brain development in infants → DHA maximizes neuron speed often at the expense of brain span.

4.3 DARWINIAN SELECTION OF THE FITTEST OILS FOR NEURONS OF DIFFERENT ANIMALS

Neurons of many animals including mammals, birds, reptiles, and aquatic insects are enriched with DHA or EPA. The nematode or roundworm *C. elegans* uses EPA instead of DHA but is able to incorporate DHA when present in the diet. However, there are major classes of animals such as the vast world of terrestrial insects and even nematode relatives of *C. elegans* that lack or contain only traces of DHA/EPA in membranes of their brains and nervous systems, instead using more saturated fatty acid chains as membrane building blocks.

Evolution of the fatty acid composition in neurons of insects is an especially interesting topic. Chemical, biochemical, physiological, and ecological studies show that EPA ($20:5_{n3}$) and arachidonic acid (ARA) ($20:4_{n6}$) play at least two essential roles in insects. ARA and EPA act as precursors for eicosanoid hormones essential in insects (Stanley, 2000). This book contains a windfall of information providing clues about a second essential role played by this class of highly unsaturated fatty acids (HUFAs). We define HUFAs here as fatty acids with four to six double bonds. Some clues found in the literature on insects and relevant to the evolution of HUFAs in membranes of neurons are as follows:

- Membranes of most terrestrial insects contain only small amounts or traces of ARA, sufficient for hormone production but not for bulk synthesis of phospholipids, with often undetectable or barely detectable levels of EPA.
- Aquatic insects, especially species adapted to cold water, generally enrich their membranes with significant levels of ARA and EPA.
- DHA is generally missing from insect membranes.
- Di-polyunsaturated chains (e.g., di-18:2) in neuron membranes of terrestrial insects likely play the same role as DHA in human neurons.

These data are consistent with ARA and EPA playing important structural roles especially in the molecular architecture of aquatic insect membranes including neuron membranes. Thus, insects represent a growing list of organisms showing a dramatic ecological bias with respect to the presence of DHA and HUFAs as membrane building blocks. Other examples include marine plants versus land plants, marine bacteria versus terrestrial bacteria, oxygenic phototrophs living in cold seawater versus those living in warm oceanic ecosystems, and humans and other large mammals where most organs and cells contain only trace amounts of DHA/EPA in contrast to the brain and nervous system. It is generally agreed that temperature plays an important role in regulating the distribution and levels of HUFAs in ectothermic animals. However, in humans other poorly understood mechanisms modulate the specificity of DHA in membranes. Data from insects support the view that unique biochemical needs of specific organs and cells in individuals and between terrestrial versus aquatic ecotypes determine when and which HUFAs are beneficial or not. For example, phospholipids in the energy-intensive tails of fireflies are highly enriched with ARA and EPA in contrast to other organs (NorAliza et al., 2001). A biochemical mechanism to explain temperature modulation of HUFA levels in membranes of aquatic insects versus terrestrial insects is developed in Chapter 12. This model seems to have broad implications for explaining how neurons selectively target DHA to or away from different classes of membranes.

Data on lipid dynamics in insects support the concept that neuron membranes are subject to Darwinian selection of the fittest fatty acid composition relevant to the neurosensory needs of different animals living in different environments. It is clear that DHA/EPA chains are not essential or beneficial building blocks for neurosensory membranes across all animals and might even be harmful under certain circumstances (see Chapter 2).

Having made the case that DHA/EPA are not universally beneficial in neurons of all animals, we next consider the environmental or ecological parameters that might have led to the selection of DHA as the fittest structure in the membranes of neurons of mammals, including humans. We begin with a brief discussion of one of the most abundant animals on earth, Antarctic krill. Neurons in krill are adapted to function near the freezing point of seawater, below 0°C (Figure 4.2). We contrast membranes of these cold-tolerant neurons with those of humans, the latter operating at 37°C. Recall that oil compositions of membranes of neurons have evolved

FIGURE 4.2 One of the largest populations of neurons on earth occurs in Antarctic krill, which spend their entire lives in cold waters near 0°C. About 550 to 825 million tons of krill inhabit Antarctic seas annually, supported by DHA/EPA made by phytoplankton. Krill oil is now marketed in capsule form and competes with fish oil for human consumption. (2174C6 Krill [*Euphausia superba*]. Photograph by Stephen Brookes, Australian Antarctic Division © Commonwealth of Australia.)

to support neuron function across animal life over a ~50°C range of temperature (i.e., –4°C to about 50°C). Ultra-cold-adapted neurons are represented by Antarctic krill (Hamner et al., 1983). These shrimp-like animals live their entire lives at temperatures around –1.3°C to about 3°C and depend on their cold-adapted neurons for food gathering, egg laying, reproduction, avoiding predators, and other life functions. Krill membranes in general contain among the highest levels of DHA/EPA found in nature, which is not surprising given their cold environment and their diet composed of DHA/EPA-rich phytoplankton. Krill are crustaceans like shrimp and lobsters. Researchers have succeeded in dissecting and analyzing the single giant neuron from lobster tail muscle, which they found is slightly more enriched with EPA than DHA (Chacko, Barnola, and Villegas, 1977). Thus, we are faced with the quandary of explaining why neurons in krill versus humans, separated by about 40°C difference in operating temperature (i.e., –2°C to 37°C), both require DHA or other HUFAs. The explanation we offer is that DHA works the same way in neurons of both krill and humans, being essential for rapid neuron function in extreme cold in krill while maximizing speed of neuron function in humans. As detailed in Chapter 5, we propose that the unique conformational dynamics of DHA chains contribute powerful membrane antifreeze properties to membranes of neurons of krill, facilitating electrical signaling that otherwise would be too slow at this low temperature to be effective. At first glance, the concept of DHA chains contributing similar antifreeze properties to krill versus human neurons seems implausible because of our warm body temperature. However, the membrane surface of human neurons including synaptosomes harbors problematic hard patches that might be overcome by "antifreeze" properties of DHA. The membrane surface is believed to consist of "islands" of hardened regions (often enriched with cholesterol) called *lipid rafts*, surrounded by liquid regions. Even at warm human body temperature, lipid rafts form and are essential (Simons and Toomre, 2000) but are also thought to obstruct or serve as a barrier against freedom of motion of membrane components important in electrophysiology. Thus, DHA phospholipids working in membranes of both ectothermic and endothermic organisms are believed to function like miniature icebreakers, freeing components trapped in lipid rafts (Diaz et al., 2002) or dissolving an excess of rafts. This function has the effect of removing navigational obstacles, essentially allowing membrane components to move with less hindrance.

Membrane surfaces are considered to exist in two physical states, a hardened one like butter and a liquid one like olive oil. Simply warming butter by increasing temperatures beyond the melting point is sufficient to switch the physical state of butter from solid to liquid. This same property applies to the oils of neuron membranes. Theoretically, the human brain could gain more speed for enhancing membrane biochemistry by melting away any obstructing, hardened regions on its membrane surfaces by simply allowing brain temperature to rise say by 10°C—from 37 to 47°C. Such a spike in temperature would melt lipid rafts but would at the same time denature key membrane components and destroy brain function and homeostasis. Note that some animals including hummingbirds are believed to gain in energy output by maintaining their body temperatures significantly above that of humans (see Chapter 12). However, the human brain can achieve sufficient membrane speed without raising the temperature, instead enriching its membranes with DHA. This strategy of facilitating membrane speed or in the case of human neurons, extreme motion, is explained by the following rule of thumb developed by membrane biologists: temperature = membrane liquidity, viscosity or fluidity = motion (especially lateral motion) or membrane speed = levels of membrane unsaturation (e.g., as contributed by DHA and other HUFAs). The main point is that the presence of DHA in membranes works like increasing temperature but without some drawbacks associated with high temperature. We estimate that a 5 percent increase in the level of DHA as total fatty acids in membrane lipids is equivalent to about a 10°C rise in temperature (see Chapter 5). We conclude that cold temperature led to the selection of DHA as antifreeze in neurons of cold-blooded animals such as krill. This same property is applicable to human neurons where the objective is to maximize speed (i.e., extreme membrane motion) above that conferred by our constant 37°C body temperature. For more background on the equivalence of DHA and membrane motion or speed, see Valentine and Valentine (2009). As discussed below, membrane speed is only one of at least three major biochemical parameters used to explain the molecular roles of DHA in neurons.

4.4 DHA MEMBRANES IN OUR BRAIN RESULT FROM A DELICATE EVOLUTIONARY BALANCING ACT

We theorize that the evolutionary history of DHA in neuron membranes is dominated by two major chemical forces surprisingly at odds—essential motion/antiviscosity properties of DHA versus DHA behaving as a mediocre permeability barrier against leakage of sodium, potassium, and protons (see Valentine, 2007). This war-within-membranes pits the need for membrane motion against the need for lack of motion (the latter of which creates a more robust permeability barrier, thus conserving energy). That is, the same chemical factors that facilitate essential motion in neuron membranes are proposed to weaken membrane permeability properties and waste energy. Or turning this idea around, reduced membrane motion saves energy. Therefore, human neurons in the choice of DHA face a dilemma. On the one hand these cells require specialized membranes with extreme motion, and on the other they benefit from energy-tight or energy-conserving membranes. We suggest that this delicate membrane-balancing act, which helps define the DHA principle at the biochemical level, plays out not only in neuron membranes but is broadly applicable to membranes of virtually all cells and organisms. We previously applied the DHA principle to explain how one form of prokaryotic life can dominate another form (Valentine, 2007). As discussed later we show that neurons, in their evolutionary quest toward extreme motion, are faced with a dilemma. In the next section we introduce how Mother Nature has partially overcome this "DHA quandary."

4.5 DHA DOES NOT WORK ALONE AND IS BLENDED WITH CHOLESTEROL AND OTHER LIPID STRUCTURES TO MAINTAIN ENERGY-EFFICIENT NEURONAL MEMBRANES

We suggest that three functions of DHA have guided its evolutionary history, leading to its current predominant position in maximizing speed and efficiency in membranes of human neurons. These include DHA's contribution to extreme membrane motion, contribution to the permeability properties of neuronal membranes, especially those of axons and synaptic vesicles, and membrane instability in the presence of oxygen (i.e., peroxidation). Permeability properties are introduced here and covered in detail in Chapters 6 and 7. Oxidative damage of DHA is covered in Chapters 8 through 10. The central theme developed here is that in the history of neurons there was great selective pressure at the level of membranes for producing as well as saving or conserving as much energy as possible. In Chapter 1 we saw a glimpse of how neurons acquire and utilize their available energy with the conclusion that energy is often in short supply. Neurons are capable of generating a great deal of energy, but most of this cellular energy is consumed as fast as it is made to fuel the electrical process carried out by these cells. We suggest that one of the biggest surprises in the field of neuron bioenergetics concerns the vast amount of energy unavoidably lost because of spontaneous leakage of electrically active ions—sodium and potassium—across neuronal membranes. In Chapters 6 and 7 we elaborate on how energy wastage always occurs in normal neuron function, but at the same time neurons strive to minimize ionic leakage across their membranes as a means to save energy. Thus, step by evolutionary step, membranes of human neurons have evolved to their current state of efficiency. The concept that tripartite biochemical rationale is behind the use of DHA in the human brain allows for brain diversity, including the Darwinian selection of brain span (see Chapters 11 through 17).

DHA is not the only essential oil needed for building efficient neuron membranes (Björkhem and Meaney, 2004). For example, the brain has the highest concentration of membrane cholesterol, based on size, of any organ in the body. The term *membrane cholesterol* is defined here to highlight cholesterol as an essential lipid blended with phospholipids to form effective membrane surfaces. In the face of much negative publicity surrounding "bad" cholesterol acting as an artery-clogger, it is not widely appreciated that cholesterol is universally required as an essential membrane building block for all animal life. This essentiality of cholesterol in neurons is linked to its beneficial qualities in modifying the molecular architecture of neuron membranes to conserve energy. This role will be discussed in detail in Chapter 6. A great deal of cholesterol is also incorporated into the myelin sheaths of axons of white matter, likely contributing multiple benefits including more energy-efficient and faster transmission of neural impulses.

The essential membrane building blocks needed to blend effective neuron membranes include other lipids besides DHA and cholesterol. For example, a rather simple change in the chemical linkage that attaches DHA and other fatty acid tails to their phospholipid head groups might be an important energy saver or energy producer in neurons. The evolution of two classes of phospholipids distinguished by their chemical linkages to their head groups is introduced here because we suspect that these structures play important roles in the energetics of human neurons (Figure 4.3). The first class of phospholipid is called *cardiolipin* (Figure 4.3a) (Schlame, 2008), named after the heart tissue where it was first isolated. Cardiolipin, introduced in Chapter 2, is unique in at least two ways. First, it has four acyl or fatty acid chains instead of two as found in most membranes. A glycerol bridge joins two conventional phospholipids, each contributing two acyl chains to form this conjoined arrangement.

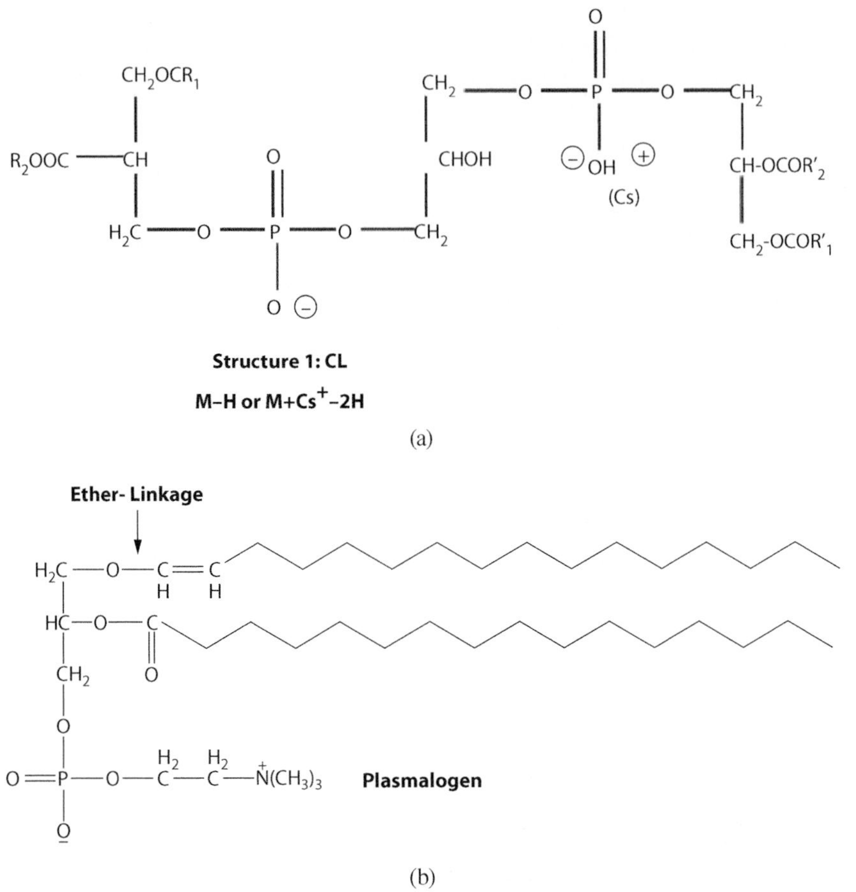

Structure 1: CL

M–H or M+Cs$^+$–2H

(a)

Ether- Linkage

Plasmalogen

(b)

FIGURE 4.3 Different chemical classes of DHA phospholipids. (a) Cardiolipin is a specialized phospholipid found in bacterial and mitochondrial membranes. This structure with four long fatty acid tails is formed by condensation of two conventional phospholipids via a glycerol bridge moiety at their head groups. (See Schlame, M. 2008, Cardiolipin Synthesis for the Assembly of Bacterial and Mitochondrial Membranes, *J. Lipid Res.* 49:1607–1620.) Cardiolipin is now believed to play multiple important biochemical roles in bioenergetics of neurons. DHA-cardiolipin, common in cold-adapted animals and certain endothermic animals including hummingbirds, is seldom found in human membranes and is discussed later as a pro-neurodegenerative molecule (Chapter 16). (Reprinted from *J. Am. Soc. Mass Spectrom*, 18, Wang, H.-Y.-J., S. N. Jackson, and A. S. Woods, Direct MALDI-MS Analysis of Cardiolipin from Rat Organs Sections, 567–577, Copyright 2007, with permission from Elsevier.) (b) Plasmalogens. At first glance this class of phospholipids found in neuron membranes and sperm tails seems to be similar to conventional phospholipids. However, a closer look shows that the chemical linkage of the outermost (*sn*-1) fatty acid chain to its head group is unique, an ether linkage in contrast to the conventional ester linkage. The ether linkage is believed to endow this class of phospholipids with energy-conserving properties important in neuron membranes. DHA usually occupies the inner or *sn*-2 position of phospholipids. (From Valentine, R., and Valentine, D., *Omega-3 Fatty Acids and the DHA Principle.* Copyright 2009. Reproduced by permission of Taylor & Francis Group, LLC, a division of Informa plc.)

Cardiolipin such as isolated from heart cells displays not only a unique tandem head arrangement but also often has four polyunsaturated tails (e.g., four 18:2 chains; $(18:2)_4$-CL). In neurons and other human cells cardiolipin is the signature phospholipid in mitochondria and is targeted specifically to the inner or energy-transducing membrane housing the electron transport chain. We argue that cardiolipin unsaturation levels in mitochondrial membranes not only govern energy efficiency by maximizing energy production (see Chapter 11) but also enhance the permeability barrier, which simultaneously reduces energy loss. It is unlikely that DHA is incorporated into cardiolipin in human neurons because this highly unsaturated structure is proposed to expose mitochondria to severe oxidative damage (Chapters 2, 11, and 17). The blending of an efficient human neuron membrane continues with a second class of phospholipid called *plasmalogens*. This structure features a unique attachment joining the second fatty tail, usually a saturated chain, to its head group in the *sn*-1 position (Figure 4.3b). Note that DHA often present in plasmalogens is joined to the head group by a conventional ester linkage. DHA plasmalogens present in high levels in membranes of axons as well as synaptic vesicles, and sperm tails are yet another example of an adaptation perhaps improving the energy efficiency of neuron membranes, and are likely part of the Darwinian selection of the most energy-efficient neuron membranes. The chemistry and biochemistry of plasmalogens as energy savers is not covered in detail here (see Valentine and Valentine, 2009).

In our previous book in which the tripartite blending code for membrane fatty acids was developed (Valentine and Valentine, 2009), we stated that higher animals, in contrast to bacteria, have not evolved methyl-branched fatty acids for membrane synthesis. We take this opportunity to correct that statement (see Christie, 2010). Kniazeva and colleagues (2004) discovered that monomethyl-branched-chain fatty acids play essential roles in the nematode *Caenorhabditis elegans* (Kniazeva et al., 2004). This class of fatty acids is widespread in membranes of bacteria and is even important in feather conditioning in birds and hair conditioning in mammals including sheep and humans (Christie, 2010). An extra methyl group attached to the second or third carbon of the methyl end of the fatty chain is typical of methyl-branched fatty acids and is believed to contribute essential energy-saving or water-repelling properties to membranes. In one sense methyl-branched fatty acids in membranes of many bacteria, and *C. elegans,* can be envisioned as a sort of "poor man's" class of cholesterol-like molecules, replacing cholesterol as a membrane constituent but at a much cheaper metabolic cost to produce. Methyl-branched chains have membrane-fluidizing properties similar to monounsaturated fatty acid chains but are far more oxidatively stable. The extra methyl group is also believed to add extra bulk to the membrane important in blocking formation of water-wires, a point discussed in Chapters 6 and 7. The mode of action of methyl-branched fatty acids in *C. elegans* is of great interest especially if these chains are incorporated into neuron membranes. In *C. elegans* branched chains in levels of 5 to 9 percent of total fatty acids are essential for growth and development. Methyl-branched fatty acids might be important in energy efficiency of neurons of *C. elegans*, perhaps mirroring the mode of action of cholesterol. Methyl-branched chains appear to impact stress tolerance in *C. elegans* (Horikawa and Sakomoto, 2009). The role of branched fatty acids in humans is also an important topic for future research.

4.6 DHA PLAYS AN ESSENTIAL ROLE IN MEMBRANE GROWTH, DEVELOPMENT, AND EFFICIENCY

We propose that structure-function of DHA working in neuron membranes can be explained by a tripartite blending code as introduced above. In developing this code we were aware of a possible fourth important biochemical function of DHA—playing essential roles in membrane growth and development. However, we chose not to discuss this area as a separate topic at the time based on data by Lesa and colleagues who found EPA/DHA to be essential at the level of the synaptosomal cycle in neurons of *C. elegans*. (Lesa et al., 2003; Marza et al., 2008). The main point here is that these researchers propose that lateral motion contributed by EPA/DHA maximizes endocytosis of synaptic vesicles, an apparent rate-limiting step in neurons firing fast bending-muscles in this worm (see Chapter 5). Their data show that EPA/DHA play essential roles in the sophisticated developmental cycle for synaptic vesicles, functions that we explain using the tripartite code. Lesa and colleagues reviewed data consistent with a unified role for DHA/EPA in synaptosomal cycling in mammals (Lesa et al., 2003; Marza et al., 2008). As shown in Figure 4.4, a significant fraction of the total surface area of a synaptic vesicle is crowded with large protein complexes. The presence of these crowded conditions is thought to hinder freedom of motion or collisions essential in rapid cycling of synaptic vesicles. Marza et al. (2008) propose that the presence of EPA maximizes motion along with rates of endocytosis. Thus, the tripartite code seems to adequately explain the role of DHA/EPA in the synaptosome cycle, a topic continued in Chapter 5.

As introduced in Chapter 2, DHA plays an important role in spermatogenesis in mice (Roqueta-Rivera et al., 2010; Stroud et al., 2009). DHA null mutant mice were constructed that deprive the testes of sufficient DHA for sperm development. The result is a dramatic roadblock in which spermatogenesis grinds to a halt. None of the substitutes for DHA tested so far (e.g., 22:4) approach the effectiveness of 22:6. These data are consistent with the view that DHA is not only essential for biochemical functions in mature sperm but is also critical for the development or differentiation of sperm (see Chapter 2). A similar role in neuron and axon development in mammals including humans is being considered by these researchers. It is

FIGURE 4.4 *(See facing page)* Membranes of synaptic vesicles are enriched with DHA. (a) Outside view of a single synaptic vesicle shows numerous proteins embedded in the membrane. (b) Vesicle sectioned in the middle exposes the vesicle lumen or interior where neurotransmitters are concentrated. (c) Model showing only one species of membrane-embedded protein, synaptobrevin, to illustrate the surface density of the most abundant protein in vesicles. DHA chains, not specifically identified in the image, are enriched in vesicle membranes and are proposed to play essential roles in speeding up collisions among membrane components (see Chapter 5 for details). (Reprinted from *Cell*, 127, S. Takamori et al., Molecular Anatomy of a Trafficking Organelle, 831–846, Copyright 2006, with permission from Elsevier.)

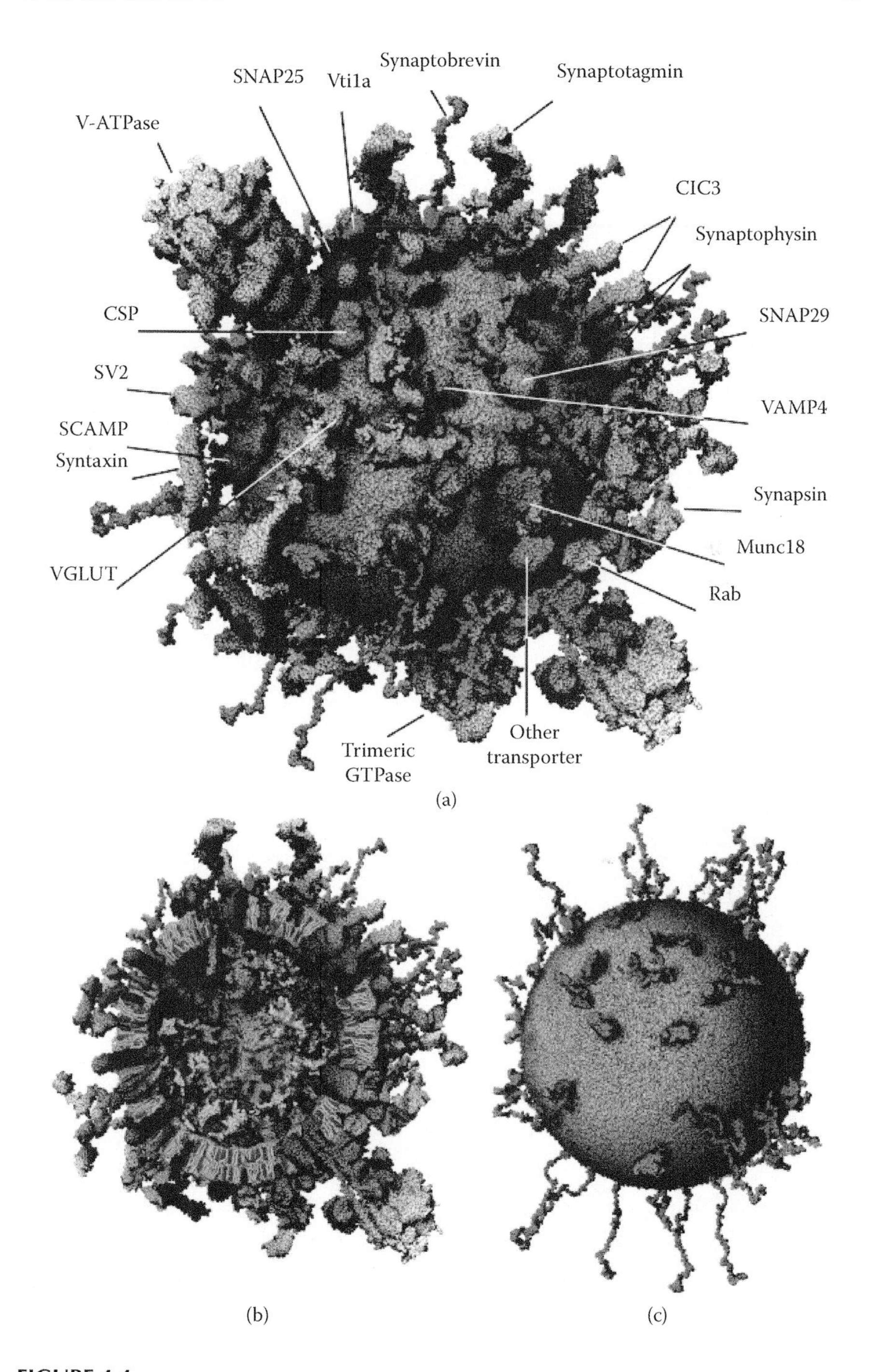
SNAP25
Vti1a
Synaptobrevin
Synaptotagmin
V-ATPase
CIC3
Synaptophysin
CSP
SNAP29
SV2
VAMP4
SCAMP
Syntaxin
Synapsin
Munc18
VGLUT
Rab
Trimeric
GTPase
Other
transporter
(a)
(b)
(c)

FIGURE 4.4

well known that DHA levels in brain tissue of mammals are almost always constant across species. One interpretation of this data is that wide fluctuations of DHA levels cannot be tolerated in the brain. One scenario that comes to mind to explain these data is that even a slight deprivation of DHA at the neuron level might exert a profound downstream effect on axon development, perhaps breaking a delicate balance between rates of new axon growth versus rates of axon senescence. Thus, even a slight shift in DHA homeostasis, as might occur with aging, could favor the senescence of axons over new growth. Recent data show that overexpressing DHA enhances axon development in fat-1 mice (He, Qu, and Cuietal, 2009). This finding confirms data from dietary studies that show that DHA stimulates neurogenesis (Kan et al., 2007).

4.7 ENERGY LIMITATION OF AN ISLAND ENVIRONMENT MIGHT HAVE SELECTED THE TINY, YET *HOMO SAPIEN*–LIKE BRAIN OF HOBBIT (*HOMO FLORESIENSIS*)

The November 2009 issue of the highly respected *Journal of Human Evolution* has 12 papers devoted to the analysis of about a dozen skeletal remains of dwarf people discovered in a cave on the island of Flores in Indonesia (Figure 4.5a and Figure 4.5b) (Morwood and Jungers, 2009). The discovery of *Homo floresiensis* in

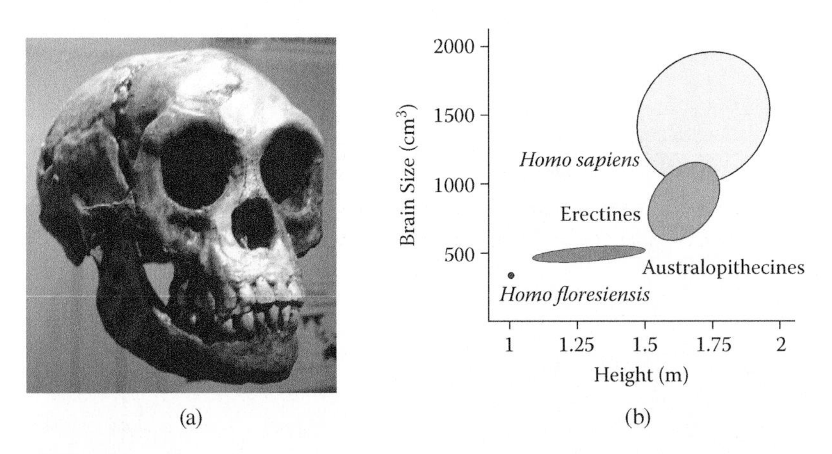

FIGURE 4.5 Brain evolution: tiny brain yet "human" behavior? (a) A skull of *Homo floresiensis*. (b) The cranial volume of this tiny hominoid, apparently a recent ancestor of humans, is a little more than one quarter (417 cc) that of modern humans. Yet *H. floresiensis* seems to have evolved relatively advanced mental capability. This discovery has fueled a good deal of debate on the importance of the environment on brain evolution and the relationship between brain size and human behavior. According to one theory, brain miniaturization is caused by the limitation in available energy occurring in an island environment. *Homo erectus* is the most familiar among the "erectines," and Lucy is the most famous of the australopithecines. Note that the brain of *H. floresiensis* is closest in size to Lucy, an ancient ancestor of man. (Parts (a) and (b) reprinted by permission from Macmillan Publishers Ltd: *Nature*, M. Mirazón Lahr, and R. Foley, Copyright 2004.) **(See color insert.)**

2003 has shocked and divided scientists studying human brain evolution. *H. floresiensis* is remarkable in that it lived until relatively recent times, likely overlapping with modern humans. It is well known that long-term isolation on islands results in eventual dwarfism of mammals, perhaps including brain size in the case of hobbit. One of the most interesting aspects of *H. floresiensis* concerns the assignment of various indicators of intelligence to such a relatively small brain (i.e., about 417 cc compared to about 1500 cc for modern humans). This tiny hominoid, about 3 feet, 6 inches tall, is believed to display advanced behaviors similar to *H. sapiens* including cooperative hunting and the use of stone tools and fire for cooking. A region of the brain associated with self-awareness is about the same size as that of modern humans. These studies open Pandora's box concerning the linkage between the environment, brain size, brain rearrangement, and intelligence. It seems likely that the dietary energy needed to power the brain of hobbit would only have been a fraction of that compared to humans, a significant selective advantage in an island environment where food might be scarce. It would be of great interest to know the life span of these tiny people, information that might provide clues about the evolution of the human brain span.

In summary, powerful forces of Darwinian selection have honed all aspects of brain-neuron functions. These include such diverse aspects as brain size, speed, efficiency, and functional brain span. We propose that DHA is the fittest oil for development of axon circuitry in the human brain. DHA-enriched bilayers are an evolutionary adaptation maximizing both speed and energy efficiency. However, neurons face a delicate balancing act in which speed and efficiency are pitted against oxidative instability. There appears to be no perfect solution: significant energy loss through futile cation cycling is likely to always occur, and chemical oxidation of DHA seems inevitable in all aerobic cells, especially at warm temperatures. Oxidative stability must be further considered as part of the tripartite balancing act thought to govern energy production or conservation and energy loss in neurons. According to a tripartite concept, energy efficiency in neurons at the membrane level involves optimizing membrane motion to speed membrane processes, minimizing cation permeability to conserve energy, and preventing oxidative damage of DHA membranes, the latter of which avoids futile energy cycling and may be a key to understanding brain span. Thus, the dietary energy cost of fueling brain function is large enough to be an important selective force in the evolution of the human brain. Finally, recent data support essential roles for DHA in the sophisticated developmental or differentiation pathway of sperm with implications for how neurons grow, develop, and senesce.

REFERENCES

Björkhem, I., and S. Meaney. 2004. Brain cholesterol: Long secret life behind a barrier. *Arterioscler. Thromb. Vasc. Biol.* 24:806–815.

Chacko, G. K., F. V. Barnola, and R. Villegas. 1977. Phospholipid and fatty acid compositions of axon and periaxonal cell plasma membranes of lobster leg nerve. *J. Neurochem.* 28:445–447.

Christie, W. W. 2010. Fatty acids: Branched-chain—Structures, occurrence and biosynthesis. Copyright © lipidlibrary.aocs.org (AOCS Lipid Library Web site: http://lipidlibrary.aocs.org/Lipids/fa_branc/index.htm).

Crawford, M. A., M. Bloom, C. L. Broadhurst et al. 1999. Evidence for the unique function of docosahexaenoic acid during the evolution of the modern hominid brain. *Lipids* 34 Suppl:S39–S47.

Cunnane, S. C., and M. A. Crawford. 2003. Survival of the fattest: Fat babies were the key to evolution of the large human brain. *Comp. Biochem. Physiol. A Mol. Integr. Physiol.* 136:17–26.

Cunnane, S., and K. Stewart, Eds. 2010. *Human Brain Evolution: The Influence of Freshwater and Marine Food Resources*. Hoboken, NJ: Wiley-Blackwell.

Diaz, O., A. Berquand, M. Dubois et al. 2002. The mechanism of docosahexaenoic acid-induced phospholipase D activation in human lymphocytes involves exclusion of the enzyme from lipid rafts. *J. Biol. Chem.* 277:39368–39378.

Faisal, A. A., J. A. White, and S. B. Laughlin. 2005. Ion-channel noise places limits on the miniaturization of the brain's wiring. *Curr. Biol.* 15:1143–1149.

Fox, D. 2011. The limits of intelligence. *Sci. Am.* 305:36–43.

Hamner, W. M., P. P. Hamner, W. Strand et al. 1983. Behavior of Antarctic krill, *Euphausia superba*: Chemoreception, feeding, schooling, and molting. *Science* 220:433–435.

He, C., X. Qu, and L. Cuietal. 2009. Improved spatial learning performance of fat-1 mice is associated with enhanced neurogenesis and neuritogenesis by docosahexaenoic acid. *Proc. Natl. Acad. Sci. USA* 106:11370–11375.

Horikawa, M., and K. Sakamoto. 2009. Fatty-acid metabolism is involved in stress-resistance mechanisms of *Caenorhabditis elegans*. *Biochem. Biophys. Res. Commun.* 390:1402–1407.

Kan, I., E. Melamed, D. Offen et al. 2007. Docosahexaenoic acid and arachidonic acid are fundamental supplements for the induction of neuronal differentiation. *J. Lipid Res.* 48:513–517.

Kniazeva, M., Q. T. Crawford, M. Seiber et al. 2004. Monomethyl branched-chain fatty acids play an essential role in *Caenorhabditis elegans* development. *PLoS Biol.* 2(9):e257 (doi:10.1371/journal.pbio.0020257).

Lesa, G. M., M. Palfreyman, D. H. Hall et al. 2003. Long chain polyunsaturated fatty acids are required for efficient neurotransmission in *C. elegans*. *J. Cell Sci.* 116:4965–4975.

Marza, E., T. Long, A. Saiardi et al. 2008. Polyunsaturated fatty acids influence synaptojanin localization to regulate synaptic vesicle recycling. *Mol. Biol. Cell* 19:833–842.

Mirazón Lahr, M., and R. Foley. 2004. Palaeoanthropology: Human evolution writ small. *Nature* 431:1043–1044.

Morwood, M., and W. L. Jungers, Eds. 2009. Paleoanthropological research at Liang Bua, Flores, Indonesia. *J. Hum. Evol.* 57:437–648.

Niven, J. E., J. C. Anderson, and S. B. Laughlin. 2007. Fly photoreceptors demonstrate energy-information trade-offs in neural coding. *PLoS Biol.* 5:e116.

Niven, J. E., and S. B. Laughlin. 2008. Energy limitation as a selective pressure on the evolution of sensory systems. *J. Exp. Biol.* 211:1792–1804.

NorAliza, A. R., J. C. Bedick, R. L. Rana et al. 2001. Arachidonic and eicosapentaenoic acids in tissues of the firefly, *Photinus pyralis* (Insecta: Coleoptera). *Comp. Biochem. Physiol. A Mol. Integr. Physiol.* 128:251–257.

Roqueta-Rivera, M., C. K. Stroud, W. M. Haschek et al. 2010. Docosahexaenoic acid supplementation fully restores fertility and spermatogenesis in male delta-6 desaturase-null mice. *J. Lipid Res.* 51:360–367.

Schlame, M. 2008. Cardiolipin synthesis for the assembly of bacterial and mitochondrial membranes. *J. Lipid Res.* 49:1607–1620.

Simons, K., and D. Toomre. 2000. Lipid rafts and signal transduction. *Nat. Rev. Mol. Cell Biol.* 1:31–39.

Stanley, D. W. 2000. *Eicosanoids in Invertebrate Signal Transduction Systems*. Princeton, NJ: Princeton University Press.

Stroud, C. K., T. Y. Nara, M. Roqueta-Rivera et al. 2009. Disruption of FADS2 gene in mice impairs male reproduction and causes dermal and intestinal ulceration. *J. Lipid Res.* 50:1870–1880.

Valentine, D. L. 2007. Adaptations to energy stress dictate the ecology and evolution of the Archaea. *Nat. Rev. Microbiol.* 5:316–323.

Valentine, R. C., and D. L. Valentine. 2009. *Omega-3 Fatty Acids and the DHA Principle.* Boca Raton, FL: Taylor & Francis.

Section II

Benefits of DHA

Darwinian selection has honed the human brain to be perhaps the fittest on Earth. Comparative evolutionary studies of brains of a hierarchy of animals reveal the history of the human brain. Several essential properties of the brains of mammals subject to evolutionary change are as follows:

- Chemical composition
- Brain size
- Organization
- Speed
- Efficiency
- Brain functional life span (brain span)

The next three chapters focus on the benefits of speed and energy efficiency for the brain. In a search for answers at the molecular level, we delve into the inner workings of specialized neuron membranes. Once inside we come face-to-face with an extraordinary fatty acid—DHA. It will be shown that the conformational dynamics of DHA chains, which change on a picosecond to nanosecond time scale, are at the heart of membrane motion. This motion enables trigger proteins in the visual cascade to collide faster (rod cells) and synaptic vesicles to cycle faster (neurons). The same conformational properties of DHA that speed up membrane biochemistry also confer energy efficiency to neurons and even sperm. We also explore the molecular architecture of neuron membranes toward understanding how DHA chains work cooperatively with other critical membrane lipids to form robust, energy-efficient structures. Finally, the behavior of DHA as a self-sealing and energy-saving molecule during growth stretching of axons, curvature stretching of synaptic vesicles, and violent tail whipping of sperm tails is explored. A model in which DHA conformational dynamics saves energy during membrane stretching is developed and generalized to other important human cells.

5 DHA Is "King Omega" for Maximizing Mental Speed

In the early 1970s scientists were first able to calibrate the physical speed of the essential light-sensing protein rhodopsin swimming or immersed in DHA-enriched membrane disks of rod cells of the eye (Liebman and Entine, 1974; Poo and Cone, 1974). These now classic studies demonstrate that rhodopsin rotates and moves laterally across the surface of membrane disks at amazing speeds (Figures 5.1a, 5.1b, and 5.1c). For biochemical studies large numbers of rod cells each containing about 1000 rhodopsin disks can be isolated from bovine eyes obtained from meat processing plants. The motional properties of integral membrane protein, rhodopsin, were calibrated using micro-lasers to rotate in a molecular swirl timed at 0.00002 seconds per turn. This remains atop the motion scale for a full swing of a membrane-bound protein and links DHA phospholipids with extreme membrane motion. Lateral motion of rhodopsin across the surface of membrane disks is also extremely fast and plays an important role in triggering the visual cascade. In this chapter we show how the unique conformational dynamics of DHA working in specialized neuron membranes is harnessed to maximize the motion of membrane components important in sensory perception and mental processes. We define *extreme speed* as a state of membrane motion in which membrane lipids, proteins, lipophilic electron carriers, and other membrane components rotate and move laterally across the bilayer at extraordinarily high speeds with DHA-enriched rhodopsin disks serving as a standard.

5.1 DHA'S TWISTY TAIL SPEEDS SENSORY AND MENTAL PROCESSES

DHA tails are akin to the Cirque de Soleil, but at the molecular level, as seen in Scott Feller's video on his Web site at Wabash College, Crawfordville, Indiana (Feller, 2008, http://lipid.wabash.edu/). A selection of images pulled from the above video shows the contortionistic nature of DHA tails (Figure 5.2). However, to fully appreciate the dynamics we encourage the reader to view the videos. The individual images in Figure 5.2a are phospholipids composed of a headgroup portion that serves as an attachment point for one or two long DHA tails. DHA in the image shown is the thicker, twisty tail in gold, and the blue straight chain is a saturated fatty acid (stearic acid). This phospholipid structure is common in rhodopsin disk membranes and neurons. The highest level of DHA reported to date in rhodopsin disks of certain rodents is 53 percent of the total fatty acids. Phospholipids spontaneously join

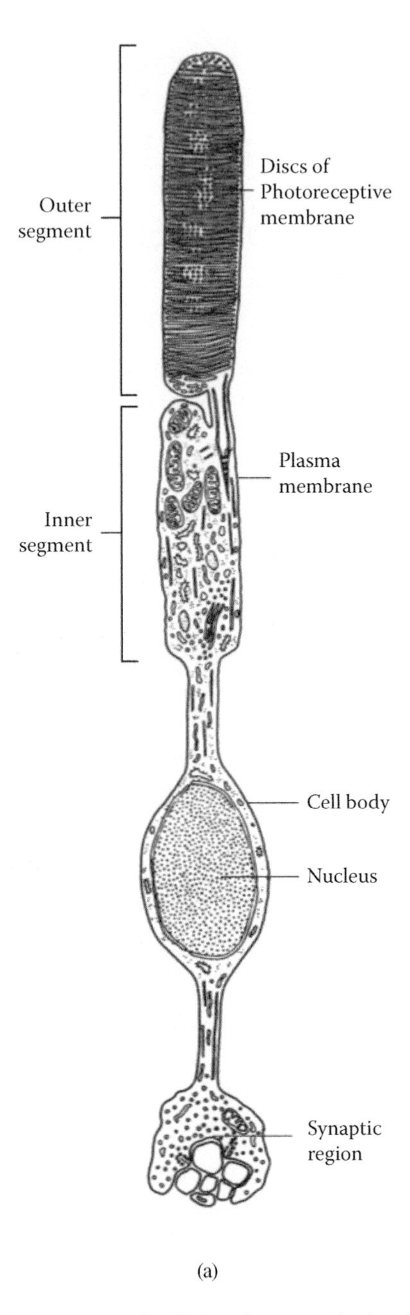

FIGURE 5.1 DHA maximizes rates of collisions between rhodopsin and its G-protein transducin triggering the visual cascade. (a) Perception of dim light occurs using strategically placed membrane disks located in the outer segment of rod cells in the eye and which act as light-sensing antennae. Approximately 1000 DHA-enriched membrane disks are present per rod cell. (From Alberts, B. et al., *Molecular Biology of the Cell*, Copyright 2008. Reproduced by permission of Taylor & Francis Group, LLC, a division of Informa plc.) *(Continued on next page)*

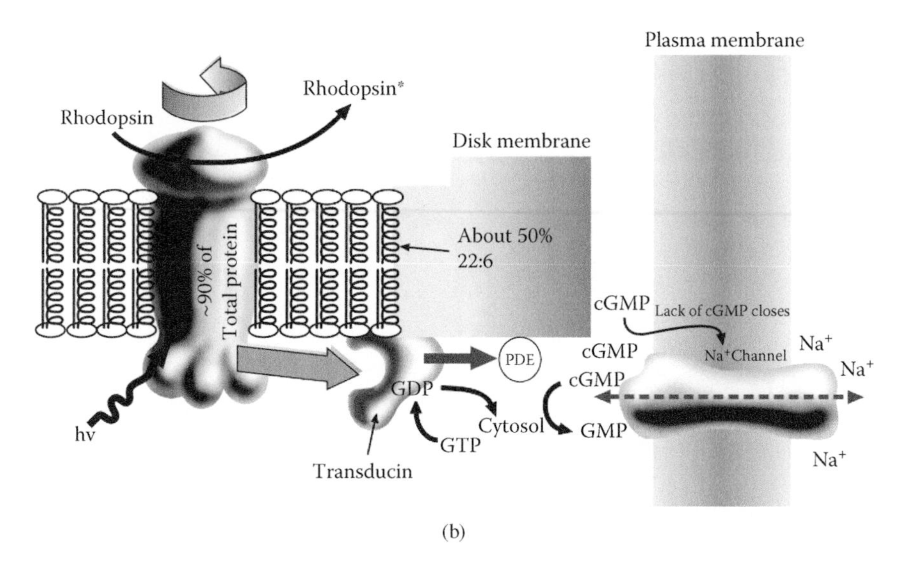

FIGURE 5.1 *(Continued)* (b) DHA-enriched membranes compared to other membrane fatty acids enable rhodopsin to rotate and move laterally at amazing speed. This facilitates collisions between rhodopsin and its G-protein, transducin, firing the visual cascade. (From Valentine, R., and Valentine, D., *Omega-3 Fatty Acids and the DHA Principle.* Copyright 2009. Reproduced by permission of Taylor & Francis Group, LLC, a division of Informa plc.) (c) A single rhodopsin molecule embedded in a disk is shown surrounded by DHA phospholipids in an image generated by chemical simulation. (Courtesy of Scott Feller.) In addition to driving extreme motion of rhodopsin, DHA phospholipids might be involved in the biochemistry of rhodopsin. (Grossfield et al., 2006.)

with each other to form the membrane surfaces of axons. Because DHA tails are in perpetual motion (Figure 5.2b) (see papers by Feller and colleagues, 2008a; Soubias and Gawrisch, 2007; Stillwell and Wassall, 2003), these chains do not stand still long enough to bind with their neighbors to form hardened oils typical of butter or lard. Thus, the contortions of DHA chains keep the membrane surface in constant motion even in the extreme cold. DHA is responsible for this extreme motion because it keeps itself, neighboring chains, and other membrane components in a perpetual state of movement (e.g., spinning and lateral movement) as seen with rhodopsin. Neurons and sensory cells harness the motion in DHA tails to boost the speed of their signals. Hence, DHA becomes a pacemaker of brain speed and without it our sensory system would likely be much slower. Examples from both chemistry and ecology help us visualize the power of membrane speed contributed by DHA chains. First, chemists have found that DHA oils resist hardening until temperatures reach the equivalent of a frigid winter night in Antarctica (i.e., lower than –50°C) (Koynova and Caffrey, 1998). This property is important in nature allowing neurons in cold-adapted animals to function at near-freezing temperatures. For example, some 500 to 800 million metric tons of tiny shrimp-like krill grow in frigid Antarctic waters and help drive this amazing food chain. Krill grow and their neurons work at temperatures that hover near 0°C year-round (Buchholz and Saborowski, 2000). Krill eat about 10 times their weight in tiny cold-adapted marine plants called *phytoplankton* that

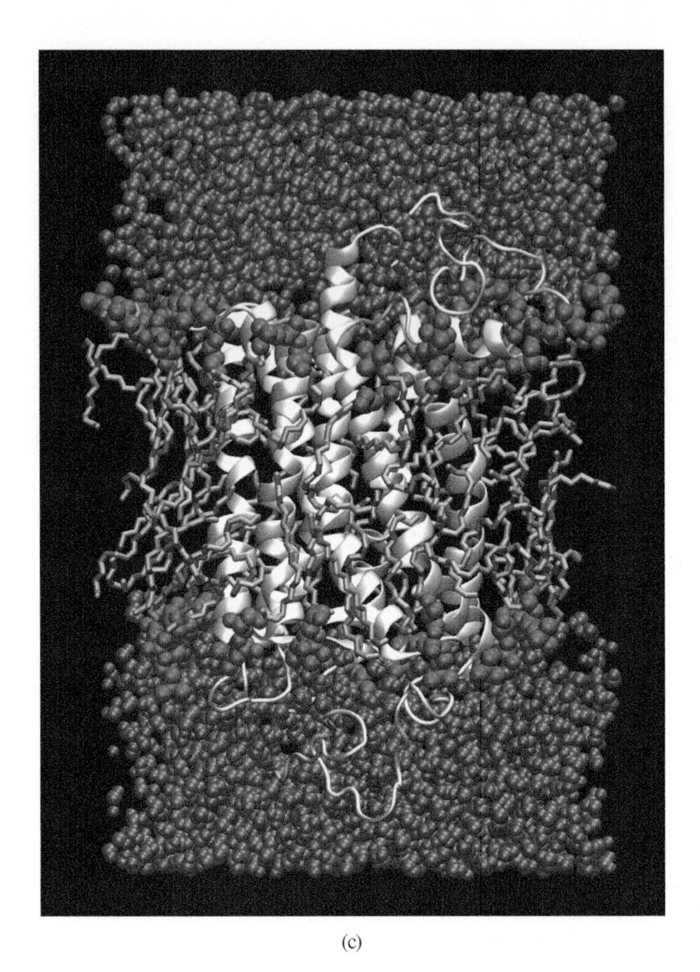

(c)

FIGURE 5.1 *(Continued)*

provide DHA/EPA for building their cold-adapted neurons. Phytoplankton, which form the base of the food chain, harness the antifreeze properties of DHA/EPA in their photosynthetic membranes to maximize rates of harvesting sunlight under Antarctic conditions. DHA/EPA flows from krill up the food chain to penguins and whales and eventually humans.

5.2 SENSORY PERCEPTION DEPENDS ON EXTREME MEMBRANE MOTION CONTRIBUTED BY DHA

DHA enables the eye to sense dim light as summarized in Figure 5.1 (Anderson, 1970). This specialized cell called a *rod cell* contains about 1000 membrane disks enriched with DHA (Figure 5.1a). These specialized membranes serve as miniature antennae for detecting rays of light (Jeffrey et al., 2001). This micro-array of light sensors is believed to be able to detect a single ray or photon of light striking the surface of the disks. Rhodopsin (Litman et al., 2001), embedded in these disks and

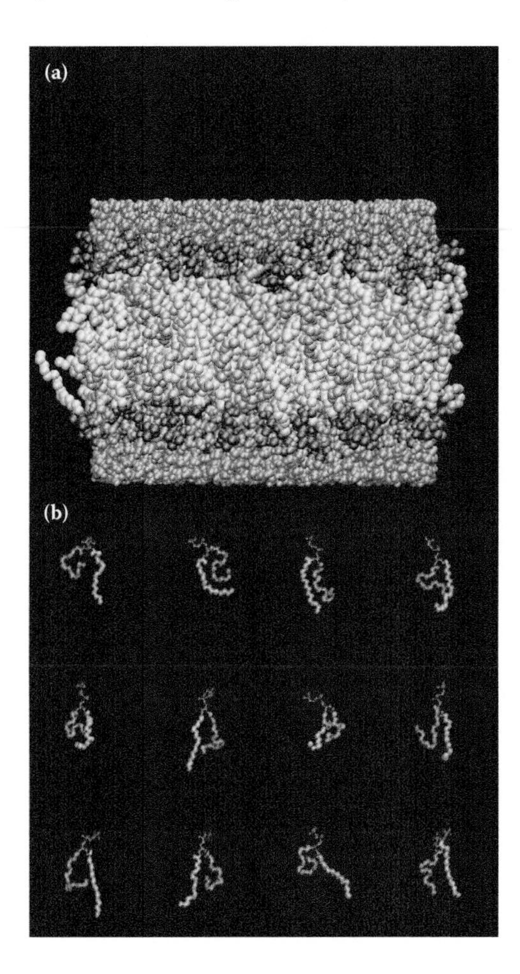

FIGURE 5.2 Dynamic conformations of DHA chains keep neuron membrane components in motion. (a) DHA phospholipids join together to form axon membranes. (b) DHA chains as membrane phospholipids shown in this set of images change shape on a picosecond to a nanosecond time scale. This property is harnessed to speed up electrical impulses in our brain. The dynamic shapes of DHA also explain its antifreeze properties in neurons of extreme, cold-adapted animals such as Antarctic krill. (Images courtesy of Scott Feller, and generated by Matthew B. Roark, both of Wabash College, Crawfordsville, Indiana.) **(See color insert.)**

making up about 90 percent of their total protein content, is the actual sensor of light (Figure 5.1b). Rhodopsin is a membrane-bound protein and becomes an effective light trapper when combined with an orange-colored carotenoid derived from carrots and other vegetables. The carotenoid is tightly bound to rhodopsin protein and serves as a chromophore able to detect light.

When light strikes a single rhodopsin molecule a structural change is induced, which in turn is sensed by a second distinct membrane-attached protein that works as a partner molecule. The amazingly rapid molecular dance step executed by rhodopsin enables rhodopsin to collide with its partner, defining the trigger reaction of vision (Figure 5.1b) (Bruckert, Chabre, and Vuong, 1992; Calvert, Govardovskii, and Krasnoperovaetal, 2001; Eldho et al., 2003; Kahlert and Hofmann, 1991; Niu and Mitchell, 2005; Niu et al., 2004; Stinson, Wiegand, and Anderson, 1991). Imagine trying to complete a dance rotation in 0.00002 seconds. A single rhodopsin molecule accomplishes its turn in this time and moves laterally across the membrane in search of its partner in an equally impressive time frame. If this dance, taking place in membranes between rhodopsin and its partner, took place in a watery environment the speed of this dance would not be very impressive. Instead, rhodopsin must do its dance in an oily world that restricts motion. This is where DHA-enriched membranes are important. In essence, DHA has ideal conformational dynamics to enable rapid rhodopsin motion. Figure 5.1c shows an image of a single rhodopsin protein molecule in white surrounded and embedded in a membrane enriched with DHA chains (gold), which are believed to maximize its motion across the surface of rhodopsin disk membranes. Development in the 1960s of micro-laser technology made possible these now classic experiments, conducted in the 1970s, that calibrated the speeds of rotation and lateral movement of rhodopsin. The dynamic conformational model of DHA developed in the early 2000s by Scott Feller and other physical chemists (Feller, 2008; Soubias and Gawrisch, 2007; Stillwell and Wassall, 2003) provides the clearest picture yet of how the extreme conformational dynamics of DHA help drive the motion of membrane components including rhodopsin to their zenith. Note that about half of the phospholipids forming the membrane surface in rhodopsin disks contain at least one chain of DHA. It is also interesting to recall that the power of DHA in driving collisions among membrane components is harnessed on a grand scale during one of the most important processes in the biological world—photosynthesis. As introduced above, photosynthetic membranes of chloroplasts of marine plants depend on DHA or similar molecules to speed collisions along the photosynthetic electron transport chain, providing energy for marine plant growth. From a global ecological perspective, it is difficult to imagine a more important role for speedy membranes than the case of photosynthetic membranes (Anderson and Andersson, 1982; Malkin and Niyogi, 2000). Thus, DHA creating extreme membrane motion is important not only for detection of dim light in rod cell membrane antennae in the eye but also in harvesting light by photosynthetic membranes of marine plants. The key point is that the conformational dynamics of DHA are essential in both classes of light-sensing membranes.

Smell is another critical sense that depends on membranes in motion. Humans are able to smell about 10,000 different odorants, and the receptors for these odorants are located in olfactory cilia, hair-like tubes growing from the end of specialized neurons called *olfactory neurons* (Malnic, Godfrey, and Buck, 2004). Figure 5.3 shows the olfactory cilia of frogs, which are surprisingly similar to humans. About 350 different membrane odor receptors are embedded in the membranes of human olfactory cilia. These receptors combine their efforts through molecular cross talk enabling us to sense thousands of different smells. Like vision, olfactory receptors must move rapidly to search out their signaling partners for triggering the olfactory sensing cascade. Once again, rapid motion of membrane components necessary for smell is contributed not by the odor receptor protein molecule but rather by the nature of the oils used to build these specialized tubular membranes (Russell,

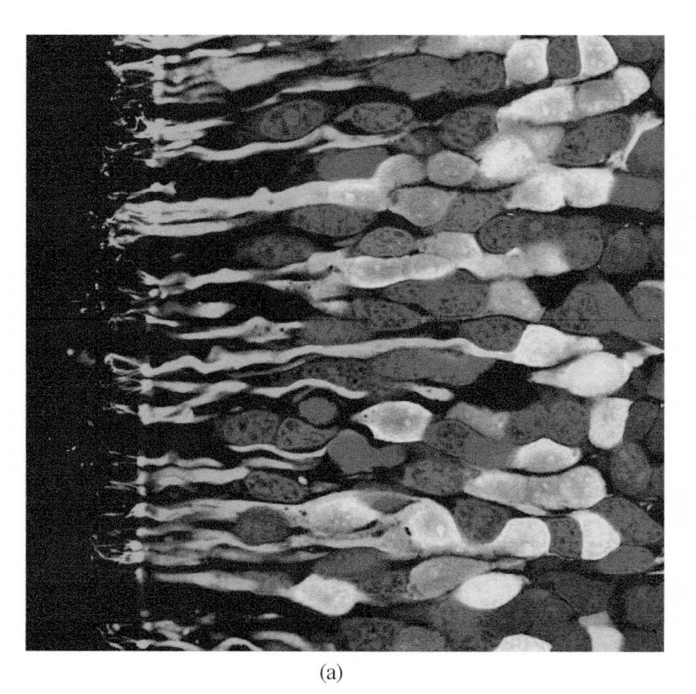

(a)

(b)

FIGURE 5.3 The sense of smell depends on rapidly moving olfactory receptors in membranes. (a) Several olfactory cilia are shown attached to the tip of each olfactory neuron in the olfactory organ of a frog. In humans about 40 million olfactory neurons producing some 200 million olfactory cilia line our nasal passage and allow us to detect about 10,000 different odors. Olfactory cilia of humans house about 350 different odor receptors, which bind odorants and then must quickly collide with another membrane-bound protein triggering the sense of smell. (b) Olfactory cilia of frogs at higher magnification. (Photos of olfactory cilia of larval *Xenopus* courtesy of Detlev Schild.)

Evans, and Dodd, 1989). We hypothesize that DHA and other fatty acids with similar dynamics are of great selective value for the evolution of sensitive and efficient olfactory neurons. Interestingly, olfactory neurons lining the nasal passage age and die in a natural cycle, and unlike neurons of the brain they are reborn from a lifetime supply of stem cells located at their base.

An elaborate signaling cascade in olfactory neurons modulates circulation of calcium ions (Ca^{++}) (Delmas, Crest, and Brown, 2004). Calcium plays a critical role in neurons as a "second messenger" for modulating the firing and transmission rates of synaptic signals. Calcium is considered a second messenger because it is directed to a specific site by a primary signal. Calcium cascades are of great importance in the brain and neurosensory system, and the main point here is not to elaborate on a detailed picture of olfaction but rather to view the cascade from a membrane perspective. Key components of the Ca^{++} cascade are intimately linked to membranes. For example, membrane proteins as key participants in the olfactory cascade collide with each other to pass on information from step to step along the cascade route. These collisions like those already discussed for vision occur between membrane-bound components and likely depend on DHA for maximizing rapid rates of interaction. One unique aspect of the Ca^{++} cascade is that membrane phospholipids working as intermediaries in the signaling process are membrane bound and depend on rapid membrane motion enabling collisions with enzymes, which convert the parent phospholipid to its hormonal form. Thus, membrane proteins as well as phospholipids and their derivatives are maintained in a state of extreme motion by DHA. A special kind of scissoring protein, a membrane-embedded enzyme called *phospholipase C*, breaks apart signaling phospholipids and releases derivatives that behave as potent modulators of biochemical processes. This implies an intimate relationship or contact between phospholipase C and its substrate within the membrane. DHA chains and cholesterol behave as opposites and together modulate rates of contact between phospholipase and its phospholipid target. Consequently, DHA governing the state of membrane motion is especially important in the case of Ca^{++} signaling essential for neuron function. The bulk fatty acid composition of the membrane contributes significantly to the state of membrane motion discussed above. However, recent data show that the ubiquitous signaling phospholipid phosphatidylinositol (PI) exists in molecular species whose fatty acid chains display different levels of unsaturation (Lee, Inoue, and Imaeetal, 2008). Different molecular species of phosphatidylinositol exhibited different bioactivities—a finding consistent with the concept that the motional state of this important signaling phospholipid is variable and dependent not only on bulk membrane motion but also on its own acyl chains. A mechanism for incorporating EPA into PI of *Caenorhabditis elegans* has been reported (Lee, Inoue, and Imaeetal, 2008).

5.3 MEMBRANE ASYMMETRY ENABLING LEAFLET-SPECIFIC MOTION IS IMPORTANT IN NEURONS

It is well known that the phospholipid classes and fatty acid compositions of the two leaflets or monolayers forming the bilayer of many membranes are strikingly different.

This difference between lipid compositions is called *membrane asymmetry*. There is considerable interest in defining the physiological or biochemical functions contributed by membrane asymmetry in specialized DHA-enriched membranes of neurons and other cells. Among human cells red blood cell membranes have been studied in detail. It has been found that different classes of phospholipids are localized in different leaflets defined as *cytosolic* or *inner membrane* (facing the cytoplasm) versus the outer leaflet in contact with the outside. In the case of axons the inner leaflet faces the high K^+ (inside) versus the outer leaflet facing high Na^+ (outside). For synaptosomes the inner leaflet faces the lumen of the vesicle, which is charged with neurotransmitters versus the outside, which faces the synaptic cytoplasm. Analysis of molecular species of phospholipids shows that DHA is often selectively enriched in certain classes of phospholipids. For example, in neurons DHA is enriched in negatively charged phospholipids (phosphatidylethanolamine [PE] and phosphatidylserine [PS]) that are often largely confined to the inner leaflet. In contrast positively charged phosphatidylcholine (PC) with less DHA generally occupies the outer leaflet of the bilayer. Membrane-bound phospholipid translocators (i.e., ATP-dependent flippases) generate and maintain membrane phospholipid asymmetry.

Membrane asymmetry, especially the localization of DHA in PE and PS present in the cytosolic leaflet, generates what might be called *leaflet-specific membrane motion*. This specialized form of membrane motion, especially lateral motion, allows numerous critical membrane proteins bound only to the cytosolic surface of the membrane to move rapidly. For example, many important signaling proteins whose mode of action involves collisions with other membrane proteins or components are attached only to the inner membrane surface. Often the attachment involves a myristate (14:0) chain covalently attached to the protein with the acyl chain acting as a membrane anchor. Other single leaflet anchors include PI, which is discussed above, and prephenate, which is synthesized via the terpenoid pathway. PI, discussed above, is present in small amounts as a membrane phospholipid building block where one of its most important roles is as a major signaling molecule. PI is activated as a potent hormone via cleavage by a membrane-bound phospholipase, which must collide with the PI substrate, an interaction enabled by DHA phospholipids. The simultaneous presence of leaflet-specific motion along with lipid rafts on one or both leaflets open Pandora's box regarding the dynamic nature and biochemical functions of different membrane surface assemblies. The role of PI-anchors in prion disease is discussed in Chapter 15.

In neurons an ATP-dependent enzyme called *flippase* maintains membrane asymmetry, while in senescing or dead neurons or dysfunctional axons-synapses an enzyme called *scramblase* (not dependent on ATP) may become activated and purposely destroy asymmetry. Scramblase works as the name implies by randomly flipping phospholipids from one leaflet to the other. In the absence of flippase activity, scramblase is believed to play a beneficial role in flipping DHA phospholipids and their dysfunctional oxidation products, normally found on the inside axon surface, to the outside where they are in a position to signal microglia and activate phagocytosis (see Chapters 8 and 17).

5.4 DHA SPEEDS UP RATES OF CYCLING OF SYNAPTIC VESICLES IMPORTANT IN FAST-FIRING NEURONS

An estimated 60 to 240 trillion synapses (Figure 5.4) are at work in neurons of the brain, and their role is to convert electrical signals terminating at the ends of axons into chemical signals traveling across the synaptic cleft to dendrites of the next neuron (Attwell and Gibb, 2005). Synapses are made of two opposing membranous regions—one found at the terminus of axons (presynaptic region) and the other located across the synaptic cleft at the beginning of the next neuron. The postsynaptic region is not shown in Figure 5.4. Dendrites relay electrical impulses from the postsynaptic region toward the head of the next neuron. The two opposing faces of synapses are separated by an aqueous region. This watery zone is the realm of neurotransmitters that are highly water-soluble molecules, which when released by synaptic vesicles diffuse rapidly across the synaptic cleft and bind to receptors in the postsynaptic region. Thus, neurotransmitters zip across, binding to receptors that fire an electrical impulse that travels along the next neuron. Because of their important roles in the brain and nervous system, an incredible amount of information is known about the nature of synapses. Synaptic vesicles (described in molecular detail in Chapter 6) store neurotransmitters for release across the synapse. In Figure 5.4 neurotransmitters are released as signaling molecules that are sensed by receptors on the postsynaptic side. Synaptic vesicles undergo an elaborate cycle including neurotransmitter loading, docking to the plasma membrane, neurotransmitter release, escape from the plasma membrane by the process of endocytosis, and finally, return to the synaptic vesicle pool. Evidence points to the synaptic vesicle cycle as a possible

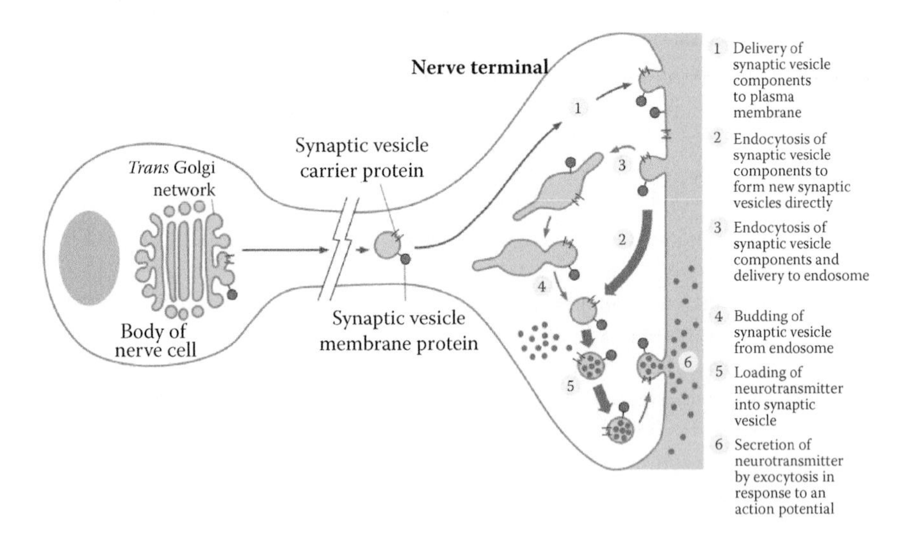

FIGURE 5.4 Formation, cycling, and function of synaptic vesicles. These tiny DHA-enriched vesicles originate from the *trans* Golgi network and are delivered specifically to the presynaptic membrane. Six steps of synaptic cycling are described in the diagram. (From Alberts, B. et al., *Molecular Biology of the Cell*, Copyright 2008. Reproduced by permission of Taylor & Francis Group, LLC, a division of Informa plc.)

rate-determining step governing the overall speed of electrical transmissions, especially in fast-firing neurons.

As shown in Figure 5.4, synaptic vesicles deliver neurotransmitters at synapses by fusing with the presynaptic membrane. This process is called *exocytosis*. The reverse of this process is called *endocytosis*. Endocytosis involves extremely rapid reformation of synaptic vesicles often using the same exocytotic vesicle membrane surface as a substrate for newly minted vesicles. It has been known for many years that this rapid pinching-off process requires numerous specific proteins that trigger and guide the endocytotic pathways. Mutational analysis has clearly demonstrated that endocytosis stalls when key proteins, including dynamin, are mutated (see Figure 5.5a,b). Figure 5.5a,b summarize data from *Drosophila*, which show

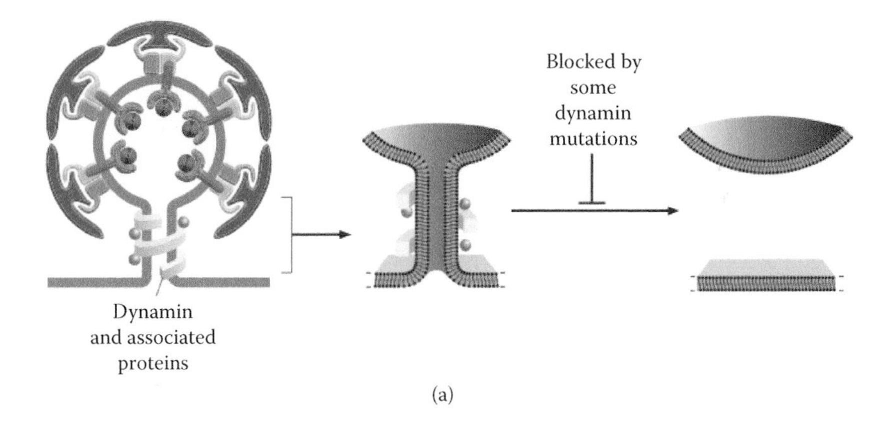

FIGURE 5.5 Omega-3 fatty acids maximize neuromuscular function of *Caenorhabditis elegans* at the level of endocytosis of synaptic vesicles. (a) How mutations in dynamin, essential for endocytosis, block the synaptic cycle in *Drosophila*. (From Alberts, B. et al., *Molecular Biology of the Cell*, Copyright 2008. Reproduced by permission of Taylor & Francis Group, LLC, a division of Informa plc.) (b) Mutations in dynamin cause accumulation of synaptosomes unable to complete endocytosis in synapses resulting in paralysis of *Drosophila*. (Reproduced with permission of the Society of Neuroscience from Koenig. J. H., and K. Ikeda, 1989, Disappearance and Reformation of Synaptic Vesicle Membrane upon Transmitter Release Observed under Reversible Blockage of Membrane Retrieval, *J. Neurosci*., 9: 3844–3860, Copyright 1989. Permission conveyed through Copyright Clearance Center, Inc.) (c) EPA and DHA maximize neuromuscular function of *C. elegans* using EPA-minus mutants. Note, purified EPA and DHA as well as aliquots taken from fish oil tablets packaged for human consumption increase body bends from about one per second to three. Researchers have found that EPA/DHA maximize synaptosomal cycling as discussed in the text. (See Lesa, G. M., M. Palfreyman, D. H. Hall et al., 2003, Long-Chain Polyunsaturated Fatty Acids Are Required for Efficient Neurotransmission in *C. elegans*, *J. Cell Sci*., 116: 4965–4975; Marza, E., T. Long, A. Saiardi et al., 2008, Polyunsaturated Fatty Acids Influence Synaptojanin Localization to Regulate Synaptic Vesicle Recycling, *Mol. Biol. Cell*, 19: 833–842.) (From Valentine, R., and Valentine, D., *Omega-3 Fatty Acids and the DHA Principle*. Copyright 2009. Reproduced by permission of Taylor & Francis Group, LLC, a division of Informa plc.) (d) *C. elegans* is a voracious feeder using powerful and fast-acting body-bending muscles for propulsion. The scale bar in the photo is 100 microns. (Photo courtesy of Dr. Ian Chin-Sang, Queen's University, Kingston, Ontario, Canada.)

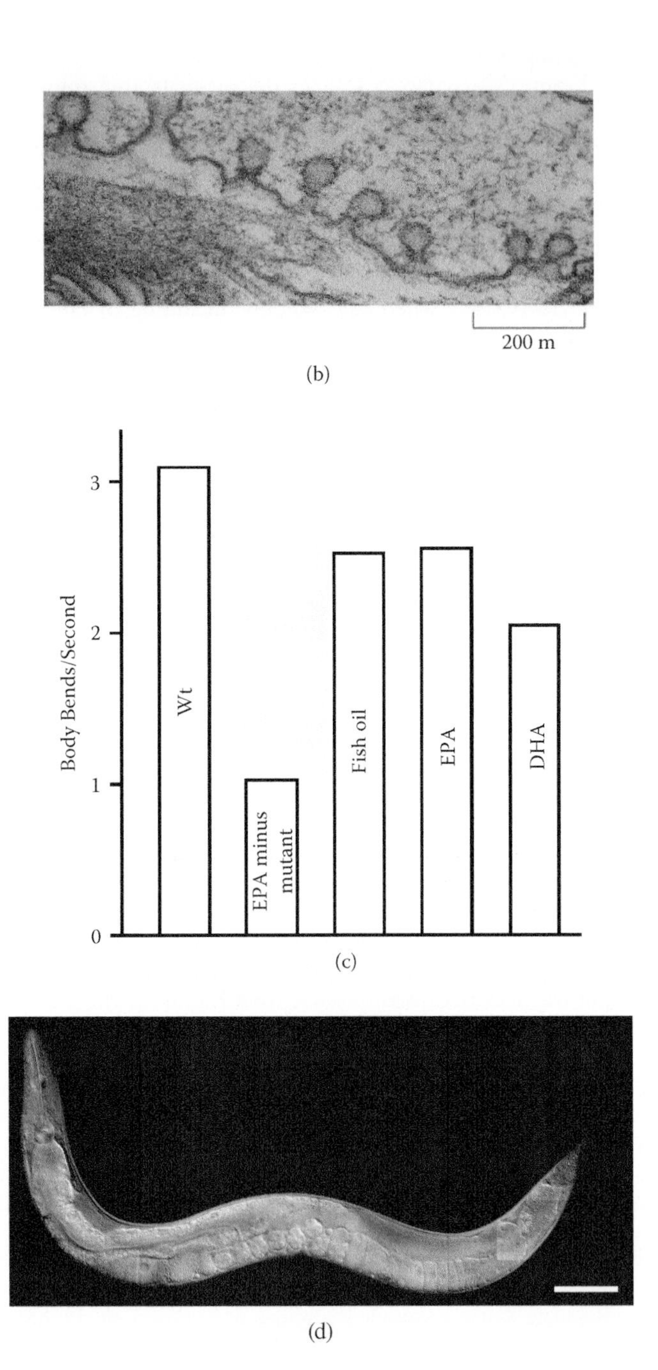

FIGURE 5.5 *(Continued)*

that mutations in dynamin, essential for synaptic cycling at the level of endocytosis, cause the accumulation of incompletely closed synaptic vesicles at synapses. This mutation causes paralysis in *Drosophila*. A similar experiment has been conducted with the nematode *C. elegans* with the exception that instead of dynamin being blocked, the synthesis of the omega-3 fatty acid EPA was blocked by a targeted mutation. EPA minus mutants displayed a phenotype in which neuromuscular activity is slowed significantly, although the worms are not paralyzed as were *Drosophila*.

As background, *C. elegans*, shown in Figure 5.5d, contains about 302 neurons compared to our 100 billion and has perhaps the best understood neurons in nature. The knowledge of the molecular biology of neurons of *C. elegans* is monumental and provides a framework for understanding the speed of synaptic cycling as a critical parameter of neuronal function and evolution in all animals. Neuronal speed in this case is defined as rates of electrical flow or impulses along neural connections ending in repetitive firing of powerful bending muscles needed for rapid movement. *C. elegans* is a voracious feeder, depending on its bending muscles for propulsion as it searches for food. This powerful muscle group allows the worm to bend at the rate of about three bends per second. It is known that the firing rate of these muscles declines sharply in worms deprived of EPA by a mutational block.

Figure 5.5c summarizes the reproduction of an earlier experiment conducted by Lesa et al. (2003) and duplicated by Luke Hillyard and Raymond Valentine in Bruce German's laboratory at the University of California–Davis. This experiment shows the stimulatory effects of either EPA, DHA, or fish oil, the latter taken from capsules from a health food store, on the neuromuscular firing rate of EPA minus mutants of *C. elegans*. The rates of muscle firing in this case can be monitored by eye using the assistance of a low-magnification microscope. Using a handheld counter, one can easily measure the rates of muscle bending. The rate measurements shown on the graph were taken by this simple method. For this experiment *C. elegans* is deprived of any available EPA by a knockout mutation in its natural pathway of synthesizing this membrane fatty acid. The mutant used in this experiment was generated in Browse's laboratory (Watts and Browse, 2002). Lack of EPA is not lethal, and mutants carrying this defective gene can be grown under laboratory conditions. One explanation is that EPA minus mutants are still able to produce di-polyunsaturated molecular species of phospholipids whose properties are similar to EPA phospholipids. However, in its natural environment it is unlikely to survive. We were able to confirm results that showed that DHA, while not normally produced by *C. elegans*, nevertheless substitutes for EPA when present in the diet. However, considerable retro-conversion of DHA back to EPA appears to occur, confirming results by Lesa et al. (2003). One explanation of why *C. elegans* converts DHA → EPA is that EPA is significantly more beneficial than DHA in neurons of *C. elegans*. Also note that aliquots of fish oil concentrate taken directly from fish oil capsules packaged for human consumption restore faster rates of muscle action (Figure 5.5c). DHA, EPA, and arachidonic acid (20:4) are all present in these capsules. Thus, studies with *C. elegans* help establish the essential neuromuscular functions of fish oils and point toward a vital role or roles in firing of fast muscles.

Recently, Marza et al. (2008) have shown that mutants of *C. elegans* unable to produce eicosapentaenoic acid (20:5) display a reduced capacity for synaptic vesicle endocytosis. The mutants have abnormally low levels of the phosphoinositide phosphatase, synaptojanin, at presynaptic release sites and accumulate the main synaptojanin substrate phosphatidylinositol 4,5 bisphosphate at these sites. These defects in vesicle endocytosis can be rescued by providing dietary arachidonic acid, which is incorporated directly into membranes or converted to EPA, suggesting that the endocytosis defect is caused by depletion of 20:4 and 20:5 in synaptosomal membranes. These data show that the genes *fat-3* (Watts and Browse, 2002) and *synaptojanin* (Harris et al., 2000) act in the same endocytic pathway at synapses. These findings suggest that 20:4 and 20:5 are required for efficient synaptic vesicle recycling, probably by modulating the localization of synaptojanin or PI (or both) at endocytotic sites on membranes of synapses. Studies with rats chronically deprived of highly unsaturated fatty acids in the diet show a decrease in synaptic vesicle numbers at the presynaptic sites of dopaminergic neurons of the frontal lobes (Zimmer et al., 2000). Kim, Chang, and Danielletal (2002) used mouse knockout mutants to demonstrate that synaptojanin is important for the cycling of synaptic vesicles in neurons subjected to prolonged stimulation. Thus, the important contribution of highly unsaturated membranes for synaptic transmission is likely a universal property. Note that mutants of *C. elegans* unable to produce 20:4 or 20:5 are still able to synthesize 18:3, which can be converted to di-18:3 molecular species of phospholipids. These phospholipids are likely used for the synthesis of membranes of synaptosomes in EPA minus mutants whose membranes are devoid of EPA. These data suggest that two additional double bonds in EPA, compared to linoleic acid (18:3), are required to maximize neuromuscular function in *C. elegans*.

A working model for the role of DHA/EPA in endocytosis of synaptic vesicles includes the following points:

- Synaptojanin has dual biochemical functions as a phosphoinositide phosphatase and as a binding or scaffolding protein for recruiting other endocytic proteins into complexes.
- Synaptojanin plays important roles in initiating or triggering formation of shallow clathrin-coated pits, an essential early stage of endocytosis of synaptosomes.
- Synaptojanin is not initially localized or enriched in membranes forming clathrin-coated pits but is recruited and moves laterally across the membrane from nearby presynaptic membrane surfaces.
- It is proposed that DHA/EPA contributes bulk membrane motion enabling extreme membrane mobility at endocytic sites. Increasing the collision rates of synaptojanin, sparsely distributed on the presynaptic membrane, with its membrane-bound inositol phosphate, substrates are proposed to enable rapid rates of synaptosomal cycling.
- *C. elegans* has evolved a mechanism for selective incorporation of EPA and other polyunsaturated fatty acids into PI presumably for the purpose of increasing the rates of collisions between PI and synaptojanin.
- The mode of action of DHA/EPA in endocytosis of synaptic vesicles is envisioned to maximize the rate of collisions among critical membrane components similar to the trigger reactions for vision and other sensory systems.

In summary, the unrivaled conformational dynamics of DHA chains are harnessed for multiple essential biochemical roles in the human brain and nervous system. For example, vision and olfaction are dependent on DHA for driving fast rates of collisions between photoreceptors or odor receptors and their respective G proteins triggering these vital sensory cascades. The exact biochemical functions of DHA in neurons are not fully understood, but recent data support a universal role in maximizing rates of synaptosomal cycling. Data from *C. elegans* suggest that EPA increases the rates of collisions between synaptojanin and its phospholipid substrate (PI), a prerequisite for closure of synaptic vesicles before their return via endocytosis to the free pool of synaptosomes (Marza et al., 2008). Thus, EPA is believed to enable faster firing or repetition rates of neural impulses in *C. elegans*. In humans, DHA likely maximizes rates of endocytosis of synaptic vesicles helping overcome a rate-limiting step in neural biochemistry.

REFERENCES

Anderson, J. M., and B. Andersson. 1982. The architecture of photosynthetic membranes: Lateral and transverse organization. *Trends Biochem. Sci.* 7:288–292.

Anderson, R. E. 1970. Lipids of ocular tissues IV. A comparison of the phospholipids from the retina of six mammalian species. *Exp. Eye Res.* 10:339–344.

Attwell, D., and A. Gibb. 2005. Neuroenergetics and the kinetic design of excitatory synapses. *Nat. Rev. Neurosci.* 6:841–849.

Bruckert, F., M. Chabre, and T. M. Vuong. 1992. Kinetic analysis of the activation of transducin by photoexcited rhodopsin. Influence of the lateral diffusion of transducin and competition of guanosine diphosphate and guanosine triphosphate for the nucleotide site. *Biophys. J.* 63:616–629.

Buchholz, F., and R. Saborowski. 2000. Metabolic and enzymatic adaptations in northern krill, *Meganyctiphanes norvegica*, and Antarctic krill, *Euphausia superba*. *Can. J. Fish. Aquat. Sci.* 57(Suppl. 3):115–129.

Calvert, P. D., V. I. Govardovskii, and N. Krasnoperovaetal. 2001. Membrane protein diffusion sets the speed of rod phototransduction. *Nature* 411:90–94.

Delmas, P., M. Crest, and D. A. Brown. 2004. Functional organization of PLC signaling microdomains in neurons. *Trends Neurosci.* 27:41–47.

Eldho, N. V., S. E. Feller, S. Tristram-Nagle et al. 2003. Polyunsaturated docosahexaenoic vs. docosapentaenoic acid—Differences in lipid matrix properties from the loss of one double bond. *J. Am. Chem. Soc.* 125:6409–6421.

Feller, S. E. 2008a. Lipid animation: 500 ps dynamics of a stearoyl docosohexaenoyl PC lipid molecule (14MB), http://lipid.wabash.edu/, Wabash College Chemistry Department, Crawfordville, IN.

Feller, S. E. 2008b. Acyl chain conformations in phospholipid bilayers: A comparative study of docosahexaenoic acid and saturated fatty acids. *Chem. Phys. Lipids* 153:76–80.

Feller, S. E., K. Gawrisch, and A. D. MacKerell, Jr. 2002. Polyunsaturated fatty acids in lipid bilayers: Intrinsic and environmental contributions to their unique physical properties. *J. Am. Chem. Soc.* 124:318–326.

Feller, S. E., and K. Gawrisch. 2005. Properties of docosahexaenoic acid-containing lipids and their influence on the function of rhodopsin. *Curr. Opin. Struct. Biol.* 15:416–422.

Feller, S. E., K. Gawrisch, and T. B. Woolf. 2003. Rhodopsin exhibits a preference for solvation by polyunsaturated docosahexaenoic acid. *J. Am. Chem. Soc.* 125:4434–4435.

Grossfield, A., S. E. Feller, and M. C. Pitman. 2006. A role for direct interactions in the modulation of rhodopsin by ω-3 polyunsaturated lipids. *Proc. Natl. Acad. Sci. USA* 103:4888–4893.

Harris, T. W., E. Hartwieg, H. R. Horvitz et al. 2000. Mutations in synaptojanin disrupt synaptic vesicle recycling. *J. Cell. Biol.* 150:589–600.

Jeffrey, B. G., H. S. Weisinger, M. Neuringer et al. 2001. The role of docosahexaenoic acid in retinal function. *Lipids* 36:859–871.

Kahlert, M., and K. P. Hofmann. 1991. Reaction rate and collisional efficiency of the rhodopsin-transducin system in intact retinal rods. *Biophys. J.* 59:375–386.

Kim, W. T., S. Chang, and L. Danielletal. 2002. Delayed reentry of recycling vesicles into the fusion-competent synaptic vesicle pool in synaptojanin 1 knockout mice. *Proc. Natl. Acad. Sci. USA* 2002. 99:17143–17148.

Koenig, J. H., and K. Ikeda. 1989. Disappearance and reformation of synaptic vesicle membrane upon transmitter release observed under reversible blockage of membrane retrieval. *J. Neurosci.* 9:3844–3860.

Koynova, R., and M. Caffrey. 1998. Phases and phase transitions of the phosphatidylcholines. *Biochim. Biophys. Acta* 1376:91–145.

Lee, H. C., T. Inoue, and R. Imaeetal. 2008. *Caenorhabditis elegans* mboa-7, a member of the MBOAT family, is required for selective incorporation of polyunsaturated fatty acids into phosphatidylinositol. *Mol. Biol. Cell.* 19:1174–1184.

Lesa, G. M., M. Palfreyman, D. H. Hall et al. 2003. Long chain polyunsaturated fatty acids are required for efficient neurotransmission in *C. elegans. J. Cell Sci.* 116:4965–4975.

Liebman, P. A., and G. Entine. 1974. Lateral diffusion of visual pigment in photoreceptor disk membranes. *Science* 185:457–459.

Litman, B. J., S.-L. Niu, A. Polozova et al. 2001. The role of docosahexaenoic acid containing phospholipids in modulating G protein-coupled signaling pathways: Visual transduction. *J. Mol. Neurosci.* 16:237–242; discussion 279–284.

Malkin, R., and K. Niyogi. 2000. Photosynthesis. In: Buchanan, B. B., W. Gruissem, and R. Jones, editors. *Biochemistry and Molecular Biology of Plants.* Rockville, MD: American Society of Plant Physiologists, pp. 568–628.

Malnic, B., P. A. Godfrey, and L. B. Buck. 2004. The human olfactory receptor gene family. *Proc. Natl. Acad. Sci. USA* 101:2584–2589.

Marza, E., T. Long, A. Saiardi et al. 2008. Polyunsaturated fatty acids influence synaptojanin localization to regulate synaptic vesicle recycling. *Mol. Biol. Cell.* 19:833–842.

Niu, S.-L., D. C. Mitchell, S.-Y. Lim et al. 2004. Reduced G protein-coupled signaling efficiency in retinal rod outer segments in response to *n*-3 fatty acid deficiency. *J. Biol. Chem.* 279:31098–31104.

Niu, S. L., and D. C. Mitchell. 2005. Effect of packing density on rhodopsin stability and function in polyunsaturated membranes. *Biophys. J.* 89:1833–1840.

Poo, M., and R. A. Cone. 1974. Lateral diffusion of rhodopsin in the photoreceptor membrane. *Nature* 247:438–441.

Russell, Y., P. Evans, and G. H. Dodd. 1989. Characterization of the total lipid and fatty acid composition of rat olfactory mucosa. *J. Lipid Res.* 30:877–884.

Soubias, O., and K. Gawrisch. 2007. Docosahexaenoyl chains isomerize on the sub-nanosecond time scale. *J. Am. Chem. Soc.* 129:6678–6679.

Stillwell, W., and S. K. Wassall. 2003. Docosahexaenoic acid: Membrane properties of a unique fatty acid. *Chem. Phys. Lipids* 126:1–27.

Stinson, A. M., R. D. Wiegand, and R. E. Anderson. 1991. Fatty acid and molecular species compositions of phospholipids and diacylglycerols from rat retinal membranes. *Exp. Eye Res.* 52:213–218.

Watts, J. L., and J. Browse. 2002. Genetic dissection of polyunsaturated fatty acid synthesis in *Caenorhabditis elegans. Proc. Natl. Acad. Sci. USA* 99:5854–5859.

Zimmer, L., S. Delpal, D. Guilloteau et al. 2000. Chronic *n*-3 polyunsaturated fatty acid deficiency alters dopamine vesicle density in the rat frontal cortex. *Neurosci. Lett.* 284:25–28.

6 DHA Improves Energy Efficiency in Neurons

We look back in time to an implausible scenario in the age of hunter-gatherers to gain an appreciation of the importance of brain energy efficiency for the evolution of our species. Imagine getting up in the morning, eating breakfast, brushing your teeth, taking the kids to arrow-making school, and then starting your day's routine of searching for enough food to feed your family. From 9:00 A.M. to noon you roam and search for enough food to satisfy the energy needs of your own brain. It is well after dark before you return home with food for the rest of your family. Because 20 to 25 percent of our ancestor's waking time is a minimum required for the gathering of brain food, it is easy to define brain efficiency in terms of the hours spent each day on feeding the voracious energy appetite of this 3-pound organ alone. If the brain were miraculously made more energy efficient; say by 33 percent, our ancestors could have knocked off looking for their own brain food at 11:00 A.M. rather than noon (i.e., more leisure time) and could have returned home before dark in time to avoid saber-toothed tigers that hunted at dusk. But this brain miracle never happened; instead brain efficiency is believed to have maxed out by the era of hunter-gatherers. Obviously, as we enter the age of obesity, finding enough food for our brain is not a high priority; but keep in mind that there are still parts of the world where it does matter.

6.1 A SIGNIFICANT AMOUNT OF BRAIN ENERGY IS USED TO MAINTAIN NEURONS READY TO FIRE DAY AND NIGHT

To explain the concept of the high cost of maintenance energy in the brain, we take a chapter from the U.S. Department of Defense. Imagine the maintenance cost of generations of antiballistic missiles and their support facilities. Developed only to intercept and explode incoming, hostile nuclear missiles, the antiballistic missile and its facility require maintenance day and night, year after year, and perhaps decade after decade before new models supersede it. Thus, the missile must always be maintained at great energy cost waiting for an event that we hope never happens. Obviously, test-firing the missile burns up a considerable amount of fuel as does firing of a neuron. However, this cost of firing is small compared to the "energy cost" associated with maintaining a state of readiness. We see neurons in the same light; for example, our brain works all night although we are aware of it only through dreams.

We believe that maintaining our 100 billion neurons and their vast circuitry in a state ready to fire electrical impulses (Figure 6.1) uses up about half of our brain food. This is likely the biggest piece of the energy pie and largely accounts for energy efficiency or, conversely, energy inefficiency in the brain. The value of about half of

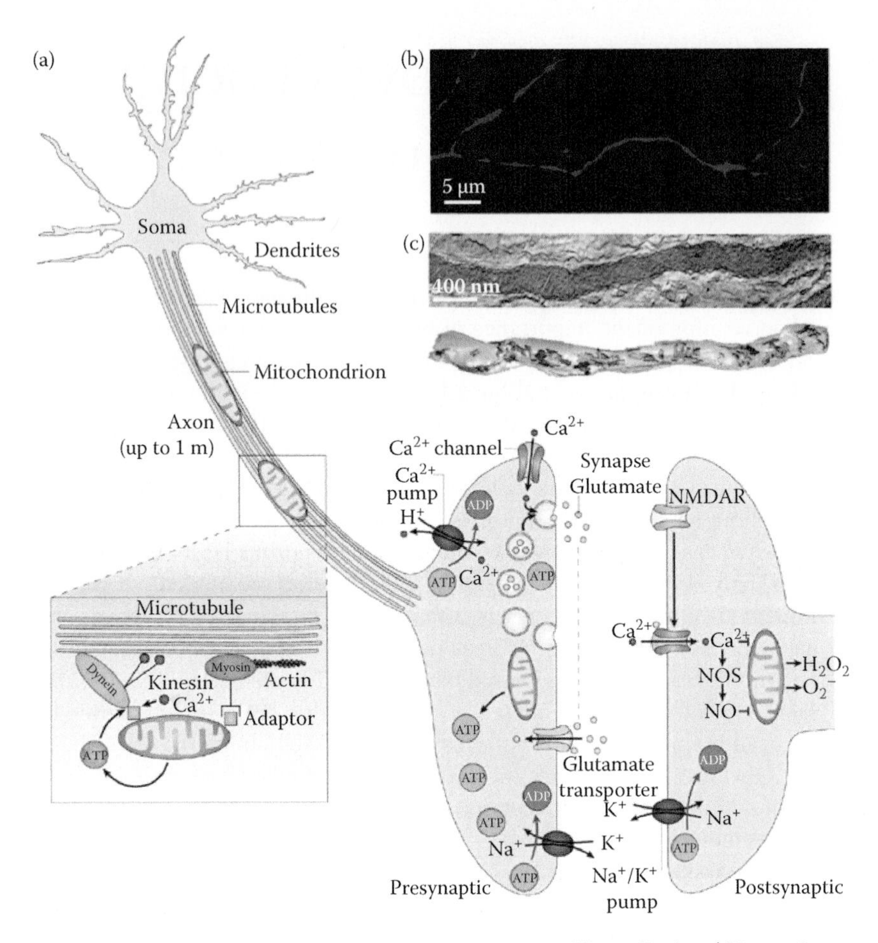

FIGURE 6.1 Neurons produce and consume a prodigious amount of energy. (a, b, c) Energy produced by mitochondria is consumed to energize numerous important processes in neurons. These include balancing ionic gradients such as sodium and potassium, energizing synaptosomes, modulating circulation of Ca^{++} ions as signaling molecules, driving nutrient uptake, and even energizing the "walking proteins" dynein and kinesin, which move cellular organelles such as mitochondria and other material for long distances back and forth through axons. (Reprinted by permission from Macmillan Publishers Ltd: *Nature Rev. Neuro.*, Knott et al., Copyright 2008.)

total brain energy comes from studies of the energy costs of maintaining ionic balance in various kinds of animal cells (Ames et al., 1992; Haines, 2001; Isler and van Schaik, 2006). Estimates for human cells run from more than 80 percent of the total energy of red blood cells devoted to ionic balance to 50 to 55 percent for cells such as neurons (see Valentine and Valentine, 2009).

Neuron membranes as well as all natural membranes tend to spontaneously leak ions such as sodium, potassium, and protons in a downhill fashion (i.e., from

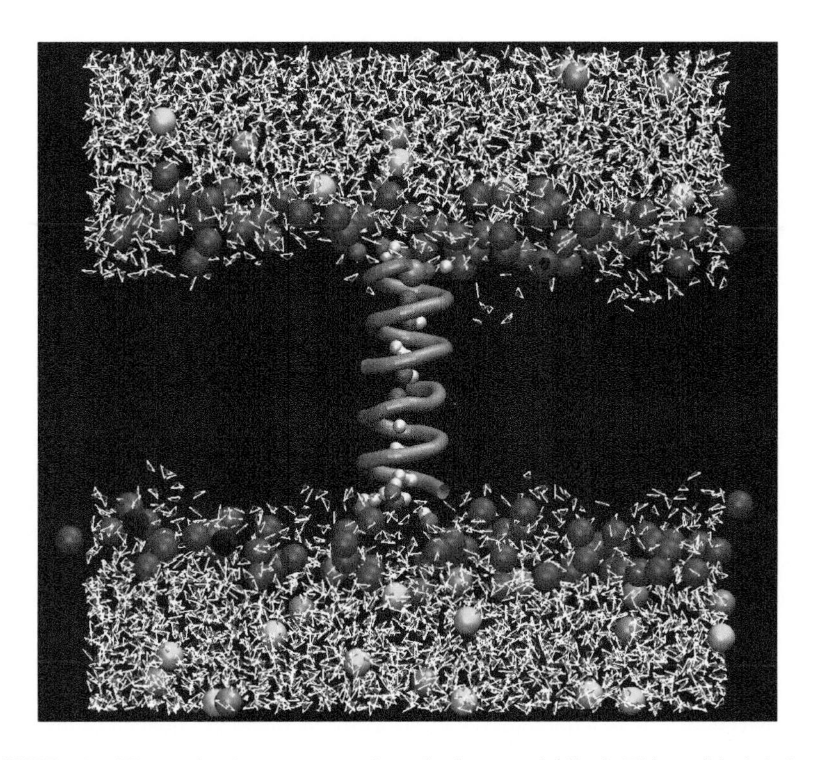

FIGURE 6.2 Water-wire theory was confirmed using gramicidin A. This antibiotic behaves as a membrane-spanning, energy-uncoupling nanotube holding 21 water-wires in single file inside its pore. Membrane water-wires are believed to occur spontaneously in all membranes and more readily in DHA membranes. Water-wire theory states that threads of water form spontaneously in DHA membranes of neurons, wasting a great deal of energy and driving up the already prodigious cost of fueling the brain. Ions including protons, sodium, and potassium move with blazing speed across water-wires and uncouple critical ion gradients. DHA, along with certain of its oxidation products, is believed to favor formation of water-wires in neuron membranes. (Image generated by chemical simulation.) (Courtesy of Serdar Kuyucak.) **(See color insert.)**

high to low concentrations such as high sodium outside of neurons compared to low inside). It is now believed that sodium, potassium, and protons (H^+) cross vast tracts of neural membrane surfaces by a most unusual path called *cation tunneling* (Andersen, Koeppe, and Roux, 2005; Nagle and Morowitz, 1978; Nichols and Deamer, 1980; Venable and Pastor, 2002). According to this process molecular threads composed of water molecules in single-file form spontaneously and momentarily across the plasma membranes of neurons allowing sodium to move from outside to inside and potassium to move along a similar wire in the opposite direction. This tunneling mechanism for cations, envisioned to occur in axons, has been convincingly proven using a naturally occurring antibiotic that kills target bacteria (Figure 6.2). The antibiotic gramicidin A behaves as an energy-uncoupling nanotube (Andersen, Koeppe, and Roux, 2005; Baştuğ, Patra, and Kuyucak, 2006; Wallace and Ravikumar, 1988) that enters and spans the membrane of a target cell. Twenty-one water molecules held single file in its core allow gramicidin A to form an amazing conduit for cations. Similar spontaneously generated water-wires forming in membranes are proposed to open a circuit for movement or flow of protons >>> sodium > potassium from high to low concentrations, breaching the permeability barrier of all natural membranes (Deamer, 1987). Protons have been found to move piggyback across a gramicidin-generated water-wire at rates of 10 million per second. This quickly bleeds all of the energy out of the target cell causing energy starvation and death, defining the role of gramicidin A as an energy-uncoupling antibiotic. Sodium moves along water-wires more slowly than protons, but because Na^+ levels surrounding neurons are extremely high compared to protons, massive losses of sodium are believed to occur due to spontaneous formation of water-wires in neuron membranes. Potassium can also leak out of neurons via a water-wire process but at a slower rate compared to Na^+. Thus, water-wires forming spontaneously in neuron membranes (Holte, Separovic, and Gawrisch, 1996; Huster et al., 1997; Infante, Kirwan, and Brenna, 2001; Stillwell et al., 1997) are believed to waste a great deal of energy and in the end are a bitter pill that must be swallowed, because this mechanism of futile cycling of cations is unavoidable. (Perhaps on another planet living computers work more energy efficiently but not on the planet Earth.) Spurious opening of Na^+ and K^+ gates or channels (i.e., when these gates open unintentionally) can also cause a futile cycle of excitatory cations. But emphasis here is on the water-wire mechanism, a form of futile energy cycling that uncouples cation electrochemical gradients. For instance, in the vast inner membranes of mitochondria, water-wires proliferate and contribute to energy stress in neurons.

6.2 TINY SYNAPTIC VESICLES SHOW HOW ENERGY EFFICIENCY IN DHA-ENRICHED MEMBRANES IS IMPROVED BY USING CHOLESTEROL TO PLUG PROTON LEAKS

One hundred thousand trillion synaptic vesicles (see Table 6.1 and Figures 6.3 and 6.4) operating in the one hundred trillion synapses in our brain cannot be wrong. Synaptic vesicles are near nano-sized membrane vesicles essential for delivering neurotransmitters at synapses (Takamori et al., 2006), which connect all neurons. The nature of DHA-enriched membranes of synaptic vesicles is explored next to show how DHA working with cholesterol improves energy efficiency. We hypothesize that cholesterol works by plugging or blocking formation of proton-conducting water-wires. Figure 6.3 shows images of individual synaptic vesicles seen by electron microscopy. Figure 6.4 shows an image of a single synaptic vesicle (Takamori et al., 2006) cut in half like a split tennis ball. Note that a protein called H^+/glutamate transporter is shown embedded in the membrane. This powerful pump concentrates the neurotransmitter glutamate inside, concurrently releasing protons (H^+) from the inside to the outside of the vesicle. The membrane portion of the vesicle formed by DHA, cholesterol, and other building blocks is depicted in Figure 6.4 as wiggly yellow tails, but individual lipid molecular species are not identified. Note that a synaptic vesicle membrane is seen packed with a variety of specialized proteins covering much of the membrane surface. The largest protein labeled V-ATPase (see bottom panel of Figure 6.3) is especially important in bioenergetics of synaptosomes, as described shortly. The tiny surface area of individual vesicle membranes can be estimated due to the relatively few fatty acids needed to build these membranes (Table 6.1). Altogether there are about 14,000 fatty acid tails in phospholipids forming the membrane surface, including an estimated 2000 DHA chains. The size of a synaptic vesicle, based on membrane surface area, is about 1/140 that of a single bacterial cell such as *Escherichia coli*. That is, an estimated 100 or so synaptic vesicles would fit inside a single bacterial cell. This detailed picture of the molecular architecture of synaptic vesicles and their membranes helps integrate ideas discussed above and later on the nature of the membrane as a permeability barrier not only in neurons but in all cells.

TABLE 6.1
Membrane Properties of Synaptic Vesicles

Physical Parameters	
Density (g/mL)	1.10
Outer diameter (nm)	41.6
Inner aqueous volume (l)	19.86×10^{-21}
Number of neurotransmitter molecules (at 150 mM)	1790
Mass (g)[a]	29.6×10^{-18}
Mass (MDa)[a]	17.8
Protein:phospholipids (w:w)	1.94
Phospholipids:cholesterol (mol:mol)	1:0.8
Transmembrane domains (number/% of surface coverage)[b]	600/20
Protein Stoichiometry (Copies/Vesicle)	
Synaptophysin	31.5
Synaptobrevin/VAMP2	69.8
VGLUT1[c]	9.0
VGLUT2[c]	14.4
Synapsins	8.3
Syntaxin 1	6.2
SNAP-25	1.8
Synaptotagmin	15.2
Rab3A	10.3
SV2	1.7
Synaptogyrin	2.0
SCAMP1	0.8
CSP	2.8
V-ATPase[d]	1.4
NSF (hexamer)	0.2

Continued

TABLE 6.1 (Continued)
Membrane Properties of Synaptic Vesicles

Membrane Lipids	
Phospholipids total (number/% of surface coverage)	6992/50.4
Phosphatidylcholine	2524
Phosphatidylethanolamine (C1-ester/C1-ether)	1621/1311
Phosphatidylserine	857
Phosphatidylinositol	132
Sphingomyelin	516
Cholesterol	5663
Hexosylceramide	108

Source: Modified from Takamori, S., M. Holt, K. Stenius et al., 2006, Molecular Anatomy of a Trafficking Organelle. *Cell,* 127: 831–846. With permission.

Note: Mean values are shown.

[a] Calculated indirectly using the vesicle number and the sum of protein and lipid masses. In agreement with this result, a mass of (26.4 ± 5.8) × 10^{-18} g was measured directly with scanning transmission electron microscopy (STEM), which corresponds to 15.9 ± 3.5 MDa.

[b] Estimate; the number calculated from the proteins included in the model amounts to 497.

[c] Corrected for the fraction of vesicles found to be positive by immunogold electron microscopy for the respective transporter.

[d] Adjusted to compensate for the loss of the V1 subunit.

Concentrating glutamate as a neurotransmitter inside synaptic vesicles requires considerable energy to fuel the glutamate uptake pumps. Before the loading of glutamate can begin, the vesicle interior must be energized or charged with proton fuel (proton motive energy). The charging of protons is carried out by a single molecule of V-ATPase embedded in the vesicle membrane. V-ATPase consumes energy as ATP to pump protons into the vesicle lumen, thus generating a proton energy gradient, high proton levels inside and low outside. ATP is provided by mitochondria operating nearby. In essence glutamate pumps located in the vesicle membrane utilize the proton gradient (high inside) to energize the influx of glutamate. Up to 5000 synaptic vesicles operate at each presynaptic region, consuming a significant amount of the energy budget of neurons.

The presence of DHA chains helping form the permeability barrier against protons in synaptic vesicle membranes would seem to be a poor choice because these chains are believed to create a mediocre molecular architecture for blocking proton leaks. Based on established rules defining membrane permeability, DHA is predicted to cause serious proton leakage across vesicle membranes. This spontaneous leakage defines a futile and wasteful energy cycle in synaptosomes because more ATP must be used to replenish lost H^+ and no useful work is accomplished. Theoretically, one way out of this quandary is to reinforce the proton barrier architecture of synaptic vesicle membranes using another readily available lipid that compensates for

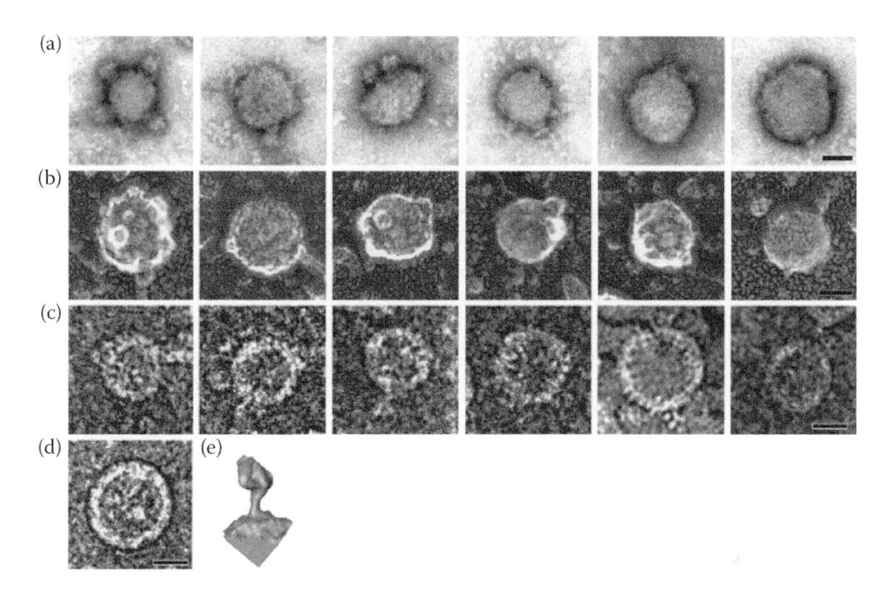

FIGURE 6.3 Synaptic vesicles imaged by electron microscopy procedures. (a) Negatively stained by uranyl acetate; scale bar = 20 nm. (b) Platinum shadowed after quick freeze/deep etching. (c) Native cryopreparation. (d) as in (c) but after pronase digestion of surface proteins. (e) Electron tomographic model of a single, large, membrane-bound protein complex, presumably V-ATPase. (Reprinted from *Cell*, 127, S. Takamori et al., Molecular Anatomy of a Trafficking Organelle, 831–846, Copyright 2006, with permission from Elsevier.)

the mediocre permeability properties of DHA and acts to bolster the permeability barrier against protons. Recent data using bacteria (Welander et al., 2009) is consistent with the theory (Haines, 2001) that cholesterol acts as an excellent proton blocker and works to tighten membrane architecture against proton leakage. This mechanism (Figure 6.5) might account for the large amount of cholesterol present along with DHA in membranes of synaptic vesicles. We suggest that the synergism between DHA and cholesterol pays off handsomely from an energy efficiency perspective and in one sense allows synaptic vesicles to have their cake and eat it too—in essence maximizing the amazing speed benefits of DHA and shoring up a weak point involving proton permeability. However, there is no free lunch because of the high metabolic energy cost of synthesizing cholesterol in the first place. On balance though, cholesterol seems to be a bargain measured in terms of energy saved by synaptic vesicles and other excitatory membranes.

Thus, DHA in combination with cholesterol is believed to significantly improve energy efficiency in synaptic vesicles. Note that energy efficiency in axonal plasma membranes in contrast to synaptic vesicles is concerned primarily with sodium and potassium permeability (Ames et al., 1992; Niven et al., 2007) rather than protons. DHA and cholesterol in combination are believed to greatly improve energy efficiency in axons because it is far easier to plug membrane energy leaks against these much larger cations compared to protons (Figure 6.5d).

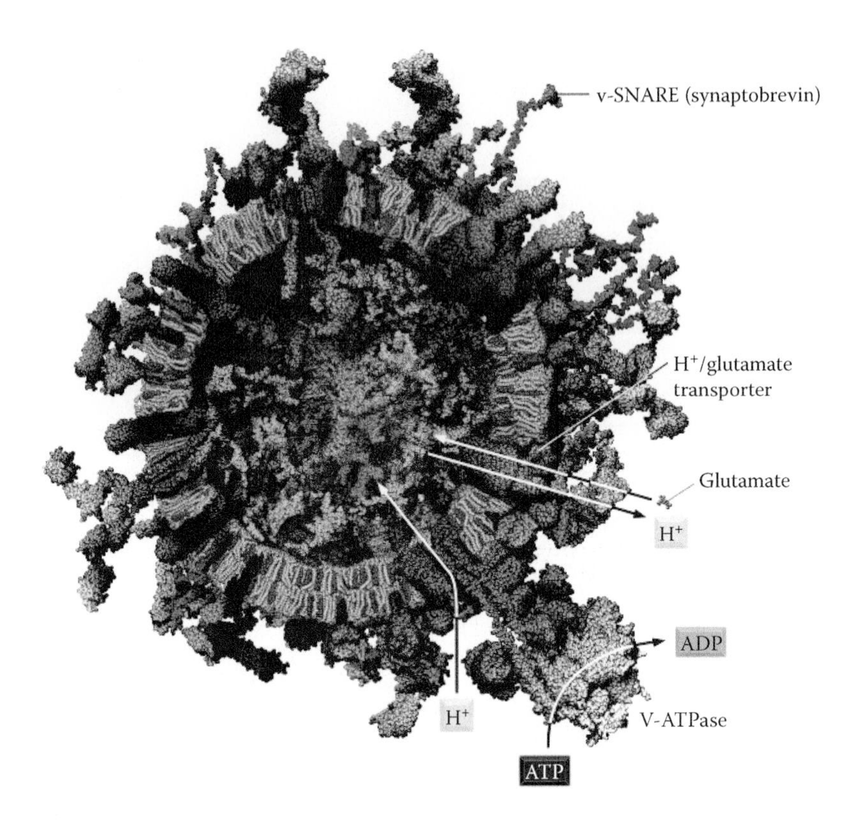

FIGURE 6.4 Molecular anatomy of a membrane of a synaptic vesicle. A synaptic vesicle cut in half and surrounded by a protein-laden bilayer (see Table 6.1) enriched with DHA and cholesterol (individual classes of membrane lipids not distinguishable in this model). Synaptic vesicles represent the best-known case in human cells, in which a DHA membrane works as a permeability barrier tight enough to prevent excessive, energy wasteful proton (H^+) leakage. High levels of cholesterol work cooperatively with DHA, presumably tightening membrane architecture against water-wires and thus minimizing proton leakage. Note that a proton energy gradient (high inside, low outside) is essential for energizing uptake pumps concentrating the critical neurotransmitter, glutamate. ATP is consumed by V-ATPase to maintain this proton gradient. (Reprinted from *Cell*, 127, S. Takamori et al., Molecular Anatomy of a Trafficking Organelle, 831–846, Copyright 2006, with permission from Elsevier.) **(See color insert.)**

6.3 "USE IT OR WASTE IT" CONCEPT OF BRAIN ENERGY EFFICIENCY

We propose that the major energy cost for neurons (Knott et al., 2008) is not the act of firing an electrical impulse but rather maintaining neurons in a state ready to fire. Time enters this picture in an unusual way in that the more times a neuron or neural circuit is usefully fired per unit time, the greater is the potential energy saving when compared on the basis of energy cost per firing. That is, the act of two firings per unit time is not much greater than firing once or not at all. The rationale is that

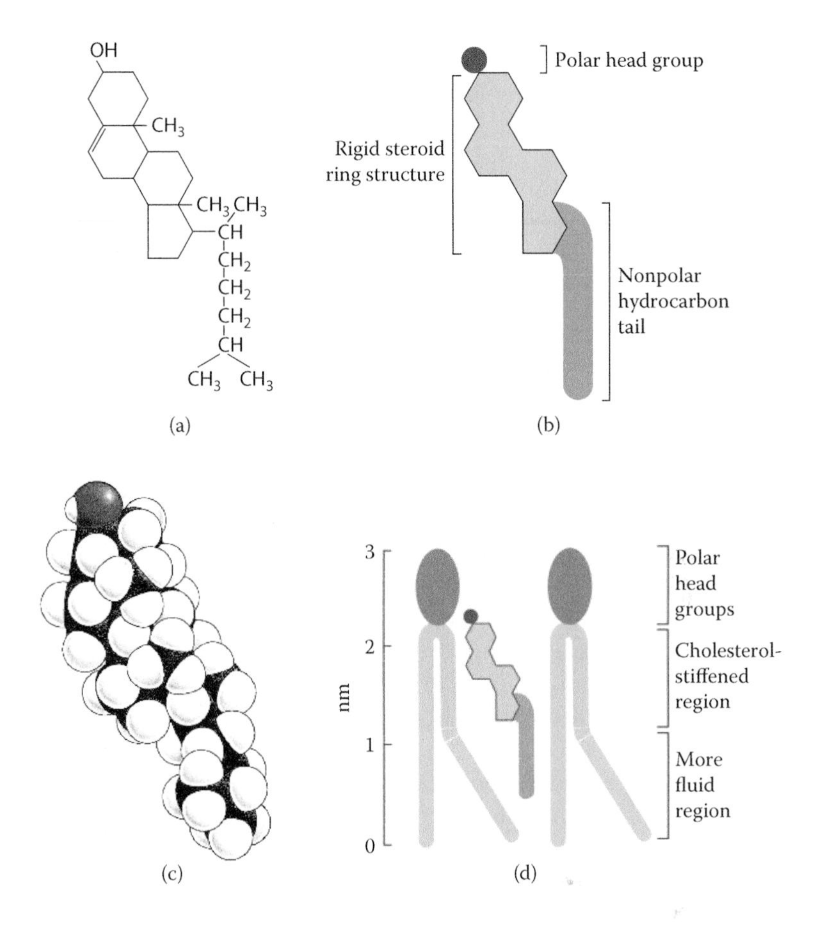

FIGURE 6.5 The structure of cholesterol. Cholesterol is represented (a) by a formula, (b) by a schematic drawing, (c) as a space-filling model, and (d) as a proton-blocker proposed to shore up membranes, including those of synaptic vesicles, against spontaneous leakage of H^+. (From Alberts, B. et al., *Molecular Biology of the Cell*, Copyright 2008. Reproduced by permission of Taylor & Francis Group, LLC, a division of Informa plc.)

energy-costly leaks of excitatory cations across neuron membranes are continuous and unavoidable and occur around the clock whether or not the neuron ever fires. We call this idea a "use it or waste it" concept of brain energy cost.

At this point, it is not known how much energy is saved by DHA working in the human brain, but we suggest the savings might be considerable. Approximately 20 to 25 percent of the food we eat is used by our brain (Martin, 1981; McNab and Eisenberg, 1989; Nilsson, 1999; Raichle and Gusnard, 2002). Here we assume that about half of the total sugar entering the brain is needed to maintain the balance of excitatory ions, mainly H^+, Na^+, and K^+. In spite of this high energy cost we suggest that DHA has evolved to maximize speed in human neurons and is the most energy efficient among a hierarchy of fatty acids. However, a synergistic relationship with cholesterol is needed to maximize energy efficiency in axons.

6.4 MYELINATION ALSO INCREASES SPEED AND ENERGY EFFICIENCY

Myelination, which takes years to complete, is such an important and metabolically expensive process that neurons surely reap multiple benefits from insulated axons. It is well known that the speed at which neurons conduct information increases some 16-fold between birth and adolescence. The vast majority of speed enhancement occurs in the first year and can be accounted for at least partially by myelination. On the other hand, virtually every measure of cognitive processing shows that children's brains perform faster as they grow older, peaking between five and eleven years of age. This disconnect between the earlier spike in myelination versus cognitive processing might be accounted for by a gain in efficiency. We define efficiency in terms of energy efficiency with increasing speed via myelination as being one mechanism to increase efficiency.

Thus, faster brains caused by myelination might be considered smarter brains (Chugani, Phelps, and Mazziotta, 1987; Hale, 1990; Müller, Ebner, and Hömberg, 1994). The more information an individual can process, store, and retrieve from his or her memory bank and analyze in a given period of time is a parameter defining greater mental capacity. Increasing brain size achieves some of these goals, but there is another way that avoids increasing brain size: increase speed. According to this idea speed is equivalent to increasing brain size. As already discussed, faster processing means that information is going to be processed more energy efficiently. Recent evidence suggests that intelligence is a summation of the relative energy efficiencies among different people's brains. Appreciation of modern techniques involving brain imaging shows that smarter subjects burn less brain energy as glucose compared to those with lower IQs while doing the same mental task (Haier et al., 1992). Another interesting study involves energy consumption of children's versus adults' brains. According to this study, children's brains are far less energy efficient than those of adults. Quantitatively, glucose use spikes rapidly from birth to about age four at which point the adult brain is roughly twice as energy efficient as that of a child. Energy efficiency increases gradually through childhood and adolescence. This pattern parallels the critical stage of brain development in which almost half of the synapses are pruned away in children's brains. This futile information flow along an excess of pathways might directly account for a lot of wasted energy. Energy efficiency can also be gained during childhood via increased myelination. Cholesterol might contribute to energy conservation working as an essential building block for formation of the myelin sheath.

Myelination begins to reverse itself after reaching a peak of roughly 100,000 miles of myelinated axons in the 20s and 30s. Thereafter, thousands of miles of myelinated axons are lost per decade. In Chapter 17 we discuss how demyelination might contribute to neurodegeneration.

6.5 WHY IS DHA FAVORED OVER EPA IN HUMAN NEURONS?

This question has stumped membrane biochemists for years, but some clues are emerging. Restated, what is the selective advantage of DHA over EPA in the evolution of the human brain, and is energy involved? We suggest distinct advantages of DHA over EPA in humans. One possible advantage is that DHA conformations in combination with cholesterol create a more robust permeability barrier compared to EPA. The bulkier size of DHA (Holte, Separovic, and Gawrisch, 1996) over EPA might account for DHA's superior membrane permeability (i.e., 22 carbons per chain for DHA versus 20 carbons for EPA), especially against futile cycling of Na^+ and K^+ in axons and H^+ in synaptosomes. However, EPA seems to be superior to DHA in neurons of certain ectothermic animals such as *Caenorhabditis elegans*. EPA levels in *C. elegans* rise sharply as temperatures drop to 15°C, suggesting that in this animal EPA is superior to DHA at temperatures near 15°C. At 15°C or colder, changes in the physical-chemical properties of EPA might improve its ability to block cations. EPA is also significantly more stable against oxidation than DHA as discussed in Chapters 8 through 10. The main point is that *C. elegans* has evolved to utilize the advantages of EPA over DHA while the opposite is true for humans. Thus, the overall needs of different animals as well as environmental conditions surrounding neurons (e.g., cold temperatures, high salinity) must be taken into account in attempting to judge whether DHA has advantages over EPA in neurons, and vice versa. In returning to the case of the human brain, selective advantages of DHA over EPA might be substantial. For example, even a 10 percent improvement by DHA over EPA at the level of the ionic permeability barrier or other advantages amounts to a significant total energy saving by the human brain.

In summary, it is assumed that a majority of energy available to neurons is used for maintaining the readiness of neurons to fire (Haines, 2001) and that the energy cost associated with the electric impulse is considerably lower than maintenance energy. If this is the case then time becomes the enemy of neural energy efficiency, and conversely more speed or rates of repetition become a friend. These relationships might still bear fruit even if the choice of DHA for the sake of speed is undermined to some extent because of its mediocre properties as a permeability barrier. Blending with a suitable amount of cholesterol or generating DHA plasmalogen phospholipids that shore up molecular defects causing cation leakage would still allow neurons to capture much of the value of DHA as the king of speed.

We suggest that DHA-mediated speed also has an energy-saving effect according to the use it or waste it idea. Recall that the role of our neurons is to conduct electricity in the form of a nerve impulse. We assume that spontaneous leaks of sodium and potassium cost human neurons as much as 50 percent of their total available energy (Haines, 2001). If this is the case then an argument can be made that if DHA speeds up impulses or decreases the time needed to rejuvenate neurons for the next firing that speed conserves energy—less energy is expended per impulse if the firing rate is high. The reason is that large amounts of energy are wasted anyway, whether the neuron is active or not. The heart of this use it or waste it concept is that the vast web of brain circuitry is always leaky against cations. Thus, judged using the energy cost of a single firing across a single neuron as a standard, the net energy cost for the brain is reduced by speeding up impulses. We suggest that energy used for the firing process of neurons is less than half of the amount of energy needed to counterbalance spontaneous leak rates. Ultimately, brainpower must have had great advantages for our ancestors to expend so much effort just to nourish this single organ. Thus, coevolution of DHA and energy efficiency of the brain likely contributed to the rise of humans.

Finally, DHA membranes of neurons are now recognized as playing important roles in energy homeostasis in these critical cells. Energy saved at the membrane level is considered of cardinal importance in the Darwinian selection of the fittest neurons for the human brain.

REFERENCES

Ames, A. III, Y.-y. Li, E. C. Heher et al. 1992. Energy metabolism of rabbit retina as related to function: High cost of Na^+ transport. *J. Neurosci.* 12:840–853.

Andersen, O. S., R. E. Koeppe II, and B. Roux. 2005. Gramicidin channels. *IEEE Trans. Nanobioscience* 4:10–20.

Baştuğ, T., S. M. Patra, and S. Kuyucak. 2006. Molecular dynamics simulations of gramicidin A in a lipid bilayer: From structure-function relations to force fields. *Chem. Phys. Lipids* 141:197–204.

Chugani, H. T., M. E. Phelps, and J. C. Mazziotta. 1987. Positron emission tomography study of human brain functional development. *Ann. Neurol.* 22:487–497.

Deamer, D. W. 1987. Proton permeation of lipid bilayers. *J. Bioenerg. Biomembr.* 19:457–479.

Haier, R. J., B. Siegel, C. Tang et al. 1992. Intelligence and changes in regional cerebral glucose metabolic rate following learning. *Intelligence* 16:415–426.

Haines, T. H. 2001. Do sterols reduce proton and sodium leaks through lipid bilayers? *Prog. Lipid Res.* 40:299–324.

Hale, S. 1990. A global developmental trend in cognitive processing speed. *Child Dev.* 61:653–663.

Holte, L. L., F. Separovic, and K. Gawrisch. 1996. Nuclear magnetic resonance investigation of hydrocarbon chain packing in bilayers of polyunsaturated phospholipids. *Lipids* 31:S199–S203.

Huster, D., A. J. Jin, K. Arnold et al. 1997. Water permeability of polyunsaturated lipid membranes measured by 17O NMR. *Biophys. J.* 73:855–864.

Infante, J. P., R. C. Kirwan, and J. T. Brenna. 2001. High levels of docosahexaenoic acid (22:6n-3)-containing phospholipids in high-frequency contraction muscles of hummingbirds and rattlesnakes. *Comp. Biochem. Physiol. B Biochem. Mol. Biol.* 130:291–298.

Isler, K., and C. P. van Schaik. 2006. Metabolic costs of brain size evolution. *Biol. Lett.* 2:557–560.

Knott, A. B., G. A. Perkins, R. Schwarzenbacher et al. 2008. Mitochondrial fragmentation in neurodegeneration. *Nat. Rev. Neuroscience* 9:505–518.

Martin, R. D. 1981. Relative brain size and basal metabolic rate in terrestrial vertebrates. *Nature* 293:57–60.

McNab, B. K., and J. F. Eisenberg. 1989. Brain size and its relation to the rate of metabolism in mammals. *Am. Nat.* 133:157–167.

Müller, K., B. Ebner, and V. Hömberg. 1994. Maturation of fastest afferent and efferent central and peripheral pathways: No evidence for a constancy of central conduction delays. *Neurosci. Lett.* 166:9–12.

Nagle, J. F., and H. J. Morowitz. 1978. Molecular mechanisms for proton transport in membranes. *Proc. Natl. Acad. Sci. USA* 75:298–302.

Nichols, J., and D. Deamer. 1980. Net proton-hydroxyl permeability of large unilamellar liposomes measured by an acid-base titration technique. *Proc. Natl. Acad. Sci. USA* 70:2038–2042.

Nilsson, G. E. 1999. The cost of a brain. *Nat. Hist.* 108:66–73.

Niven, J. E., J. C. Anderson, and S. B. Laughlin. 2007. Fly photoreceptors demonstrate energy-information trade-offs in neural coding. *PLoS Biol.* 5:3116.

Raichle, M. E., and D. A. Gusnard. 2002. Appraising the brain's energy budget. *Proc. Natl. Acad. Sci. USA* 99:10237–10239.

Stillwell, W., L. J. Jenski, F. T. Crump et al. 1997. Effect of docosahexaenoic acid on mouse mitochondrial membrane properties. *Lipids* 32:497–506.

Takamori, S., M. Holt, K. Stenius et al. 2006. Molecular anatomy of a trafficking organelle. *Cell* 127:831–846.

Valentine, R. C., and D. L. Valentine. 2009. *Omega-3 Fatty Acids and the DHA Principle*. Boca Raton, FL: Taylor & Francis.

Venable, R. M., and R. W. Pastor. 2002. Molecular dynamics simulations of water-wires in a lipid bilayer and water/octane model systems. *J. Chem. Phys.*116:2663–2664.

Wallace, B. A., and K. Ravikumar. 1988. The gramicidin pore: Crystal structure of a cesium complex. *Science* 241:182–187.

Welander, P. V., R. C. Hunter, L. Zhang et al. 2009. Hopanoids play a role in membrane integrity and pH homeostasis in *Rhodopseudomonas palustris* TIE-1. *J. Bacteriol.* 191:6145–6156.

7 Wild Conformational Dynamics of DHA Might Save Energy by Sealing Stretch-Induced Defects as Fast as They Occur in Membranes

The remarkable conformational dynamics of DHA are full of surprises and in this chapter we describe a new dimension involving membranes routinely subjected to stretching and bending. Numerous cells and their membranes in the human body are subjected to repeated cycles of stretching often induced by mechanical stress. We define the importance of mechanically stressed membranes here using selected case histories featuring different cells known to be subjected to mechanical stress. Our thick skull protects neurons in the brain from everyday sources of mechanical stress. However, during brain trauma caused by contact sports and war injuries, axons are stretched to a pathological level as discussed in Chapter 16. Axons also stretch to achieve their amazingly fast growth rates. Membranes of synaptic vesicles can be envisioned to be subjected to stretching imposed by their small size and the marked curvature of their membranes. Other membranes likely subject to stretching include sperm tails, cilia, heart muscle cells, and even red blood cells. We propose that the conformational dynamics of DHA and other polyunsaturated fatty acids contribute self-sealing powers to membranes during stretching, in essence conserving ion gradients and saving energy.

7.1 MECHANICAL STRESS MODEL OF SPERM TAILS

Membranes subjected to stretching are widespread in cells in nature ranging from bacteria to human sperm. Next, we explore the case history of mechanical stress induced by violent tail whipping in tail membranes of sperm, which we view as a surrogate for axons (Figure 7.1). The tails of sperm have likely evolved several mechanisms to withstand cycles of mechanical stress induced by powerful tail whipping essential in the race to fertilize the egg. For example, nine long, stiff, keratin-based rods seem strategically located around the axoneme, perhaps as structural support against fracturing of the tail (Figure 7.1c). We propose that the keratin rods handle

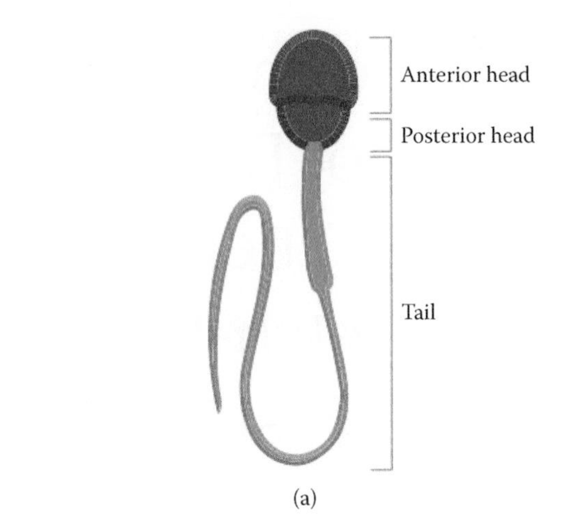

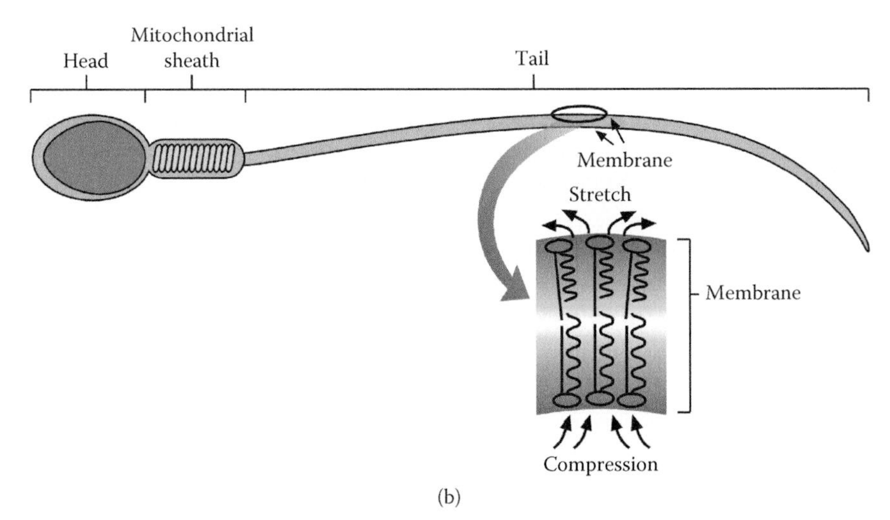

FIGURE 7.1 DHA-enriched tail membranes of sperm like axons are excitatory and provide clues about the roles of DHA in neurons. (a) Three separate domains define the membrane surrounding a sperm cell. (From Alberts, B. et al., *Molecular Biology of the Cell*, Copyright 2008. Reproduced by permission of Taylor & Francis Group, LLC, a division of Informa plc.) (b) DHA concentrated in the tail domain likely changes conformations on a sub-nanosecond time scale. We propose that these wild conformational changes allow these chains to instantaneously plug any permeability defects generated by violent tail whipping. (From Valentine, R., and Valentine, D., *Omega-3 Fatty Acids and the DHA Principle*. Copyright 2009. Reproduced by permission of Taylor & Francis Group, LLC, a division of Informa plc.) (c) Drawing of a cross section of a mammalian sperm tail showing the protein machinery for motion (axoneme) in the center surrounded by nine dense fibers proposed to stabilize the long thin tail against breaking from the strain of flagellar whipping. (From Alberts, B. et al., *Molecular Biology of the Cell*, Copyright 1994. Reproduced by permission of Taylor & Francis Group, LLC, a division of Informa plc.)

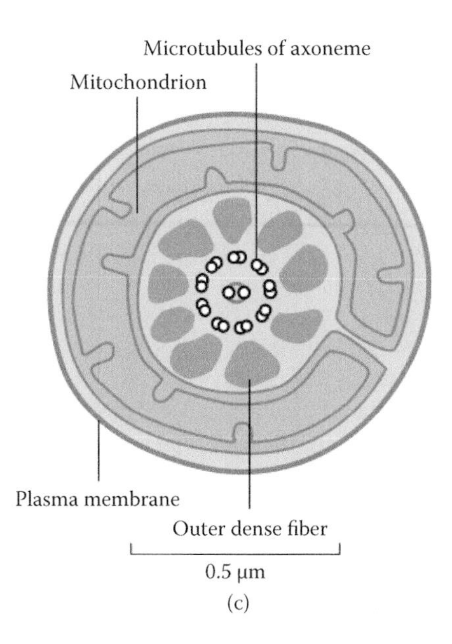

FIGURE 7.1 *(Continued)*

the brunt of mechanical stress generated by tail whipping but that this is not the only mechanism. We suggest that the DHA-enriched membrane surrounding the sperm tail is also subjected to repeated cycles of mechanical stretching strong enough to damage the integrity of membrane architecture. We hypothesize that mechanical stretching of tail membranes is an inevitable and important enough consequence of sperm motility to require evolution of a membrane-based protective mechanism. We suggest that DHA has unique physical-chemical properties capable of helping seal stretch-induced membrane defects along the tail as fast as they occur. We develop a model for the bioenergetics of sperm in which available energy is rate limiting for motility during fertilization. According to this model, DHA again working with cholesterol saves energy by instantaneously sealing stretch-induced defects that uncouple cation gradients and destroy excitatory properties of the tail. Implications of this model for neurons and other cells are discussed below.

The shape of sperm cells with a head region packed with DNA connected to a long whip-like tail is familiar to most readers. Sperm membranes surround the entire cell, tail and all, and are separated into three membrane domains, including distinct top and bottom head regions, in addition to the tail (Figure 7.1a). Among the three domains, only the membrane of the tail is significantly enriched with DHA (Connor et al., 1998). That is, DHA is localized primarily in the tail membrane domain, which surrounds the enzymatic and structural machinery needed to develop thrust or motility. Surprisingly, destruction of the integrity of the tail membrane using detergents does not completely inhibit motion, which can be restored by adding "energizing" metabolites, including ATP and Mg^{2+}. However, these membrane-defective sperm wander aimlessly and have obviously lost critical functions attributable to the membrane.

Sperm cells are unusual in spending a crucial part of their life cycle traveling as "free-living" cells in the female reproductive tract where conditions are fundamentally different from those inside the body. For example, sperm face O_2 deprivation and fluctuations of pH that would kill many cells, including neurons. In addition, during the long transit to the egg, sperm face uncertain levels of energy substrates and variable levels of cations, essential for their excitatory nature. It is now clear that sperm have adapted to their free-living environment both biochemically and structurally. Specifically, sperm appear to have evolved DHA membranes as an adaptation to the free-living state allowing them to successfully traverse the hostile environment of the female reproductive tract. Considering the roles of DHA membranes of sperm from an ecological perspective has helped illuminate in molecular terms why sperm tail membranes are enriched with DHA. Understanding this adaptation in sperm has implications for neurons and other cells whose membranes may be subject to some form of mechanical stress.

The journey of sperm in fertilizing the egg is more dependent on energy-rich substrates, nutrients, and other metabolites provided by the female than previously suspected (Miki et al., 2004; Mukai and Okuno, 2004; Turner, 2006). For example, a portion and perhaps even a majority of the energy for powering sperm movement during fertilization has been traced back to sugars present in the female reproductive tract, not energy stored in the sperm cell. The current picture is that sperm moving in the relatively low O_2 environment of the female reproductive tract require glucose, which these cells actively ferment as a source of ATP for energizing motility. Glycolytic enzymes are strategically located throughout the length of the tail region and in proximity to the machinery driving flagellar beating. In contrast, sperm such as that of sea urchin functioning in a highly aerated environment of seawater depend on a single mitochondrion in combination with a unique creatine phosphate shuttle system for energizing motion along the full length of the long flagellum. Note that human sperm have 25 or more mitochondria. The definitive finding on the essential role of glycolysis, rather than mitochondria, as a source of ATP for motility during mammalian fertilization, involves knockout mutations in mice of key glycolytic enzymes resulting in infertility. How is bioenergetics of sperm related to DHA structure-function?

Mammalian sperm are enriched in DHA, primarily in tail membranes represented by the tail membrane domain that covers the core machinery for driving flagellar motion (Figure 7.1a). Violent tail whipping characteristic of human sperm is mechanically stressful (Figure 7.1b), and sperm tails seem fragile enough to literally break off. As introduced above, the structural integrity of sperm tails seems to be reinforced against breakage by the presence of nine outer dense fibers located between the central axoneme or contractile region and the DHA-enriched membrane (Figure 7.1c). These fibers composed mainly of keratin are stiff and noncontractile and might prevent breakage. Tail membranes are known to be excitatory in nature (Navarro, Kirichok, and Clapham, 2007). The biochemistry of electrophysiology is itself an energy-intensive process involving the energizing of pumps needed to maintain cation homeostasis—high Na^+ outside and high K^+ inside—along the relatively large surface area of the tail membrane. As introduced above, the view of bioenergetics of human sperm motility has changed dramatically from ample ATP produced by mitochondria to a paucity of energy available from glycolysis during the voyage to fertilize the egg. This is consistent with a concept of sperm in which available energy might become rate limiting for motion during transit in the female reproductive tract, thus elevating the importance of membrane-based mechanisms of energy conservation. Recall that any energy saved at the membrane level effectively makes more energy available for motility. We suggest that membrane architecture and the roles of DHA as a permeability barrier against energy wastage caused by futile cycling of $Na^+/K^+/H^+$ need to be considered in the ecology and bioenergetics of sperm. High levels of cholesterol and DHA plasmalogens in tail membranes can also be explained in the context of improved energy-conserving properties.

As excitatory membranes, sperm tails face a novel stress. That is, tail membrane lipids are likely subject to mechanical stress caused by violent cycles of tail whipping (Figure 7.1b). We hypothesize that the molecular architecture of tail membranes has evolved to compensate for defects in permeability generated by mechanical stress. We further hypothesize that tightening the permeability barrier against $Na^+/K^+/H^+$ as an energy-saving mechanism represents an important adaptation of human sperm necessary for male fertility. The conformational dynamics of DHA, both flexible and space filling, seem ideally suited for avoiding energy stress in sperm.

7.2 DYNAMIC SPACE-FILLING CONFORMATIONS OF DHA

The spring-like or helical shape model once envisioned for DHA chains has now given way to a dynamic conformational model (Feller, 2008). A DHA chain is now viewed as a highly flexible molecule with rapid transitions between large numbers of conformers on the time scale from picoseconds to hundreds of nanoseconds (Soubias and Gawrisch, 2007). This unusual flexibility has been explained on the basis of low barriers to torsional rotation about C-C bonds linking *cis*-double bonds with the methylene carbons between them. In this section we explore how the flexibility of DHA chains might be harnessed to enhance the bioenergetic properties of mechano-sensitive membranes such as sperm tails and axons. The compression or compactness of the DHA chain is of interest from the standpoint of sperm tail function (Figure 7.1) or any mechano-sensitive membrane including axons undergoing mechanical stress during brain trauma or stretch growth. The intimate relationship between conformation-function of DHA chains seems to stand out in the case of mechanically stressed membranes.

One of the least understood aspects of DHA function in mechanically stretched membranes involves molecular roles of these chains in cation permeability (Tabarean, Juranka, and Morris, 1999). This is important for both electrophysiology and energy conservation. From first principles, permeability properties of stretched membranes enriched with DHA must involve conformations of the DHA chain. We hypothesize that DHA chains are best suited among a hierarchy of unsaturated fatty acids in containing sufficient bulk in the form of their dynamic conformations to rapidly fill any molecular voids, pores, or defects occurring within their intramembrane molecular domains (Figure 7.1b). In essence, what we are proposing is that the dynamic conformation of DHA chains allows these molecules to rapidly readjust their conformations when the membrane surface is mechanically stressed as occurs during cycles of flagellar whipping (Figure 7.1b), stretch growth of axons, and so forth. In the case of sperm this molecular model assumes the enhanced formation of water-wires (see Chapter 6) at "stretch points" along the tail caused by cyclical stretching/compression of the bilayer. DHA is envisioned to rapidly fill these defects so as to inhibit the formation of water-wires. Once again the greatest potential energy saving in sperm likely involves membrane adaptations against futile cycling of the excitatory cations Na^+ and K^+. Recently it has been shown that proton extrusion from human spermatozoa is mediated by flagellar voltage-gated H^+ channels (Hv1) (Lishko et al., 2010). These authors point out that maintaining a proton gradient across the flagellar membrane up to 1000-fold higher H^+ levels (inside than outside) plays an important role in activating sperm traveling in the female reproductive tract. We propose that the DHA- and cholesterol-enriched tail membrane architecture has evolved to prevent excessive proton leakage caused by flagellar beating. The role of cholesterol in preventing futile proton cycling in DHA-enriched membranes of synaptic vesicles is discussed in Chapter 6 and has implications for proton permeability of flagellar membranes.

7.3 AXON MEMBRANES ARE SUBJECT TO STRETCHING DURING NORMAL GROWTH AND DURING BRAIN TRAUMA

Researchers have reported that axons of cultured human neurons have a remarkable capacity for stretch (Figure 7.2) (Pfister et al., 2004; Smith, 2009) with no primary axotomy (severing of an axon) observed with strains below 65 percent of the length of the axon (Smith et al., 1999). In addition, axons exhibit what is termed *delayed elasticity*. In other words, stretching that causes a temporary deformation allows the axon to return to the original orientation and morphology even when internal axonal damage is sustained. Severe stretching triggers a cascade of events leading to more structural changes and metabolic dysfunctions (see Chapter 16). One of the earliest changes involves abnormal influx of Na^+ through mechano-sensitive or stretch sodium channels or perhaps through water-wires. Using an animal optic nerve stretch model it was shown that stretch below 4 mm did not cause any morphological change in axons, whereas a stretch of 5 mm altered visual-evoked potentials differing significantly from controls. Increasing stretch led to more severe axon dysfunction. Thus, it appears that any stretch occurring from ordinary activities such as running or walking might mildly stretch axons but not at a level causing any permanent damage. However, it is predicted that because of the massive surface area of axons even mild stretching increases cationic leakage across the axon bilayers, and this is another source of energy stress.

Axons are also subject to stretch exerted by growth (Huang et al., 2008; Smith, 2009). As humans grow and expand their nervous system, new axon growth occurs at seemingly impossible rates. A process called *stretch growth* is responsible and is perhaps the most remarkable axon growth mechanism. The extremely rapid expansion of the nervous system appears to be directed purely by mechanical forces on neuron tracts. Thus, axon stretch is a natural part of the growth of these DHA-rich membranes. It has been found that during stretch growth axons retain their ability to

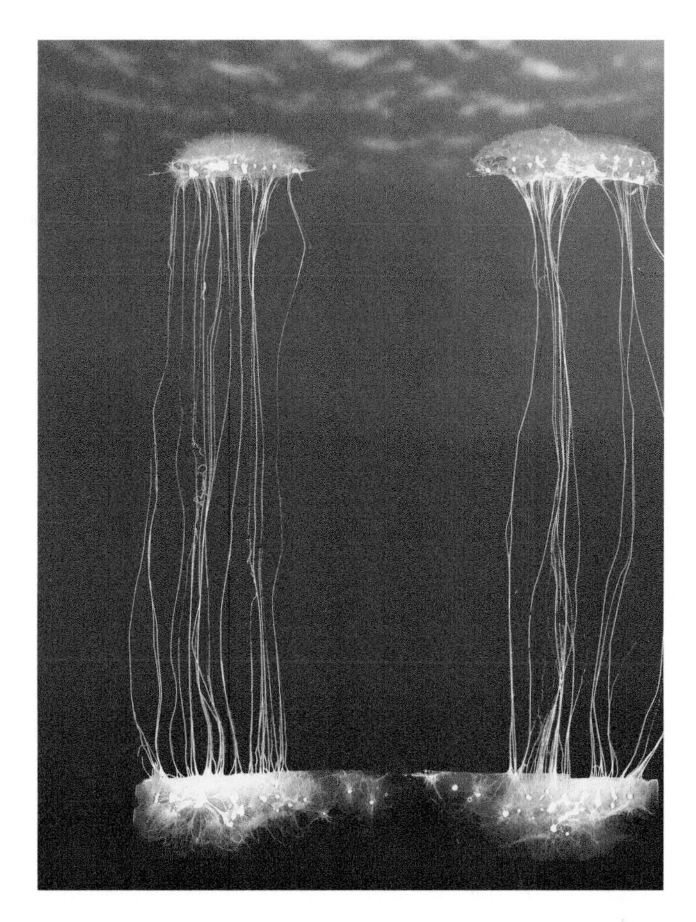

FIGURE 7.2 Extreme stretch growth of axon tracts. Before the initiation of mechanical stretching, the two populations of neurons shown in this photo were adjacent, and the bridging axons were only about 100 μm long. Axons grew to 5 cm long during progressive separation of neuron populations. Excitatory properties were found to persist in stretched axons. See text for proposed roles of DNA in axon growth and Chapter 16 for further details. (Reprinted from *Prog. Neurobiol.*, 89, Smith, D. H., Stretch Growth of Integrated Axon Tracts: Extremes and Exploitations, 231–239. Copyright 2009, with permission from Elsevier.)

transmit electrical signals (Pfister et al., 2006), an indication that their permeability properties and excitatory nature remain functional even in the face of mechanical stress. These data are consistent with DHA playing space-filling and energy-saving roles in axons similar to sperm tails.

Axons contain a vast membrane surface represented by synaptic vesicles (Chapter 6) whose small size is envisioned to generate curvature stress. Note that the 14,000 lipid chains forming the bilayer of a synaptic vesicle include about 2000 DHA chains and 5600 cholesterol molecules—lipids, which might be recruited as a mechanism against proton leakage caused by spontaneous mechanical stress.

Synaptic vesicle membranes exhibit asymmetry in that each of the two half bilayers can have a distinct lipid composition. Whereas the exact chemical composition of each leaflet is not known, the finished bilayer is responsible for the difficult task of forming membrane architecture robust enough to contain the smallest and most leaky or permeable cation—protons—required to energize uptake of neurotransmitters. Clearly, the membrane role as the proton permeability barrier in synaptosomes is likely fulfilled relatively efficiently and includes the need to plug water-wires caused by uneven stretching of the outside leaflet with an opposite effect on the inside leaflet. To see this effect on a larger scale, take a closed zipper such as on your coat and first encircle a small segment of the zipper around one finger, forming a loop. Remove your finger using the other hand to maintain the circular shape and apply pressure with the free hand in an attempt to shorten the circumference of the loop. Watch as the outside edge of the zipper teeth on the loop begins to spread apart. The point is that phospholipids forming the surface of synaptic membranes might stretch in a similar manner, perhaps needing DHA working with cholesterol to plug membrane defects against leakage of protons.

7.4 TURGOR (OSMOTIC PRESSURE) IS LIKELY REQUIRED FOR STRETCH GROWTH OF AXONS

Potassium ion is a fundamentally important excitatory cation whose razor-sharp levels in neurons are essential for a generation of neural impulses. High inside levels of K^+ also play a role as the bulk osmoregulant in maintaining neuron water balance and are thus key players in the osmoregulation of neurons. Potassium ion is considered as a compatible bulk osmoregulant, and its level inside neurons always more or less matches the high level of sodium ions surrounding neurons (about one third the seawater level of Na^+). Thus, K^+ is the major contributor of osmotic pressure or turgor pressure in neurons but not the only one. As mentioned above, a fast-growing axon maintains its electrophysiological properties. We doubt that K^+ could play triple roles—excitatory cation, bulk osmoregulant, and fine-tunable osmoregulant driving axon growth—all at the same time.

Medulla cells of the kidney face high salinity during the concentration of urine. Studies of osmoregulation of these cells provide a clue toward how turgor pressure for growth of axons might be fine-tuned. Note that kidney medulla cells are often surrounded by osmotic pressure up to five times higher than found in serum. The powerful organic osmoregulant and osmoprotectant glycine betaine is concentrated in medulla cells to prevent these essential kidney cells from osmotically bursting under such high salinity levels. Now, the osmotic environment surrounding axons is roughly that of serum, but nevertheless maintaining a positive inside turgor pressure is critical for growth and maintaining the water balance of neurons. Relatively small amounts of betaine or a similar molecule in combination with their specific osmotically modulated membrane-based pumps could theoretically generate and fine-tune turgor pressure vital for the growth of axons. The point is that unlike kidney cells, where high levels of organic osmoprotectants are needed, a relatively small amount of an available organic osmoregulant should suffice for adjusting osmoregulation inside axons to match their growth. The metabolic price to neurons of synthesis of the putative organic osmoregulant for axon stretching would of course have to be considered along with the cost of synthesis and fueling of yet another uptake pump for osmoregulants.

Because axon growth is explosive, especially during rapid brain development in infants, this is a good place to look for any roles of glycine betaine-type molecules—perhaps even preformed in mother's milk. Taurine seems to fit the requirements as follows:

- In 1984, taurine was approved by the U.S. Food and Drug Administration (FDA) as a supplement in baby formula.
- Taurine is abundant in human milk but not cow's milk.
- Levels are high in the brain/eye of newborns (i.e., twice as high in newborns as in adults).
- Evidence points to roles in brain development and osmoregulation.

The capacity to synthesize taurine is limited in humans especially in early infancy. Taurine levels are high *in utero* and are maintained in breast-fed babies, but they become depleted within a few weeks in infants fed formula without added taurine. Animal studies using cats or primates show that taurine depletion can cause serious degeneration of the eye. In humans, taurine deficiency is linked to seizures and reduction of neural excitability. There are large gaps in understanding the molecular biology of taurine in the human brain. It has been proposed that this well-established molecule is both an osmoregulant and osmoprotective agent (Bourque, 2008) perhaps used to modulate turgor pressure and the growth of axons. This focuses attention on the importance of osmoregulated taurine pumps in stretch growth in neurons, a topic discussed in more detail in Chapter 10. We also raise the possibility that taurine by virtue of its zwitterion chemical structure might directly reduce membrane permeability by entering the membrane and perhaps blocking leakage of K^+ or Na^+ across the membrane. The same mechanism might also apply to glycine betaine whose three terminal methyl groups seem ideal to add bulk to prevent water-wires from forming in membranes. This putative direct energy-saving role of organic osmoprotective molecules is an intriguing area for future research. It is well known that there is a great deal of cross talk among various kinds of stresses, such as faced by axons. Thus, it would be logical that energy stress and osmotic stress were interlinked.

7.5 STRETCH-INDUCED DEFECTS LIKELY OCCUR IN MEMBRANES OF CILIA, RED BLOOD CELLS, AND HEART MUSCLE CELLS

A significant proportion of total membranes in human cells are likely subjected to some degree of stretching. Untold numbers of tiny, hair-like cilia extend from the surface of many human cells and display constant motion. For example, huge numbers of cilia ($>10^9/cm^2$) sweep layers of mucus and other trapped particles of dust, dead cells, and debris from the respiratory tract up toward the mouth to be swallowed or spit out. Specialized olfactory cilia sprout from the ends of neurons lining the nasal passage. Olfactory cilia, like sperm tails, are subjected to bending motion.

The beat of a cilium such as that on an epithelial cell from the human respiratory tract is a miniature version of the powerful breaststroke used by Michael Phelps for winning at the Olympics. A fast power stroke drives fluid over the surface of the cell. This is followed by a slow recovery stroke. Each cycle typically takes about 0.1 to 0.2 seconds and generates a force perpendicular to the axis of the axoneme. The bending apparatus is found in the core of the cilium similar to the arrangement with sperm. A membrane surrounds the micro-tubular core assembly (Kaneshiro et al., 1979). Note that cilia are much shorter than the tails of sperm and thus are not likely subjected to the same level of mechanical stress.

Cardiac muscle cells are subject to mechanical stress as a natural part of their pumping or muscular action. Recall that heart rates across mammals vary more than 100-fold (Hulbert and Else, 1999) and more importantly, that DHA levels in heart muscles increase in a roughly linear fashion with increasing heartbeats. Based on this set of data Hulbert and colleagues have proposed that DHA membranes help modulate respiratory rates in fast muscles (see Chapter 12). The case of hummingbird flight muscle cells is especially interesting. Heartbeats of more than 20 beats per second have been recorded. Approximately one-third of the total volume of flight muscle cells is filled by mitochondria, many of which seem to be compressed tightly between powerful muscle sheaths. This raises the possibility that the extremely rapid firing of these strong muscles might subject membranes in these tightly packed mitochondria to continuous cycles of mechanical stress, especially during flight. Movies taken of individual mammalian cardiac cells show what appear to be continuous cycles of stretching, shrinking, and contortion of cellular shape likely causing membrane stretching. We suggest that the physical pumping action characteristic of individual cardiac cells might also subject the plasma membrane of heart muscle cells to mechanical stress.

Circulating red blood cells must dramatically downsize and change shape before entering the narrow passageways of capillaries. Membranes of red cells contain significant levels of 22:6, especially when seafood is a routine part of the diet. Human red cell membranes also are composed of a significant level of molecular species of diacyl phospholipids with two unsaturated chains (Connor et al., 1997). This phospholipid structure might represent a membrane-based adaptation of red cells against mechanical stress.

In summary, essentially all membranes in humans are subjected to some degree of stretching often linked to sophisticated systems of osmoregulation that maintain a slight positive turgor pressure inside versus outside the cell. Certain cells and membranes often face much greater stretching forces including violent tail whipping of sperm and stretch growth of axons. We have developed our version of a membrane-stretching model for neurons around sperm tails. The heart of the sperm-axon model resides in harnessing the wild conformational dynamics of DHA chains as a space-filling sealant against stretch-induced membrane defects. According to this model, membrane stretching results in futile cycling of critical excitatory cations, especially Na^+ and K^+ ions. Replenishing Na^+/K^+ balance costs energy, and the self-sealing properties of DHA are proposed to save energy. Such energy savings are thus equivalent to making more energy available for neuron function.

Axon growth depends on powerful membrane stretching matched with a membrane architecture that preserves excitatory biochemical properties while growth is occurring. We hypothesize that DHA in axon membranes plays an energy-saving role similar to that in sperm tails.

REFERENCES

Bourque, C. W. 2008. Central mechanisms of osmosensation and systemic osmoregulation. *Nat. Rev. Neurosci.* 9:519–531.

Connor, W. E., D. S. Lin, and G. Thomasetal. 1997. Abnormal phospholipid molecular species of erythrocytes in sickle cell anemia. *J. Lipid Res.* 38:2516–2528.

Connor, W. E., D. S. Lin, D. P. Wolfe et al. 1998. Uneven distribution of desmosterol and docosahexaenoic acid in the heads and tails of monkey sperm. *J. Lipid Res.* 39:1404–1411.

Feller, S. E. 2008. Acyl chain conformations in phospholipid bilayers: A comparative study of docosahexaenoic acid and saturated fatty acids. *Chem. Phys. Lipids* 153:76–80.

Huang, J. H., E. L. Zager, J. Zhang et al. 2008. Harvested human neurons engineered as live nervous tissue constructs: Implications for transplantation. Laboratory investigation. *J. Neurosurg.* 108:343–347.

Hulbert, A. J., and P. L. Else. 1999. Membranes as possible pacemakers of metabolism. *J. Theor. Biol.* 199:257–274.

Kaneshiro, E. S., L. S. Beischel, S. J. Merkel et al. 1979. The fatty acid composition of *Paramecium aurelia* cells and cilia: Changes with culture age. *J. Protozool.* 26:147–158.

Lishko, P. V., I. L. Botchkina, A. Fedorenko et al. 2010. Acid extrusion from human spermatozoa is mediated by flagellar voltage-gated proton channel. *Cell* 140:327–337.

Miki, K., W. Qu, E. H. Goulding et al. 2004. Glyceraldehyde 3-phosphate dehydrogenase-S, a sperm-specific glycolytic enzyme, is required for sperm motility and male fertility. *Proc. Natl. Acad. Sci. USA* 101:16501–16506.

Mukai, C., and M. Okuno. 2004. Glycolysis plays a major role for adenosine triphosphate supplementation in mouse sperm flagellar movement. *Biol. Reprod.* 71:540–547.

Navarro, B., Y. Kirichok, and D. E. Clapham. 2007. KSper, a pH-sensitive K+ current that controls sperm membrane potential. *Proc. Natl. Acad. Sci. USA* 104:7688–7692.

Pfister, B. J., A. Iwata, D. F. Meaney et al. 2004. Extreme stretch growth of integrated axons. *J. Neurosci.* 24:7978–7983.

Pfister, B. J., A. Iwata, A. G. Taylor et al. 2006. Development of transplantable nervous tissue constructs comprised of stretch-grown axons. *J. Neurosci. Methods* 153:95–103.

Smith, D. H. 2009. Stretch growth of integrated axon tracts: Extremes and exploitations. *Prog. Neurobiol.* 89:231–239.

Smith, D. H., J. A. Wolf, T. A. Lusardi et al. 1999. High tolerance and delayed elastic response of cultured axons to dynamic stretch injury. *J. Neurosci.* 19:4263–4269.

Soubias, O., and K. Gawrisch. 2007. Docosahexaenoyl chains isomerize on the sub-nanosecond time scale. *J. Am. Chem. Soc.* 129:6678–6679.

Tabarean, I. V., P. Juranka, and C. E. Morris. 1999. Membrane stretch affects gating modes of a skeletal muscle sodium channel. *Biophys. J.* 77:758–774.

Turner, R. M. 2006. Moving to the beat: A review of mammalian sperm motility regulation. *Reprod. Fertil. Dev.* 18:25–38.

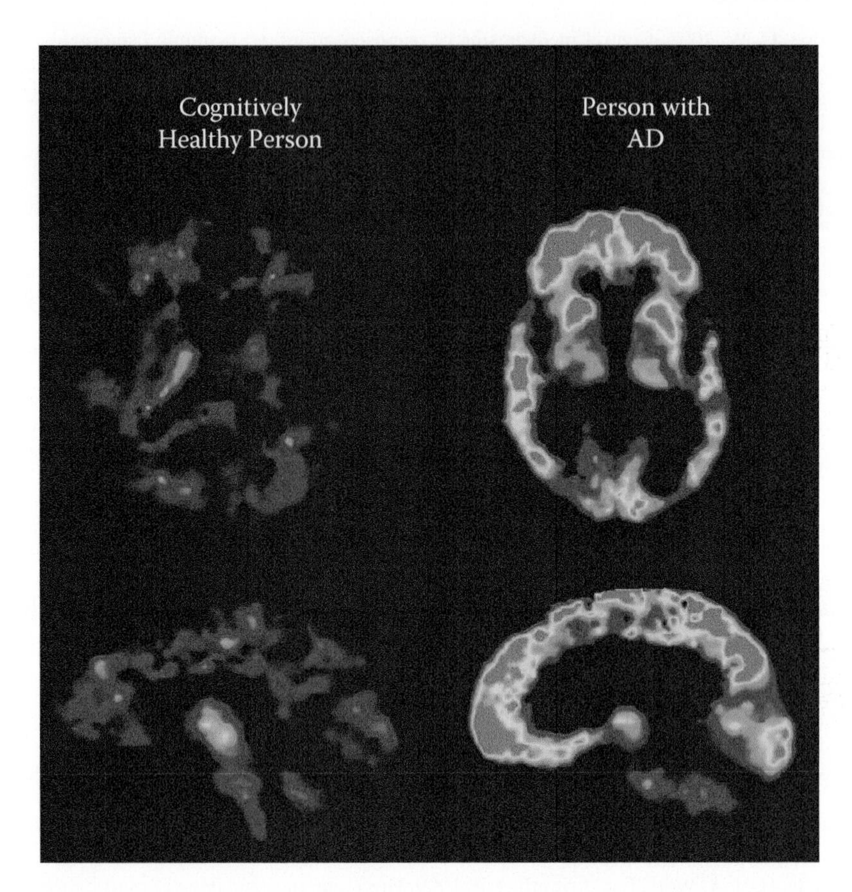

FIGURE I.3 Positron emission tomography (PET) scans of a healthy brain on the left and the brain of a patient with Alzheimer's disease (AD) on the right show high levels of amyloid plaque (red color) accumulating in advanced cases of AD. A marker dye PiB is injected to detect regions of amyloid accumulation, which are extensive in this case. (Image courtesy of the National Institute on Aging/National Institutes of Health.)

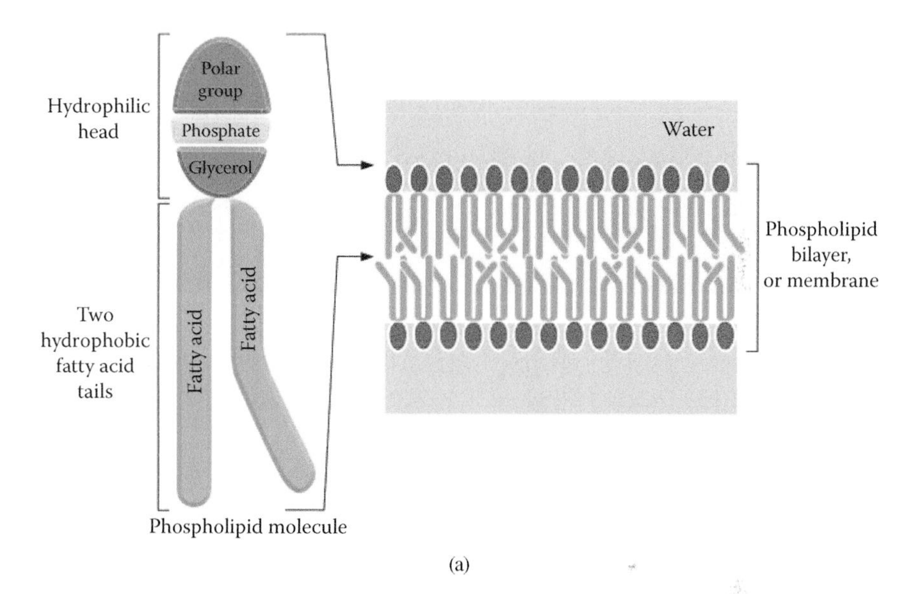

FIGURE 1.2 Phospholipids are the building blocks for axon membranes. (a) Phospholipid nomenclature. In an aqueous environment, the hydrophobic tails of phospholipids pack together to exclude water. Here they have formed a bilayer with the hydrophilic head of each phospholipid facing the water. Lipid bilayers are the basis for most cell membranes. (From Alberts, B. et al., *Molecular Biology of the Cell*, Copyright 2008. Reproduced by permission of Taylor & Francis Group, LLC, a division of Informa plc.) (b) Image of a DHA phospholipid molecule such as found in neuron membranes, as generated by chemical simulation. Gold chains are DHA; green chains are stearic acid (18:0). (Image courtesy of Scott Feller.) (c) Model of an axon-membrane-embedded K^+ channel or gate (yellow ribbon) surrounded by DHA chains (red). Potassium gates play a crucial role in electrophysiology of neurons and other cells. (Courtesy Igor Vorobyov, Scott Feller, and Toby W. Allen, unpublished communication.)

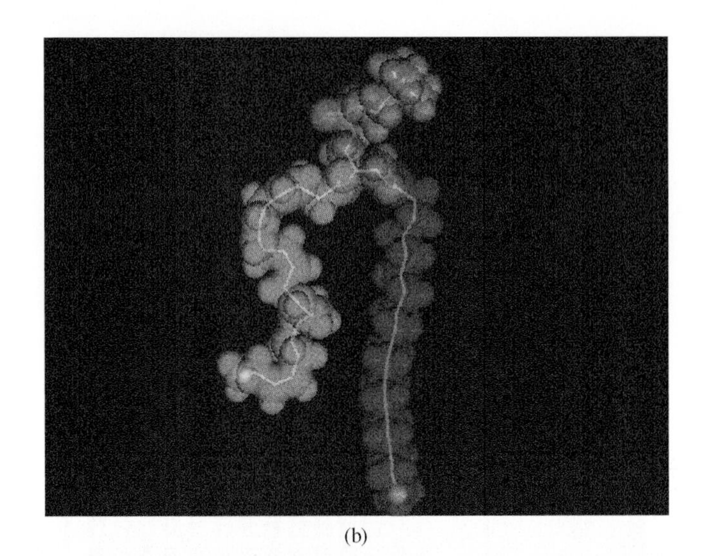

(b)

(c)

FIGURE 1.2 *(Continued)*

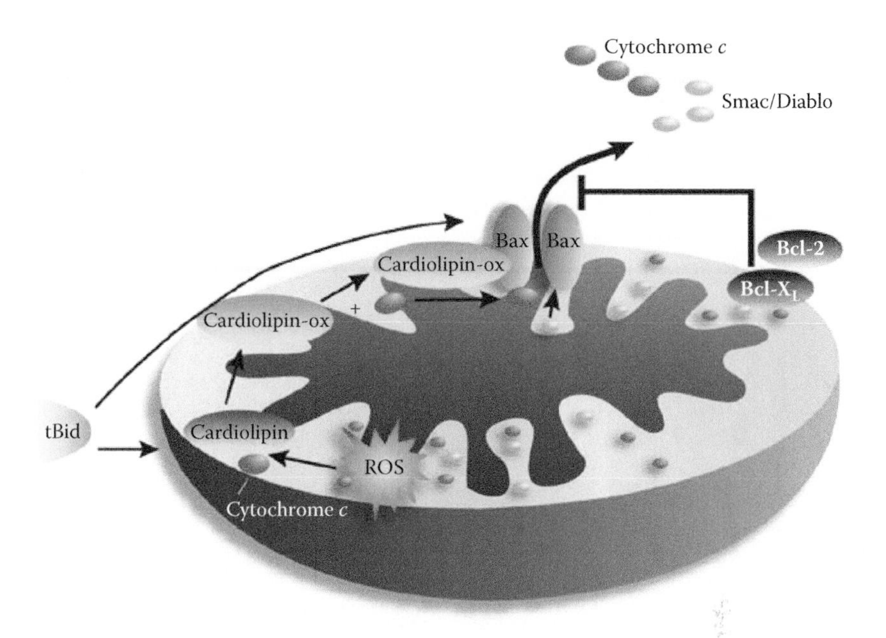

FIGURE 2.3 DHA may exert its anticancer effects following incorporation into mitochondrial cardiolipin of senescing colonic epithelial cells. DHA incorporated into cardiolipin of mitochondrial membranes is proposed to intensify apoptosis, effectively ridding the colon of potentially precancerous cells. (Cardiolipin ox, oxidized cardiolipin; tBid, 16 kDa truncated peptide of Bid, which is a pro-apoptotic effector protein. Bax is a pore-forming protein in the outer mitochondrial membrane. Bcl or Bclx are proteins that inhibit pore formation. Smac and Diablo, like cytochrome C, are pro-apoptotic proteins that are released into the cytoplasm through Bax.) (Reprinted by permission from Macmillan Publishers Ltd: *Nature Chem. Biol.*, Orrenius, S. and B. Zhivotovsky, Cardiolipin Oxidation Sets Cytochrome c free, 1:188–189, © 2005.)

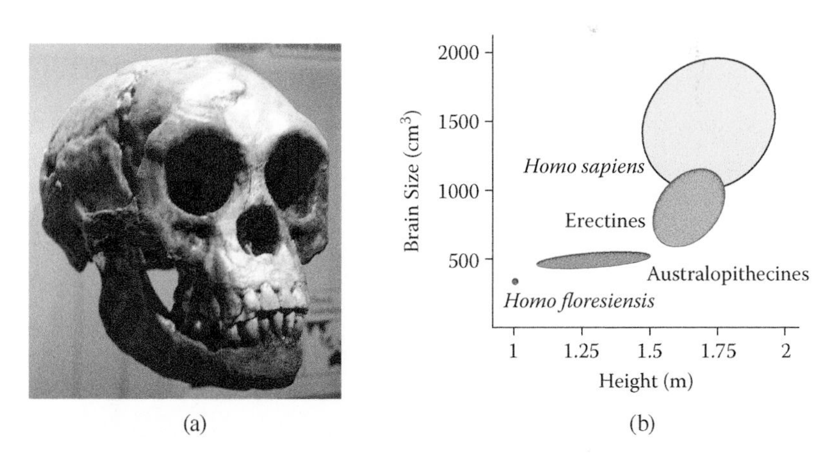

FIGURE 4.5 Brain evolution: tiny brain yet "human" behavior? (a) A skull of *Homo floresiensis.* (b) The cranial volume of this tiny hominoid, apparently a recent ancestor of humans, is a little more than one quarter (417 cc) that of modern humans. Yet *H. floresiensis* seems to have evolved relatively advanced mental capability. This discovery has fueled a good deal of debate on the importance of the environment on brain evolution and the relationship between brain size and human behavior. According to one theory, brain miniaturization is caused by the limitation in available energy occurring in an island environment. *Homo erectus* is the most familiar among the "erectines," and Lucy is the most famous of the australopithecines. Note that the brain of *H. floresiensis* is closest in size to Lucy, an ancient ancestor of man. (Parts (a) and (b) reprinted by permission from Macmillan Publishers Ltd: *Nature*, M. Mirazón Lahr, and R. Foley, Copyright 2004.)

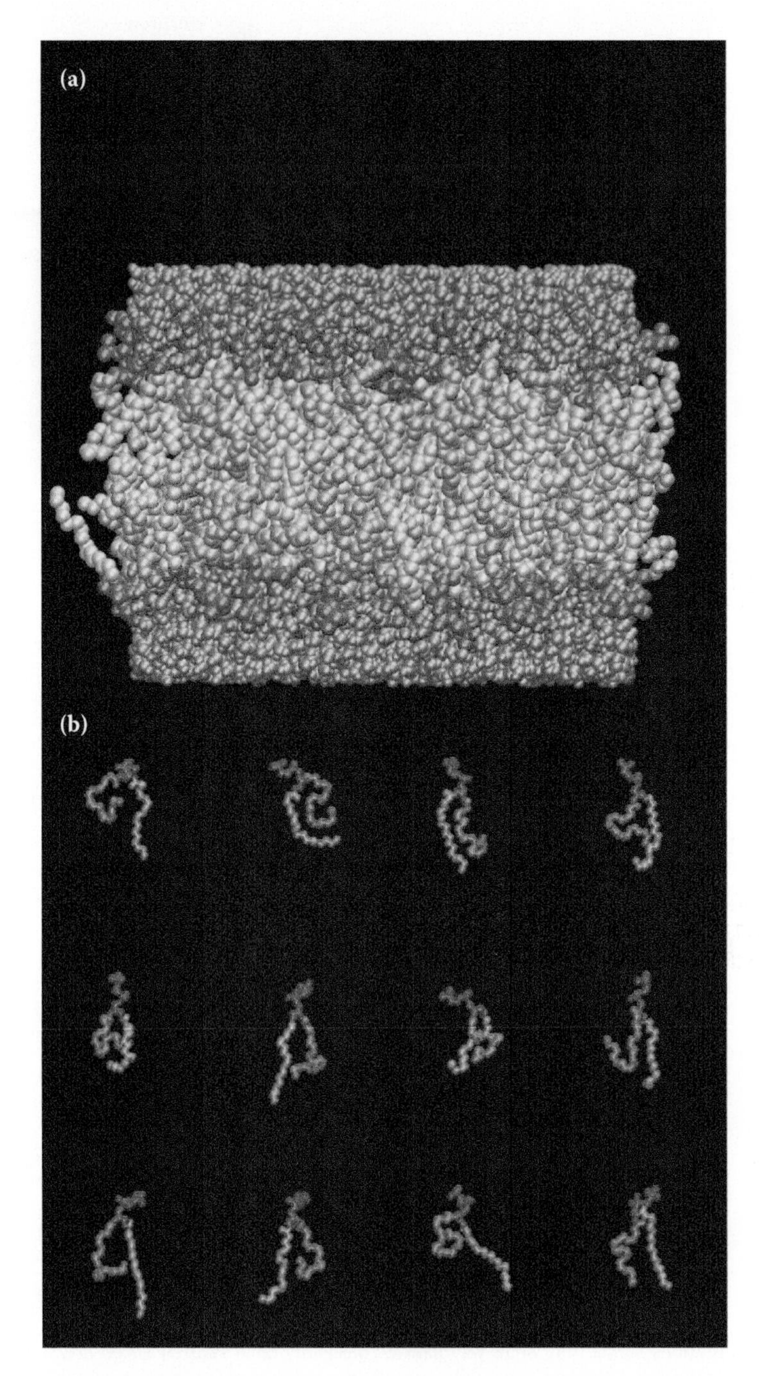

FIGURE 5.2 Dynamic conformations of DHA chains keep neuron membrane components in motion. (a) DHA phospholipids join together to form axon membranes. (b) DHA chains as membrane phospholipids shown in this set of images change shape on a picosecond to a nanosecond time scale. This property is harnessed to speed up electrical impulses in our brain. The dynamic shapes of DHA also explain its antifreeze properties in neurons of extreme, cold-adapted animals such as Antarctic krill. (Images courtesy of Scott Feller, and generated by Matthew B. Roark, both of Wabash College, Crawfordsville, Indiana.)

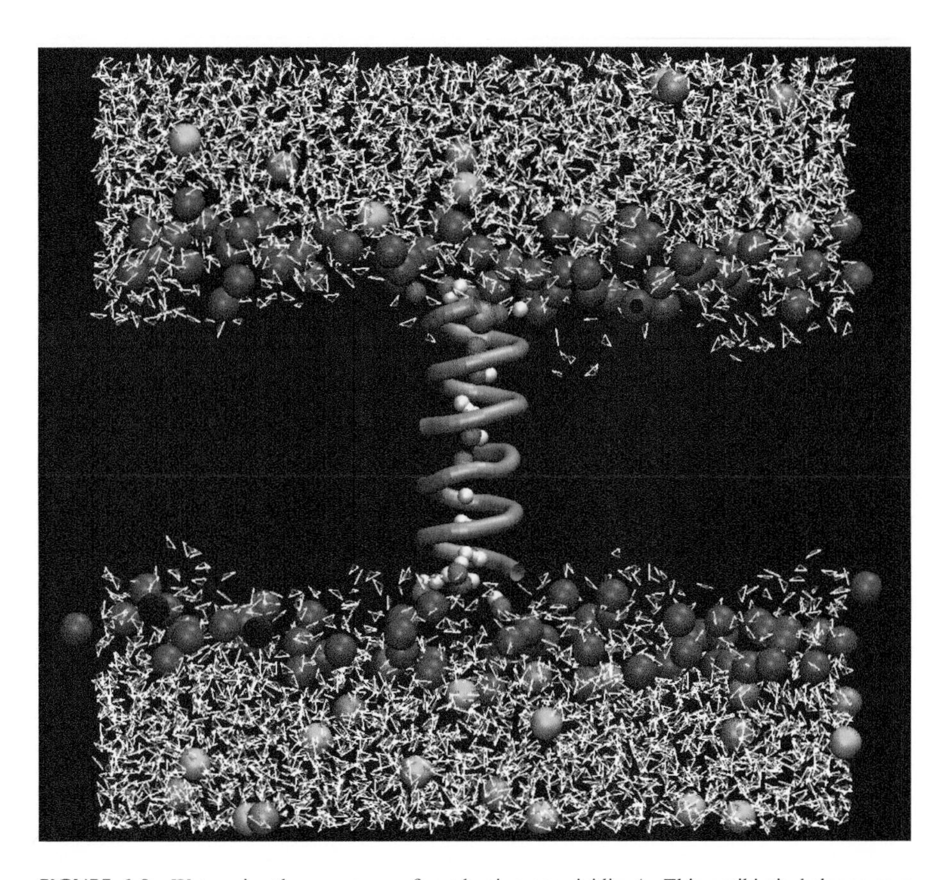

FIGURE 6.2 Water-wire theory was confirmed using gramicidin A. This antibiotic behaves as a membrane-spanning, energy-uncoupling nanotube holding 21 water-wires in single file inside its pore. Membrane water-wires are believed to occur spontaneously in all membranes and more readily in DHA membranes. Water-wire theory states that threads of water form spontaneously in DHA membranes of neurons, wasting a great deal of energy and driving up the already prodigious cost of fueling the brain. Ions including protons, sodium, and potassium move with blazing speed across water-wires and uncouple critical ion gradients. DHA, along with certain of its oxidation products, is believed to favor formation of water-wires in neuron membranes. (Image generated by chemical simulation.) (Courtesy of Serdar Kuyucak.)

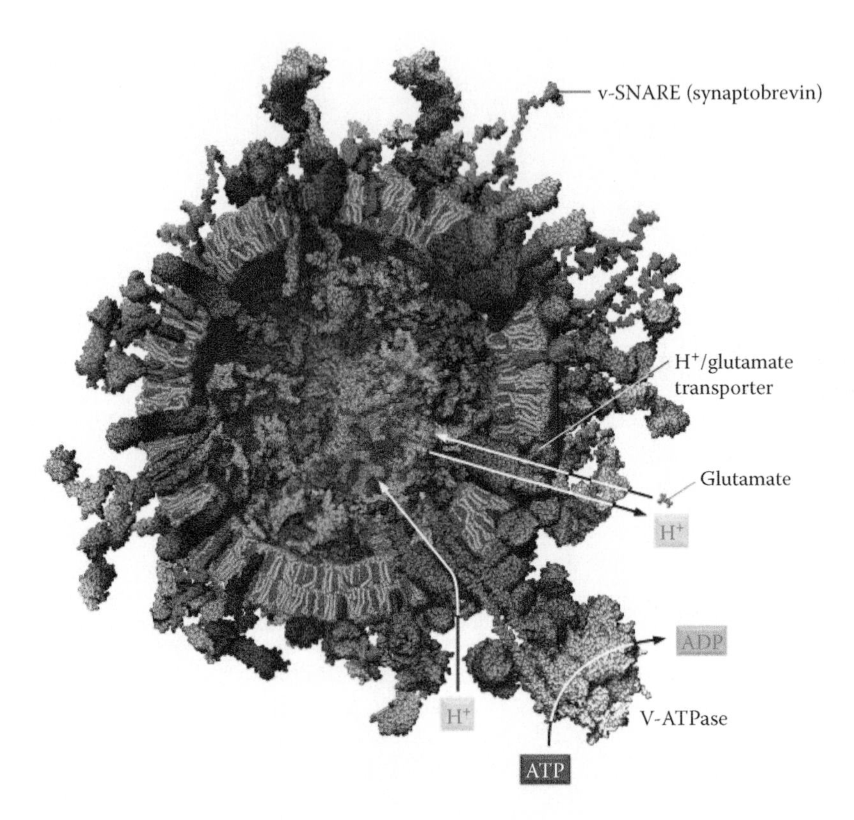

FIGURE 6.4 Molecular anatomy of a membrane of a synaptic vesicle. A synaptic vesicle cut in half and surrounded by a protein-laden bilayer (see Table 6.1) enriched with DHA and cholesterol (individual classes of membrane lipids not distinguishable in this model). Synaptic vesicles represent the best-known case in human cells, in which a DHA membrane works as a permeability barrier tight enough to prevent excessive, energy wasteful proton (H^+) leakage. High levels of cholesterol work cooperatively with DHA, presumably tightening membrane architecture against water-wires and thus minimizing proton leakage. Note that a proton energy gradient (high inside, low outside) is essential for energizing uptake pumps concentrating the critical neurotransmitter, glutamate. ATP is consumed by V-ATPase to maintain this proton gradient. (Reprinted from *Cell*, 127, S. Takamori et al., Molecular Anatomy of a Trafficking Organelle, 831–846, Copyright 2006, with permission from Elsevier.)

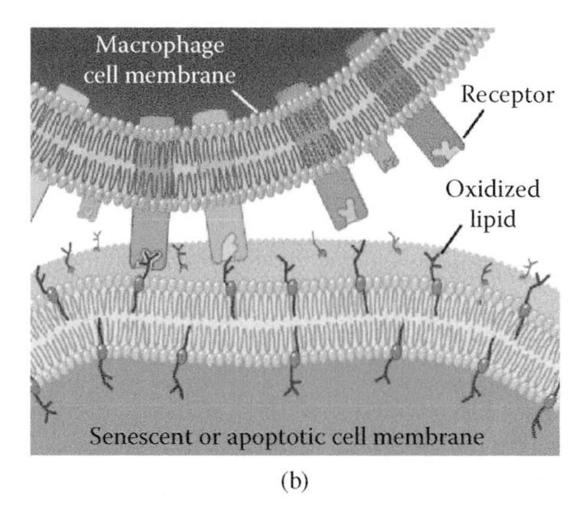

FIGURE 8.2(b) Lipid whisker model of oxidized cell membranes. Truncated DHA phospholipids as signaling molecules for phagocytes. Bioactivities being assigned to lipid whiskers range from recognition as membrane-surface signaling molecules by receptors of phagocytes to generation of a "hole" or defect in the membrane architecture that allows formation of energy-uncoupling water-wires. (Copyright 2008 The American Society for Biochemistry and Molecular Biology.)

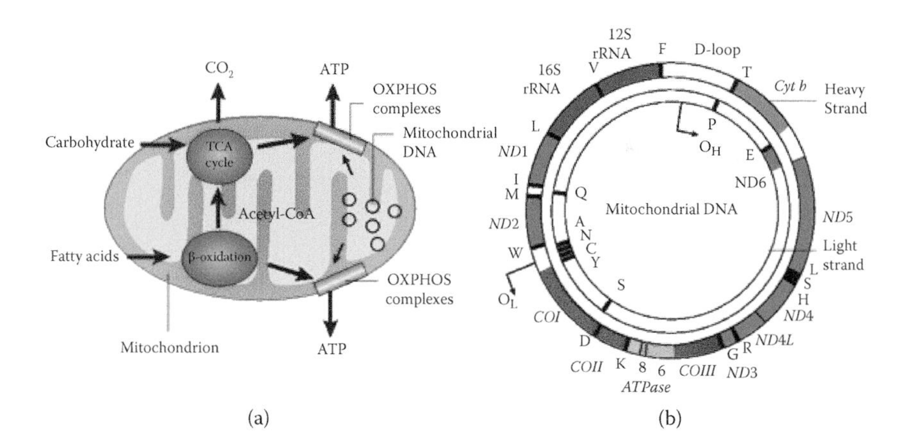

FIGURE 11.3 Human mitochondrial DNA (mtDNA) is a circular double helix encoding essential energy-producing genes. The genes that encode the subunits of complex I (ND1 through ND6 and ND4L) are shown in blue; the terminal complex, cytochrome *c* oxidase (COI through COIII) is shown in red; cytochrome *b* of complex III is shown in green; and the subunits of ATP synthase (ATPase 6 and 8) are shown in yellow. RNA genes are also listed (purple and black slashes). According to the revised mitochondrial theory of aging, mutations occurring during replication of mtDNA decrease energy production, causing aging. (Reprinted by permission from Macmillan Publishers Ltd: *Nat. Rev. Genet.*, R. W. Taylor, and D. M. Turnbull, Copyright 2005.)

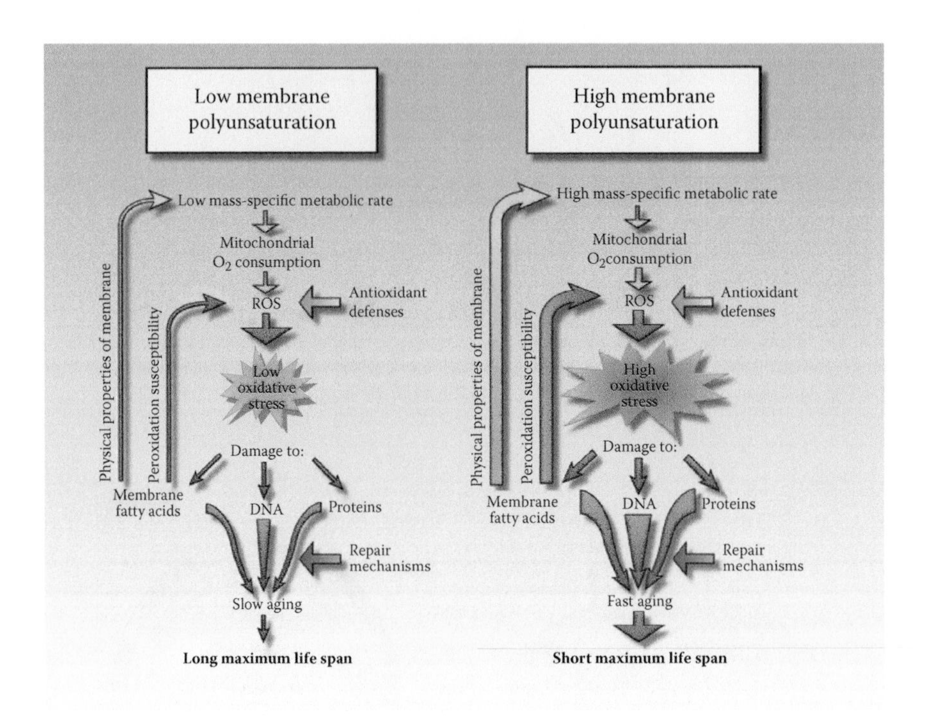

FIGURE 12.3 Peroxidation of unsaturated membranes is proposed as a mechanism to explain the membrane pacemaker theory of aging. The original membrane pacemaker model proposed by Hulbert and Pamplona states that increased levels of unsaturated chains in membranes, especially DHA or other polyunsaturated molecules, elevate rates of membrane peroxidation, causing aging. In this model peroxidation of polyunsaturated phospholipids boosts the pool of reactive oxygen species (ROS) to directly drive mutations in mitochondrial DNA. This mechanism is a derivative of the famous mitochondrial theory of aging featuring ROS as mutagen-generating mutations in mtDNA. This concept now requires some modification as discussed in the text. (Reproduced by permission of the American Physiological Society, Hulbert et al., 2007, *Physiol. Rev.*, 87: 1175–1213.)

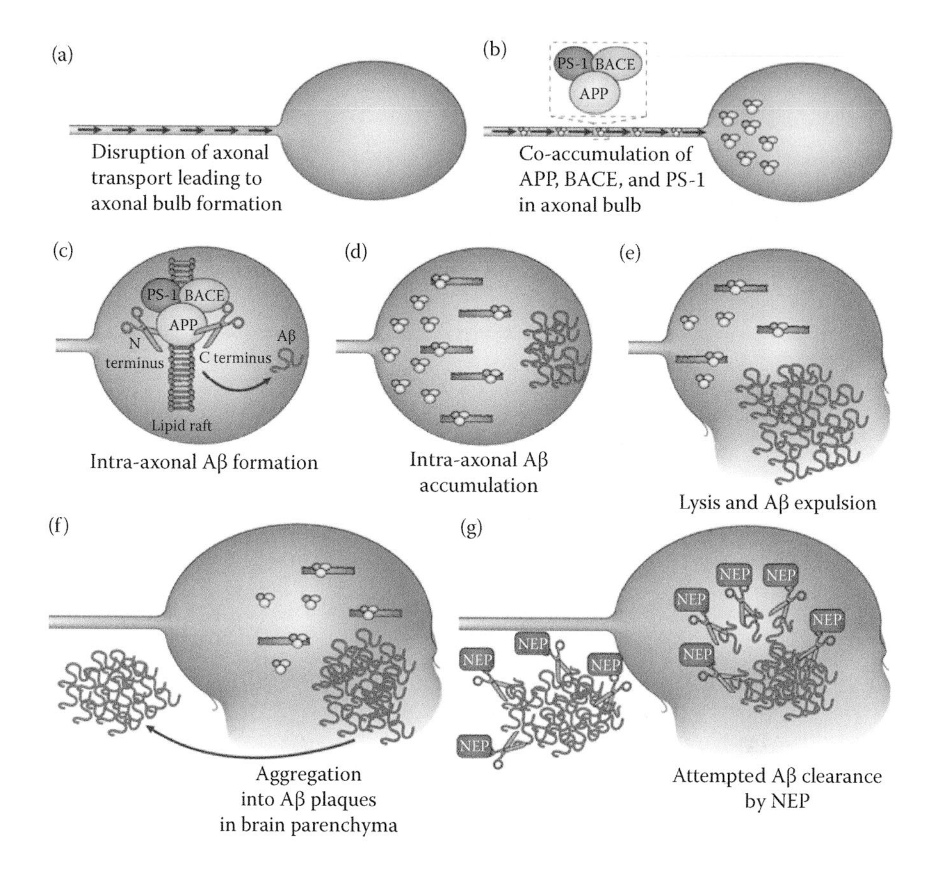

FIGURE 16.2 How mechanical damage triggers axons to produce amyloid plaque characteristic of Alzheimer's disease. Axonal transport is blocked or hindered in injured axons resulting in swelling at their disconnected terminals and accumulation of multiple misplaced proteins. Axonal bulbs contain the enzymes necessary for cleavage of amyloid precursor protein (APP) to amyloid-ß, including presenilin (PS-1), and ß-site, APP-cleaving enzyme (BACE). The enzyme that clears Aß, neprilysin (NEP), also accumulates in axonal bulbs, but the ability of this enzyme to clear amyloid-β might be overwhelmed by excessive production of this dysfunctionally produced peptide following trauma. Thus, localized brain damage caused by brain trauma is proposed to tip the balance between buildup and clearance of toxic peptides. (Reprinted by permission from Macmillan Publishers Ltd: *Nature Rev. Neurosci.*, Johnson, V. E., W. Stewart, and D. H. Smith, Traumatic Brain Injury and Amyloid-β Pathology: A Link to Alzheimer's Disease? *Nat. Rev. Neurosci.*, 11: 361–370, Copyright 2010.)

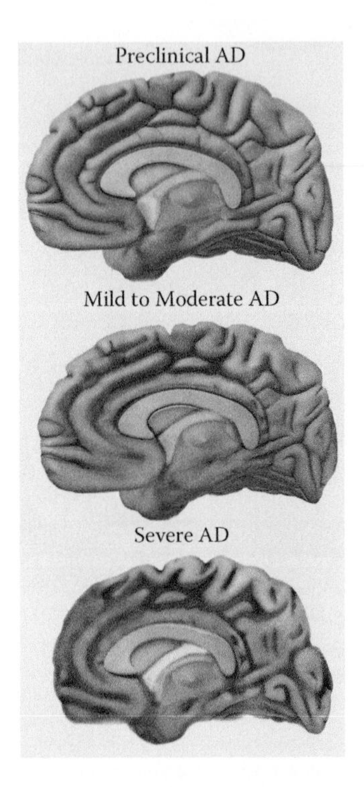

FIGURE 17.1 "Rotten apple in a barrel" concept of Alzheimer's disease. Three stages of brain damage leading to symptoms are shown—early, intermediate, and late. Massive neuron destruction and brain shrinkage occur during late states of Alzheimer's disease. (Image courtesy of the National Institute on Aging/National Institutes of Health.)

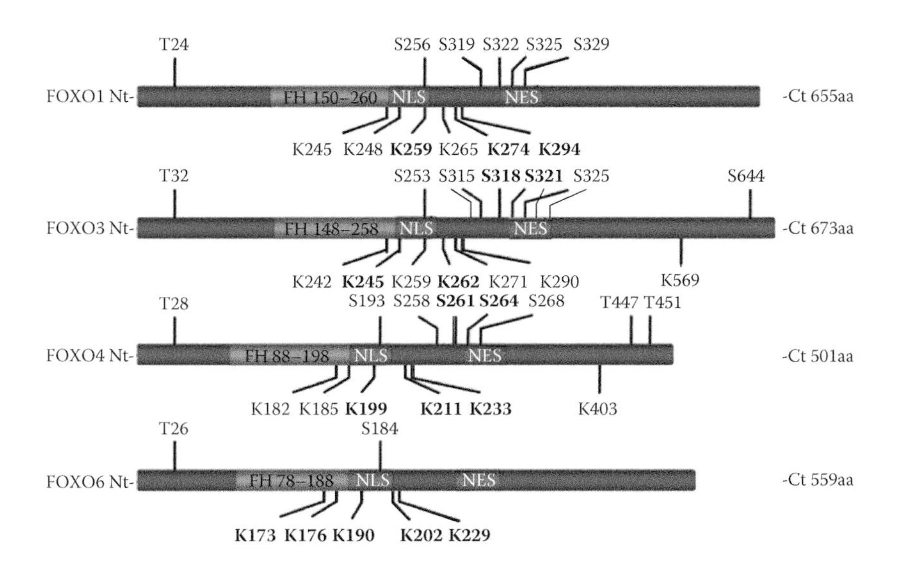

FIGURE 17.4 FOXO transcription factors bind to DNA and sense and respond to stress signals. Environmental signals are transduced, resulting in changes in patterns of phosphorylation and acetylation of specific amino acid residues along the protein chain of these transcription factors. These changing patterns alter nuclear localization, FOXO degradation, DNA-binding ability, transcriptional activity of stress tolerance genes, and protein-protein interactions. Akt sites (black); SGK, serum and glucocorticoid inducible kinase (black); IKKb, IkB kinase b (orange); JNK, Jun N-terminal kinase (green); DYRK, dual-specificity tyrosine (Y) phosphorylation-regulated kinase (red); CK1, casein kinase 1 (purple); acetylation sites (blue); FH, Forkhead domain; NLS, nuclear localization signal; NES, nuclear export sequence. Stress-induced phosphorylation sites of FOXO3 (not shown) occur at Ser90, Ser284, Ser294, Ser300, Ser413, Ser425, Thr427, and Ser574. (Reprinted by permission from Macmillan Publishers Ltd: *Oncogene*, 24: 7410–7425, Greer, E. L., and A. Brunet, FOXO Transcription Factors at the Interface between Longevity and Tumor Suppression, Copyright 2005.)

Section III

Risks of DHA

The risks associated with DHA are rarely discussed in the scientific literature, which is overwhelmed by an avalanche of claims of numerous health benefits of DHA. Interestingly, when using microorganisms as models, the risks of DHA are far easier to document than its benefits. We justify the risks of DHA through a hypothesis: with great benefit comes great risk—with evolution serving as judge and jury. The case history of molecular oxygen can be used to illustrate the concept of benefit versus risk among molecules essential for human life. Few would argue against the essentiality of oxygen. For example, a few minutes of oxygen deprivation is enough to destroy our brain. The same is true for risks of oxygen or more specifically its derivatives: reactive oxygen species. These molecules are blamed for many ailments including cancer, aging, and neurodegeneration. Thus, the concept that vital molecules have a duality of impacts is so well established among biochemists as to have become trivial.

How can molecules of DHA present in mother's milk as essential brain building blocks for infant brain development become dangerous later in life? In this section we search for answers and are beginning to find them in the conformational dynamics that make DHA so valuable. That is, the chemical properties of DHA chains conferring benefits versus risks are impossible to separate. According to the DHA principle developed in our previous monograph, DHA is always dangerous to cells living in the presence of oxygen, and to leverage DHA's great benefit, cell membranes must be continuously protected or cleansed against oxidative damage. Without such protection or as protection wanes, benefits give rise to severe risks, accentuated by time, and can help trigger cellular death. Perhaps the best evidence for this idea in humans is the fact that DHA is present in relatively low levels in most cells with the exception of a few specialized human cells, notably neurons. Here we begin to define the nature of the risks associated with DHA chains working in cellular membranes and follow up in Sections IV and V with how DHA is linked to aging, cancer, and dementia. Novel mechanisms to protect DHA-enriched membranes of neurons against oxidative stress are also discussed in this section.

8 Lipid Whisker Model and Water-Wire Theory of Energy Uncoupling in Neurons

In the highly oxygenated environment surrounding neurons, DHA chains are degraded, yielding a variety of oxidation products. Chemical oxidation occurs when oxygen attacks the multiple double bonds of DHA (Figure 8.1a). Warm temperatures (i.e., 37°C) and traces of metals such as iron and copper facilitate oxidation (Frankel, 2005). One of the most intriguing classes of oxidation products is named *truncated* or *chain-shortened derivatives* (Figure 8.1b). A DHA chain at the point of attack by O_2 or its reactive oxygen species (ROS) derivatives can be severed or truncated, yielding a chain-shortened fragment of the DHA chain still attached to its phospholipid head group. The truncated phospholipid usually contains a second long-chain, saturated fatty acid resistant to oxidation. The long acyl tail of the saturated chain firmly anchors the damaged phospholipid to the membrane (Gugiu et al., 2006). It has been proposed that the water-loving properties of chain-shortened DHA cause this stubby chain to exit the oily membrane interior where an intact DHA chain is normally located; the chain instead moves to a location in the aqueous layer above the membrane surface (Greenberg et al., 2008). According to the lipid whisker model proposed by these authors, oxidatively truncated chains stick out of the membrane surface like tiny molecular hairs or whiskers. The whisker model is important for understanding how DHA membranes of neurons age, as follows:

- Lipid whiskers are reporters of oxidative damage to membranes.
- Membrane repair lipases can recognize and access these hairs as being defective and clear them from the surface (Farooqui et al., 1997; Leslie, 1997).
- Lipid whiskers are exposed to receptors of phagocytic cells, which are triggered into a state of phagocytosis for recycling old membranes and returning useful building blocks such as DHA.
- Lipid whiskers are predicted to mark the location of molecular defects in the membrane surface, altering critical parameters such as membrane lateral motion and permeability and likely generating energy-uncoupling structures including water-wires.
- Whiskers are markers for aging or unhealthy neurons.

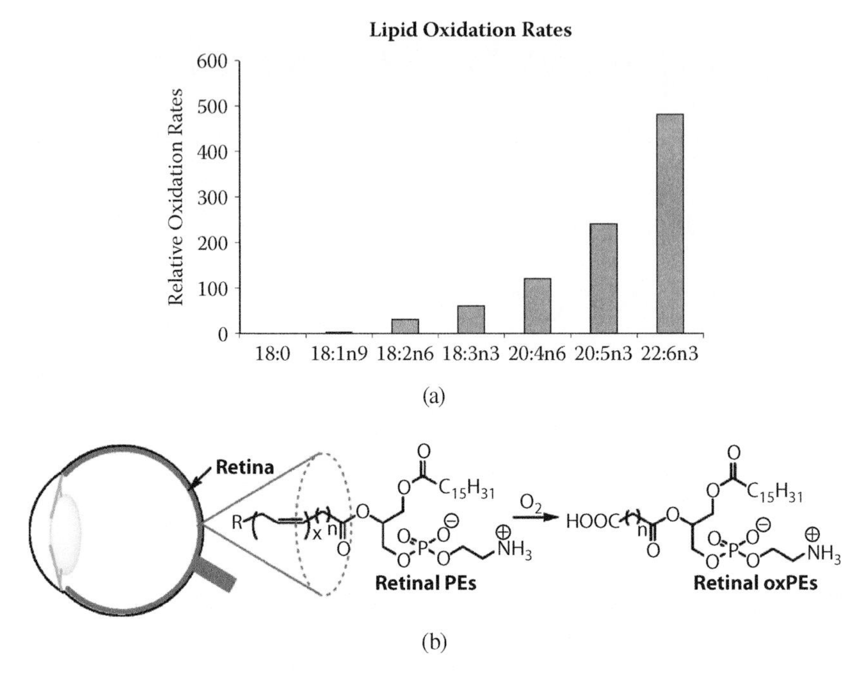

FIGURE 8.1 DHA-enriched membranes are oxidatively unstable and yield energy-uncoupling by-products. (a) Lipid peroxidation rates depend to a large degree on the number of double bonds in the fatty acid chains of phospholipids. DHA is the most readily oxidized fatty acid in nature, and wherever O_2 is present yields chemical breakdown products that create defective membranes and are also toxic to the cell. (From Valentine, R., and Valentine, D., *Omega-3 Fatty Acids and the DHA Principle.* Copyright 2009. Reproduced by permission of Taylor & Francis Group, LLC, a division of Informa plc.) (b) Truncated DHA phospholipids in the retina are reporters of oxidative damage. Oxidatively cleaved DHA phospholipids are thought to form in all DHA membranes in the presence of O_2. During the oxidation reaction a dramatic chain shortening or cleavage can occur and results in defective phospholipids that are proposed to act as powerful energy uncouplers and signals for phagocytosis. (Reprinted with permission from Gugiu et al., *Chem. Res. Toxicol.* Copyright 2006 American Chemical Society.)

8.1 LIPID WHISKER MODEL APPLIED TO NEURONS

The whisker model of oxidatively truncated DHA phospholipids was published in 2008 (Greenberg et al., 2008) following many years of pioneering research. These data support a model of oxidatively damaged DHA membranes in which truncation or cleavage of DHA chains occurs, causing the chain-shortened phospholipid to partially flip out of the plane of the membrane (Figure 8.2a). This novel structure generates a lipid whisker. Note that DHA and arachidonic acid (AA) are enriched in neuron membranes and synaptosomes, which according to the whisker model would sprout whiskers as membrane damage occurs and the truncated chains flip out of their position in a membrane leaflet into the aqueous phase. As already mentioned and discussed in more detail in Chapter 10, oxidation of neuron membranes

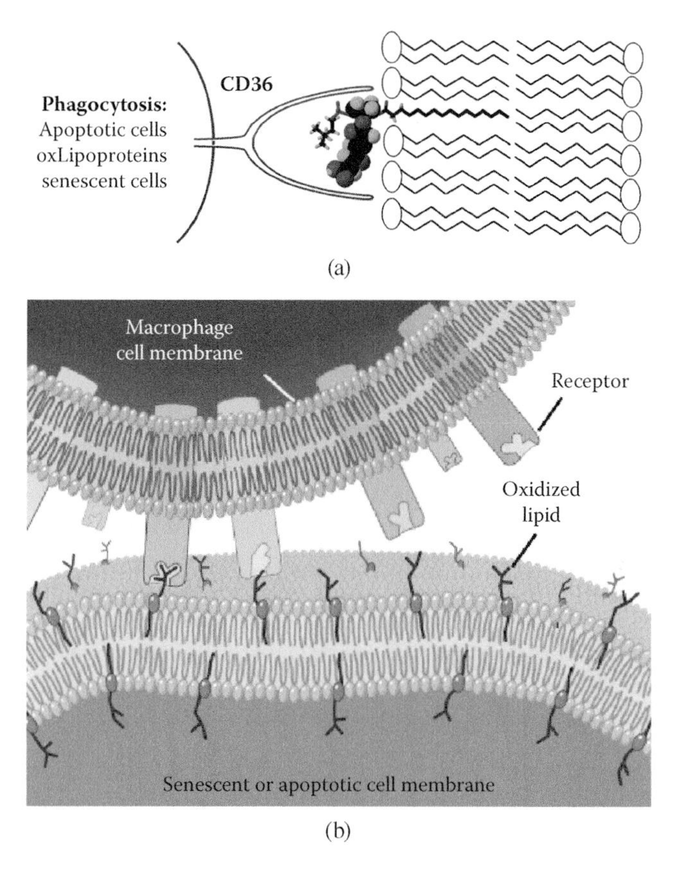

FIGURE 8.2 Lipid whisker model of oxidized cell membranes. (a) The physical-chemical properties of truncated phospholipids cause this normally membrane-embedded building block to emerge out of the oily bilayer extending as a molecular whisker into the aqueous phase. (Reprinted with permission from Li, X. et al. Conformation of an Endogenous Ligand in a Membrane Bilayer for the Macrophage Scavenger Receptor CD36, *Biochemistry*, 46: 5009–5017. Copyright 2007 American Chemical Society.) (b) Truncated DHA phospholipids as signaling molecules for phagocytes. Bioactivities being assigned to lipid whiskers range from recognition as membrane-surface signaling molecules by receptors of phagocytes to generation of a "hole" or defect in the membrane architecture that allows formation of energy-uncoupling water-wires. (Copyright 2008 The American Society for Biochemistry and Molecular Biology.) **(b: See color insert.)**

is inevitable because of conditions that exist surrounding and within neurons. Numerous physiological pathways are believed to lead to conditions favoring neuron membrane damage. As discussed in Section IV, major winds of change are now rattling the field of oxidative stress because of data that diminish the role of ROS as being directly responsible for mutations in mitochondrial DNA (mtDNA) causing aging (Chapter 11). The whisker model does not specify or depend on a particular source or class of reactive oxygen species causing membrane damage, leaving open the possibility that DHA membranes foster and even prime oxidative damage to themselves. The model posits that wherever oxidative damage occurs in unsaturated membrane phospholipids, the chemistry of truncated chains will cause these chains to sprout whiskers on the membrane surface. Rules of lipid peroxidation predict that DHA-enriched membranes of neurons and sperm and rod cells will tend to generate lipid whiskers at the greatest rates (Figure 8.1a). However, polyunsaturated membrane phospholipids composed of fatty acid chains with two to three double bonds are also expected to sprout significant levels of lipid whiskers.

The first clues that novel conformations of oxidized lipids might protrude into the aqueous phase came in studies involving receptors located on the surface of phagocytic cells that seemed to sense a lipid molecule as a signal on the membrane surface (Li et al., 2007). As shown in Figure 8.2b, lipid whiskers are now proposed to act as signaling molecules toward activation of phagocytosis. In other words, lipid whiskers are believed to send signals to phagocytic cells to begin the degradation of any membrane surface tagged by this mechanism as being unable to maintain critical membrane functionality—an unhealthy state. The concept of a "functionally unhealthy state" requires explanation. First, phagocytosis and apoptosis are considered a normal part of the cell cycle and are essential for removing cells, organelles, or cell structures that have become unhealthy or are a danger to the whole organism. Generally, oxidatively truncated DHA phospholipids forming lipid whiskers that invite phagocytosis can be viewed as an important part of the overall defense network against oxidative damage. These cells that have sustained severe oxidative damage of their membranes are targeted for removal. We suggest that such an elaborate system of handling damaged membranes has evolved at least in part as a defense not only against oxidative stress but against energy stress as well. We also suggest it is the presence of lipid whiskers, not only their mechanism of origin, which has led to the necessity to activate the phagocytic process shown in Figure 8.2b.

The authors of the lipid whisker model make the important points that the molecular architecture of membranes and the conformational dynamics of individual phospholipids are likely far more complex than previously anticipated. This led them to propose that the lipid whisker model might be a relatively universal mechanism encompassing critical events in aging and senescence, inflammation, and apoptosis. For some years these researchers have also had neurodegeneration squarely in their sights, and we continue with this theme in later chapters.

8.2 OXIDATIVELY TRUNCATED PHOSPHOLIPIDS ACT DIRECTLY AS SIGNALING MOLECULES FOR ACTIVATING PHAGOCYTOSIS

Recently an important test of the lipid whisker model has been performed using receptor proteins on the membrane surface of macrophages as a probe to calibrate critical regions of bioactivity of the putative lipid whiskers (Gao et al., 2010). According to the 2008 rendition of the lipid whisker model (Greenberg et al., 2008; Hazen, 2008), receptors on phagocytes might gain access to and bind specifically to the negatively charged head group of phospholipids as well as recognize the terminal carboxyl group of the truncated chain, the latter also negatively charged. That is, two points of recognition are opened on the truncated phospholipid by virtue of the flipping of the stub of the former DHA chain into the aqueous phase (Figure 8.2b).

Studies on the structural basis for the recognition of oxidized phospholipids by Gao et al. (2010) show that high-affinity scavenger receptors (e.g., CD36) recognize and bind to three regions of truncated phospholipids—the long *sn*-1 hydrophobic (e.g., saturated) chain, the *sn*-3 hydrophilic phosphocholine or phosphatidic acid group, and the polar-truncated *sn*-2 tail. All three are essential for high-affinity binding. A terminal, negatively charged carboxylate at the *sn*-2 position satisfies binding at this location.

One of the questions raised by this study concerns how class B receptors such as CD36 access the outer membrane leaflet to recognize or bind to a long-chain, saturated fatty acid such as a stearic acid (18:0) chain. One can imagine a hydrophobic loop or region of the receptor penetrating deeply enough into the leaflet interior to recognize the C-18 moiety. On the other hand, once the receptor makes a two-point binding to the lipid whisker then the acyl chain acting as anchor might be pulled further into the aqueous phase by microglia. A third alternative is that the rate of flipping of this defective membrane phospholipid might be greatly accelerated once oxidative truncation occurs, in essence momentarily exposing all three critical binding regions to its receptor. Lipid whiskers as truncation products of DHA phospholipids usually residing in the cytosolic leaflet of the membrane can be flipped to the outer leaflet by the mode of action of scramblase (see Chapter 5).

8.3 LIPID WHISKERS MIGHT GENERATE ENERGY-UNCOUPLING WATER-WIRES

DHA based on its permeability properties alone is considered a poor choice for neuron membranes whose efficiency and function depend on maintaining razor-sharp ion gradients. The nature of DHA as a mediocre permeability barrier against critical neuronal cations—protons, sodium, and potassium—is based on physical-chemical data summarized as follows:

- Water content in the interior of the bilayer is known to increase with increasing chain unsaturation.
- A linear relationship between DHA content and proton permeability has been demonstrated.
- DHA has been shown to increase H^+ permeability of mitochondrial membranes.
- DHA might allow water-wires to form because of loose lipid packing at the aqueous interface.
- DHA decreases membrane thickness favoring water-wire formation (Figure 8.3).

One of the simplest explanations of these data/properties is that water-wires are believed to form more readily in DHA-enriched bilayers. This concept is introduced in Chapters 4 and 6 where biochemical studies show that DHA membranes seem

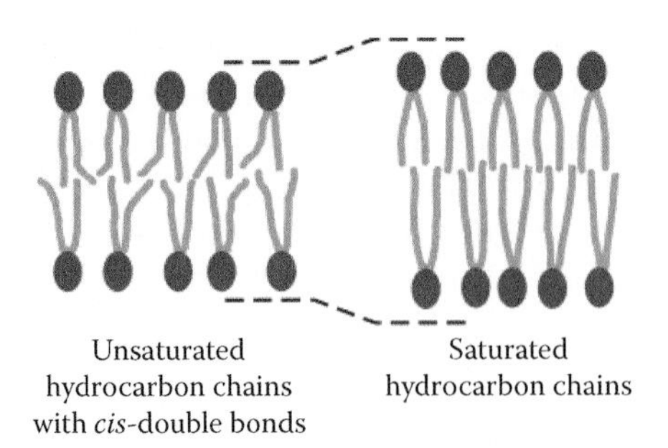

FIGURE 8.3 DHA is known to decrease membrane thickness favoring formation of water-wires. As a rule of thumb each *cis* double bond shortens the fatty acid by a length equivalent to a single carbon-carbon bond length. (From Alberts, B. et al., *Molecular Biology of the Cell*, Copyright 2008. Reproduced by permission of Taylor & Francis Group, LLC, a division of Informa plc.)

to require a partnership with cholesterol in forming membrane architecture tight enough to block futile cycling of protons. Other ecological evidence shows that DHA is readily converted into a powerful energy uncoupling conformation by even a few degrees increase in temperature (reviewed by Valentine and Valentine, 2009). Water-wires forming in DHA-enriched membranes of phytoplankton are believed to act as reporters of the harmful impact of global warming on marine productivity.

There is growing evidence that neuron membranes, and all membranes for that matter, are inherently leaky or chemically permeable especially for protons, the smallest and most elusive of essential excitatory ions in the order $H^+ >>> Na^+ > K^+$ (Figure 8.4). Neuron membranes are far less permeable to sodium ions than to protons. However, even though neuron membranes are several orders of magnitude less permeable to Na^+ than to H^+, the high levels of Na^+ surrounding axons can rapidly drive Na^+ into the cell, wasting a great deal of energy.

Many years ago studies of the electro-chemical properties of water showed that protons somehow tunnel through ice at an amazing speed (see Chapter 4). About 50 years ago, the concept of proton tunneling via water-wires was first applied to explain proton permeability properties of membranes. However, it was not until the early 1980s that proton tunneling, thought to be mediated by traces of water spontaneously entering the lipid region of membranes, was shown to occur in chemically defined membrane vesicles. Perhaps the biggest breakthrough in this field came from an unlikely source—studies of the molecular mode of action of energy uncoupling antibiotics used to fight diseases. These studies, which were already discussed in Chapter 6, show that when about 21 molecules of water enter single-file into the core of a membrane-spanning, nanotube-forming antibiotic called *gramicidin A*, a circuit is opened for conducting protons. Several mechanisms for proton energy uncoupling are summarized in Figure 8.4. It is now clear that it is the structure of chains or aggregates of water held in the pore of gramicidin A that create proton-conducting water-wires. Water-wires are proposed to form spontaneously in virtually all natural membranes (Valentine, 2007).

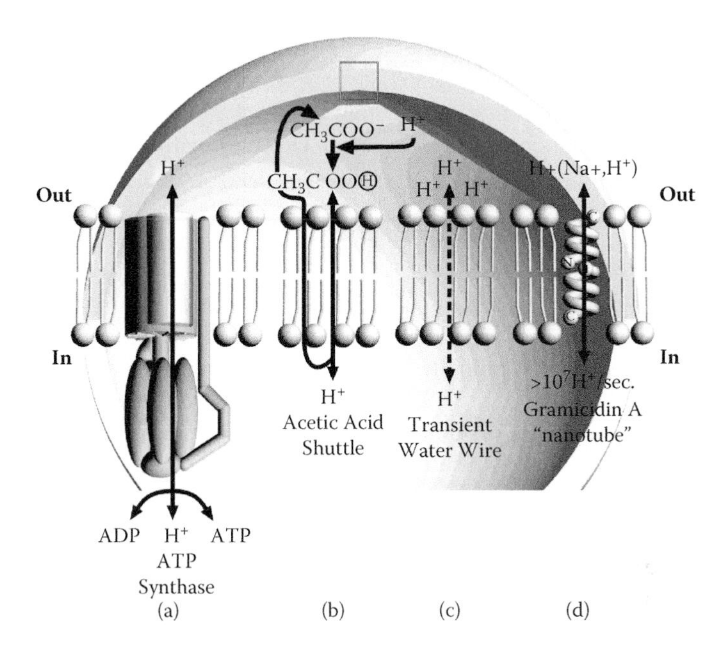

FIGURE 8.4 Water-wire theory applicable to neurons was first developed in bacteria. Four mechanisms, studied in bacteria, for energy uncoupling are from left to right: (a) proton leakage through defective proteins, such as mutated ATP synthase; (b) organic acids acting as energy uncouplers by shuttling protons across the bilayer; (c) transient water-wires forming spontaneously across the lipid portion of the bilayer; and (d) gramicidin A behaving as an energy-uncoupling antibiotic. Besides these possible mechanisms, mitochondria can purposely drain their proton energy gradients by opening a proton gate called the *mitochondrial permeation pore*. Also, certain classes of mitochondria including those in brown fat cells have evolved specific uncoupling systems for the purpose of producing heat for maintaining temperature homeostasis. In plasma membranes of neurons, mitochondrial inner membranes and membranes of axons and synaptic vesicles seem especially sensitive as targets for oxidative generation of energy-uncoupling water-wires. (From Valentine, R., and Valentine, D., *Omega-3 Fatty Acids and the DHA Principle*. Copyright 2009. Reproduced by permission of Taylor & Francis Group, LLC, a division of Informa plc.)

The antimicrobial activity of gramicidin A was first discovered during the era of massive screening of soil-born organisms for antibiotics which, in nature, serve as chemical warfare agents used to gain competitive advantage against neighboring cells. The bacterium *Bacillus brevis* synthesizes and exports gramicidin A as a helical polypeptide monomer composed of 15 amino acid residues, which inserts into the membrane of a Gram-positive target cell. A single monomer spans the distance of only one of the two leaflets forming the bilayer, requiring head-to-head dimerization with a second monomer to enable a full-length nanotube. The core of each nanotube contains approximately 21 waters in single file (i.e., water-wire), a number sufficient to span across the membrane allowing "downhill" (i.e., high outside/low inside) proton flow. This catastrophic form of proton uncoupling short-circuits proton electrochemical gradients—in essence bleeding the cell of energy. The mature nanotubes formed by two monomers joined at their amino terminal ends are unstable within the bilayer and are constantly forming and disassociating, such that the average open time for a nanotube is about 1 second. However, with a large electrochemical gradient, one water-wire can transport about 20,000 protons per open channel each millisecond. Thus, water-wires are an exceptionally fast mechanism capable of draining neuron cation gradients, but energy uncoupling is not triggered until a water-wire spanning across the bilayer is generated.

We now return to lipid whiskers as a potential source of energy uncoupling in neurons. The authors of the lipid whisker model did not discuss the possible impact of whiskers on bioenergetic properties of membranes. To gain insight we focus on a single lipid whisker (see Figure 8.2a) and attempt to envision its dynamic nature. A lipid whisker anchored by its saturated chain in a lipid raft would be expected to hold the damaged phospholipid relatively tight in the membrane. However, we would expect a truncated phospholipid in the liquid region of a neuron membrane to be strongly influenced by its truncated and water-soluble tail. The authors of the whisker model considered this property, and they provide evidence that supports their model. The gist is that the attraction of the shortened chain to water is sufficient to rearrange, flip, and hold the truncated chain into the aqueous phase. There is room in their model for a truncated phospholipid molecule to exhibit its dynamic nature by continuously flip-flopping in and out of the bilayer. We propose that this dynamic state occurs like a tiny piston plunging in and out of the membrane. The point is this dynamic action might cause localized membrane alterations or thinning (Beranova et al., 2010) that perturbs membrane internal water status and permeability, thus altering bioenergetic properties of the bilayer. Data summarizing the energy uncoupling properties of naturally occurring chain-shortened phospholipids are summarized in Chapter 13 (also see Valentine and Valentine, 2009). The main point is that oxidatively damaged DHA phospholipids in neuron membranes are expected to behave as potent uncouplers of cation gradients. This would foster energy stress in neurons. Even a few lipid whiskers escaping repair might generate enough water-wires to create catastrophic energy stress, especially in a neuron already weakened by aging.

In summary, any balanced picture regarding the benefits and risks of DHA in neuronal membranes must take into account chemical oxidation of these oils and the bioactivity and potential toxicity of oxidation products. Lipid peroxidation damages the integrity of the membrane and has the power to trigger neuron death. Damaged DHA chains remaining in the membrane are believed to behave directly as phagocytic signaling molecules and indirectly as apoptotic signals. We propose that lipid whiskers in neuron membranes act directly as powerful energy uncouplers contributing to energy stress, which acts as a signal for apoptosis. The paucity of DHA in most human cells is consistent with a toxicity mechanism to explain the absence of this otherwise essential fatty acid. Finally, we reiterate the concept that the benefits of DHA in neurons must be considerable to compensate for the potential toxicity described here. Discussion on oxidized DHA phospholipids as trigger molecules for Alzheimer's disease is continued in Chapter 17.

REFERENCES

Beranova, L., L. Cwiklik, P. Jurkiewicz et al. 2010. Oxidation changes physical properties of phospholipid bilayers: Fluorescence spectroscopy and molecular simulations. *Langmuir* 26:6140–6144.

Farooqui, A. A., H.-S. Yang, T. A. Rosenberger et al. 1997. Phospholipase A2 and its role in brain tissue. *J. Neurochem.* 69:889–901.

Frankel, E. N. 2005. *Lipid Oxidation.* Oily Press Lipid Library. Vol. 18. Bridgwater, UK: P. J. Barnes & Associates.

Gao, D., M. Z. Ashraf, N. S. Kar et al. 2010. Structural basis for the recognition of oxidized phospholipids in oxidized low density lipoproteins by class B scavenger receptors CD36 and SR-BI. *J. Biol. Chem.* 285:4447–4454.

Greenberg, M. E., X. M. Li, B. G. Gugiu et al. 2008. The lipid whisker model of the structure of oxidized cell membranes. *J. Biol. Chem.* 283:2385–2396.

Gugiu, B. G., C. A. Mesaros, M. Sun et al. 2006. Identification of oxidatively truncated ethanolamine phospholipids in retina and their generation from polyunsaturated phosphatidylethanolamines. *Chem. Res. Toxicol.* 19:262–271.

Hazen, S. L. 2008. Oxidized phospholipids as endogenous pattern recognition ligands in innate immunity. *J. Biol. Chem.* 283:15527–15531.

Leslie, C. C. 1997. Properties and regulation of cytosolic phospholipase A2. *J. Biol. Chem.* 272:16709–16712.

Li, X. M., R. G. Salomon, J. Qin et al. 2007. Conformation of an endogenous ligand in a membrane bilayer for the macrophage scavenger receptor CD36. *Biochemistry* 46:5009–5017.

Valentine, D. L. 2007. Adaptations to energy stress dictate the ecology and evolution of the Archaea. *Nat. Rev. Microbiol.* 5:316–323.

Valentine, R. C., and D. L. Valentine. 2009. *Omega-3 Fatty Acids and the DHA Principle.* Boca Raton, FL: Taylor & Francis Group.

9 Neurons Boost Their Energy State by Using Powerful Antioxidant Systems to Protect Their Membranes against Damage

Specialized DHA-enriched membranes of axons and synaptosomes as well as polyunsaturated membranes of neuron mitochondria are major substrates or targets for chemical oxidation. It is now clear that neurons have evolved numerous robust mechanisms for protecting their membranes against lipid peroxidation (see Chapter 10). Antioxidants are one of the most important lines of defense against membrane defects caused by oxidation. According to conventional theories of membrane peroxidation, DHA and polyunsaturated fatty acid chains in neuron membranes are first attacked by reactive oxygen species generating more reactive oxygen species in a snowballing chemical reaction (see Chapter 14). Chemists classify this as a chain reaction that, if uncontrolled, can destroy cellular function. Antioxidants, including ascorbic acid, emphasized here along with vitamin E, work together in neurons by snuffing out free radicals that feed a chain reaction. Reactive oxygen species are generated as a natural by-product of electron transport in neuron mitochondria, by lipid peroxidation of polyunsaturated membranes and during oxidative bursts of microglia as disease-fighting cells in the brain. It is not possible to completely stop production of reactive oxygen species in neurons, and it is not possible to develop a perfect shield against the toxic effects of these free radicals. These data gave rise to the popular oxidative stress theory of aging in which reactive oxygen species escaping cellular defenses directly attack mitochondrial DNA (mtDNA), creating mutations followed by energy stress that causes aging. According to this model, oxidatively derived mutations in mtDNA are envisioned to be the primary cause of energy deficits driving the process of aging.

The revised mitochondrial theory of aging (see Chapter 11) states that errors of mtDNA replication rather than oxidative stress are the direct cause of energy deficiency in aging mitochondria. This has led to a revised model of oxidative stress in aging and disease (Murphy et al., 2011). According to the revised membrane pacemaker model developed in Chapter 12, oxidatively damaged membranes directly waste energy as the result of defects of membrane molecular architecture, which weaken the permeability barrier against protons and excitatory cations. Thus, we propose that antioxidants in neurons are important energy savers. The essential roles of antioxidants in the brain, especially vitamin C, are discussed next.

9.1 THE BRAIN HOARDS HIGH LEVELS OF VITAMIN C: CASE HISTORY OF SCURVY

In the era of great sailing ships, seamen on long voyages became lethargic due to effects of a sometimes-fatal disease called *scurvy* and were often beaten for being laggards. Common sailors contracted scurvy, while officers usually remained healthy. With an epidemic of scurvy depopulating their navy, the British Admiralty became alarmed enough to investigate the cause of scurvy. Because officers were often fed special meals including fresh fruits such as citrus, it soon became apparent that adding lemons to the fare of common seamen solved the problem. Vitamin C (ascorbic acid) concentrates were eventually found to replace whole lemons. Many nutritionists consider this dietary solution to a serious human disease as the beginning of the modern era of vitamin research.

During the period that scurvy was the scourge of seamen, autopsies of sailors who died of scurvy revealed wholesale destruction of many organs while the brain appeared healthy. The initial interpretation of this finding is that vitamin C, while being essential for, say, a healthy liver, is not important in the brain. It is now known that the opposite is true; the brain is the last organ to lose its vitamin C content because this vitamin is so important that the brain has evolved robust mechanisms to hoard ascorbate even at the expense of other organs. It is now known that the highest levels of ascorbate in the body occur in the brain. The unique mechanisms used to shuttle and concentrate vitamin C in neurons are discussed next.

9.2 MECHANISMS FOR CONCENTRATING ASCORBATE IN NEURONS ARE KNOWN IN DETAIL

We trace the route of an ascorbate molecule carried in blood plasma into the cytoplasm of a neuron (Figure 9.1) (see Harrison and May, 2009). Figure 9.1a shows that brain tissue contains relatively high levels of vitamin C consistent with a specialized role of this antioxidant in protecting neurons. A unique two-step mechanism involving cerebrospinal fluid (CSF) is involved in concentrating ascorbate in the brain (Figure 9.1b). The first step involves transport of ascorbate from blood plasma into the CSF bathing the brain. This "backdoor" entry involves transport across the epithelium of the specialized brain region called the *choroid plexus*. High levels of the major ascorbate transporter (SVCT2) are embedded in the membranes of epithelial cells, and its role is to concentrate ascorbate from the plasma into the cytoplasm of the epithelial cell. Transport is energized by a sodium gradient (high outside and low inside). Ascorbate is next effluxed by an unknown mechanism from epithelial cells into the CSF. This step results in a fourfold step-up in ascorbate levels in the CSF—from about 40 to 60 μM to 160 μM.

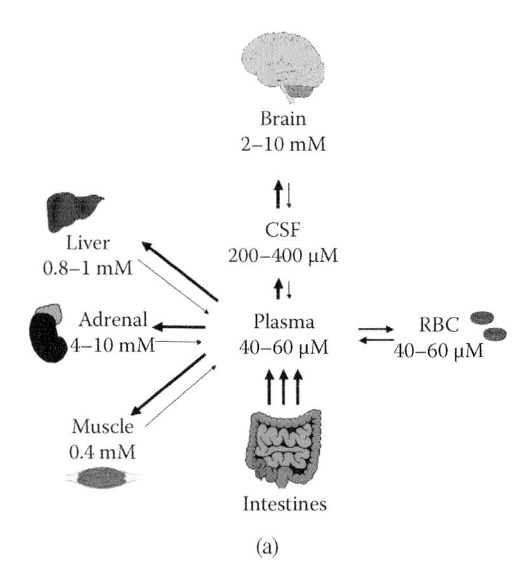

FIGURE 9.1 High levels of ascorbate (vitamin C) are needed in the brain to protect neuron membranes and save energy. (a) Ascorbate levels in various organs and cells vary greatly. The high levels of this antioxidant in brain tissue likely mean that the brain has specialized needs for antioxidants. (b) Concentration of ascorbic acid from blood plasma into neurons involves the powerful Na^+-dependent transporter SVCT2. Symbols are ASC (ascorbate), CSF (cerebrospinal fluid), and Glut (glucose transporter). Note that neurons in contrast to astrocytes in the brain require high levels of ascorbate. These data are consistent with the view that neurons need high levels of antioxidants including vitamin C to protect their DHA-enriched membranes against lipid peroxidation. (c) Chemistry of ascorbic acid as antioxidant. See text for details. (Reprinted from *Free Radic. Biol. Med.*, 46, Harrison, F. E., and J. M. May, Vitamin C Function in the Brain: Vital Role of the Ascorbate Transporter SVCT2, 719–730. Copyright 2009, with permission from Elsevier.)

From the CSF, ascorbate diffuses into the brain interstitium where the powerful ascorbate transporter SVCT2 present in neuron membranes once again concentrates this antioxidant in neurons at levels of 2 to 10 mM. Note that neurons contain the highest levels of vitamin C in the body. Ascorbate can leak slowly from neurons, but the levels in neurons are maintained from the large supply in the CSF. Data from studies of knockout mutants targeting the major ascorbate transporter in mice (Sotiriou et al., 2002) have confirmed this pathway. Knockout mutants of ascorbate transporter result in undetectable levels in brain tissue and are lethal in mice.

9.3 BRIEF CHEMISTRY OF ASCORBATE AS AN ANTIOXIDANT

Vitamin C plays many roles in humans, but its best-known function is as a water-soluble antioxidant. As discussed later ascorbate works synergistically with the membrane-specific antioxidant α-tocopherol or vitamin E. Vitamin C also works cooperatively with another important water-soluble antioxidant, glutathione (GSH).

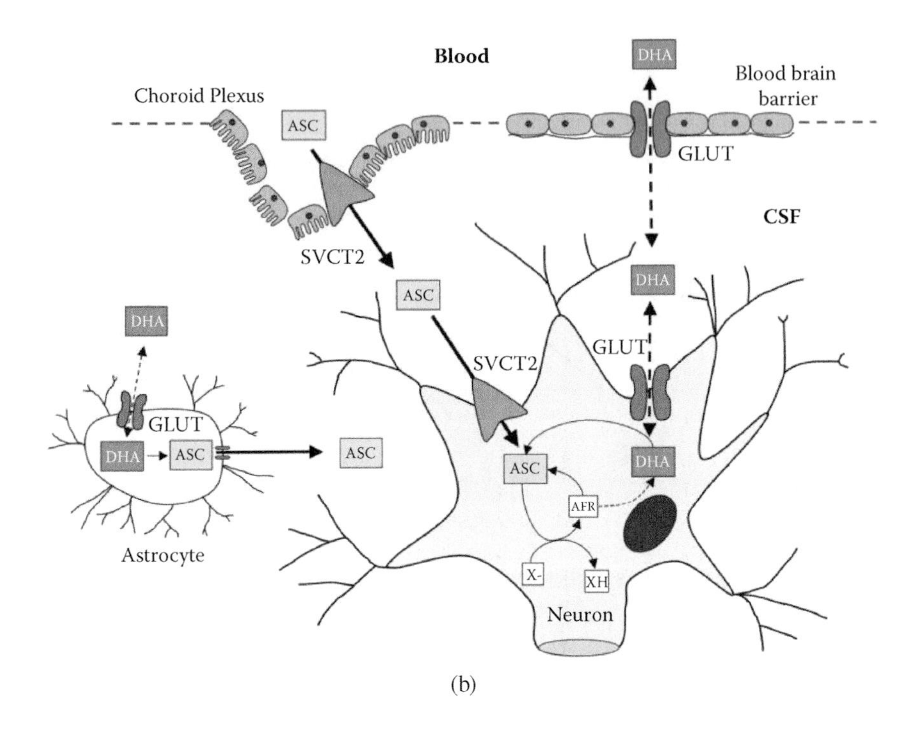

(b)

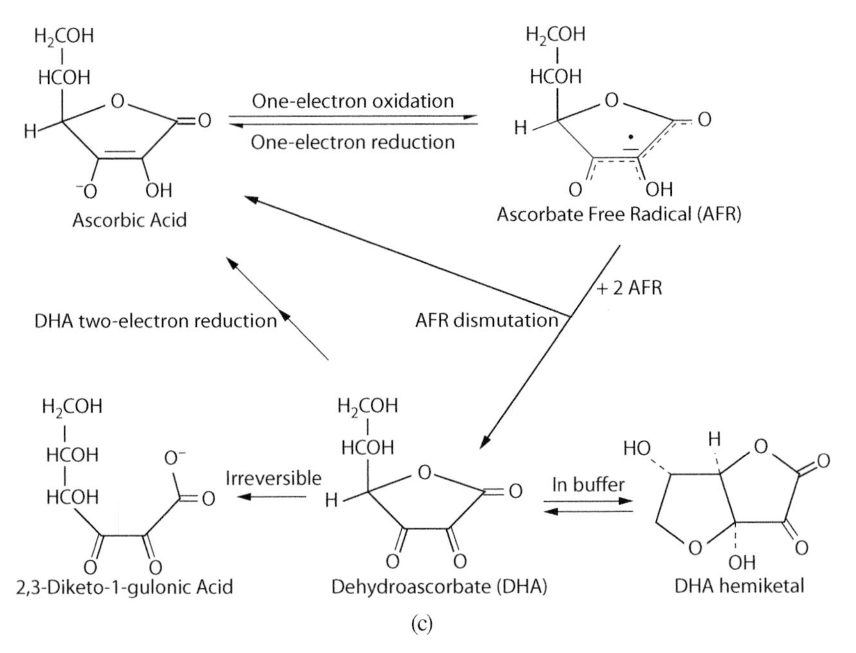

(c)

FIGURE 9.1 *(Continued)*

As shown in Figure 9.1c, ascorbate readily loses a single electron generating the ascorbate free radical (AFR). Loss of a second electron generates dehydroascorbate, also called DHA but not to be confused with the fatty acid of the same name. Dehydroascorbic acid is unstable, being degraded irreversibly to 2,3-diketo-L-gulonic acid and reversibly to its hemiketal derivative. Physiologically important two-electron donors such as NADH and NADPH along with their specific dehydrogenases donate electrons converting dehydroascorbate back to ascorbate. Another important antioxidant, glutathione, chemically donates electrons in this reaction. Note that one AFR readily reacts with another AFR yielding non-free radical products (i.e., ascorbate and dehydroascorbate). This chemistry in which AFR is formed and reacts with itself can be generalized to explain how a powerful chain reaction occurs in highly unsaturated neuronal membranes. Electron delocalization around carbon atoms in fatty acid chains such as DHA (22:6) can essentially starve the carbon atom of a full complement of its electrons. This is how free radicals form along the 22:6 chain, exposing a bonding opportunity with a free radical in the vicinity (see Chapter 14). For example, an oxidative free radical generated from molecular oxygen is highly reactive with the fatty acid radical. In essence, the oxidation of the fatty acid DHA is a chain process occurring through and breeding more free radicals. Lipid peroxidation results in formation of DHA hydroperoxides in neurons, and this chemical species acts as a catalyst to create more free radicals. This chain reaction can be prevented or stopped by antioxidants such as vitamins C and E, which act by preferentially destroying toxic free radicals. Fatty acid peroxides and various classes of reactive oxygen species (e.g., H_2O_2) are not themselves free radicals but instead are dangerous because these molecules are a chemical step away from becoming free radicals. As discussed in Chapter 10, iron and copper must be tightly sequestered in brain tissue because these metals catalyze conversion of otherwise harmless reactive oxygen species to their toxic free radical forms. One quick way to kill a neuron in the human brain is to ingest a free radical-generating neurotoxin capable of passing across the blood–brain barrier. This horrible result occurred accidentally several times among drug abusers in the late 1970s and 1980s, as discussed in Chapter 14.

9.4 VITAMIN E, A MEMBRANE-BASED ANTIOXIDANT, WORKS SYNERGISTICALLY WITH VITAMIN C TO PROTECT NEURAL MEMBRANES

Lipophilic or membrane-loving antioxidants such as vitamin E or molecules with similar antioxidant activity are routinely used in the food industry to extend the shelf life of cooking oils and by Mother Nature to protect the DHA-enriched membranes of neurons. The extremely high concentrations of vitamin C in neurons intensify its effect and its key synergistic role with vitamin E. DHA-enriched membranes of axons and synaptosomes stand out among membranes of most other human cells as being especially sensitive to oxidative damage. We propose that evolution of the sophisticated system for maintaining extraordinarily high levels of vitamin C in neurons is closely linked to the lifelong necessity to protect these specialized membranes against oxidation. As discussed in Chapter 11, neuron mitochondrial membranes in humans likely lack DHA, but their polyunsaturated membranes, especially polyunsaturated molecular species of mitochondrial cardiolipin (see Chapter 11), still require powerful antioxidant protection. In all of its many functions in cells, ascorbate functions as one electron acceptor. As an antioxidant, ascorbate acts directly to scavenge oxygen-based radical species generated during normal mitochondrial metabolism or other sources including membrane peroxidation. Superoxide is a major diffusible by-product of neuronal mitochondrial metabolism, and high concentrations of ascorbate in neurons are expected to scavenge most of this radical. Ascorbate concentrated in the cytoplasm of neurons can also recycle vitamin E, acting as an antioxidant for protecting not only DHA chains in membranes but also the numerous membrane proteins embedded in these membranes, which are also targets of oxidative damage (see Chapter 10). Vitamin E (α-tocopherol) intercepts reactive oxygen species entering the lipid portion of the membrane, a reaction generating the α-tocopherol radical as a by-product. Ascorbate has been shown to recycle α-tocopherol in lipid bilayers. This reaction involves the one-electron reduction of the spent vitamin E radical back to its active form α-tocopherol. Thus, even though the lipophilic chemistry of α-tocopherol confines it to the lipid region of membranes, ascorbate present in the aqueous phase has the ability to somehow reach oxidized vitamin E as a substrate. There are several possible mechanisms to explain this important step. Molecules of vitamin E are expected to be highly mobile in neuronal membranes, perhaps often reaching the surface of the bilayer where they might become substrates for reduction by ascorbate. Alternatively, ascorbate at such high levels might penetrate into the lipid portion of the bilayer more often than currently appreciated. Ascorbate in high concentrations is also expected to bind any free iron and copper ions near the membrane surface, sequestering these metals and thus preventing lipid peroxidation. In various model studies using cultured cells and brain slices, ascorbate has been shown to directly prevent lipid peroxidation generated by various oxidizing agents. Ascorbate as an antioxidant against lipid peroxidation is most effective in combination with α-tocopherol. Because of numerous artifacts associated with antioxidant studies *in vitro,* more emphasis is now being given to studies using whole animals. The synergy between vitamins E and C is seen from nutritional data from guinea pigs (Hill et al., 2003). In this study guinea pigs were first subjected to a state of moderate deficiency of vitamin E. After a 2-week period analysis showed a decrease of 65 percent of vitamin E in plasma versus 32 percent in brain tissue. These animals appeared normal and actually gained some weight. Superimposing a deficiency of vitamin C in their diet had a dramatic effect. Most of the animals died within 24 hours. Upon autopsy many organs appeared with little damage. However, accelerated apoptosis and degeneration occurs in the pons and long motor tracts of the spinal cord. In this animal model significant death of neurons occurs when the synergy between vitamins E and C is disturbed.

Thus, the synergy between vitamins E and C is especially important in the brain and nervous system. As discussed above, a knockout mutation in mice of the major ascorbate transporter for concentrating vitamin C in neurons is lethal. Pups die shortly after birth and have diffuse cerebral hemorrhage (Sotiriou et al., 2002). Recently, hemorrhage was found in the cortex and in the brain stem (Harrison et al., 2010b). Generalized apoptosis is observed, and the basement membrane in the fetal brain is disrupted. Whereas the brain and nervous system are especially vulnerable to low ascorbate levels, other tissues including the placenta show elevated levels of various markers of oxidative stress. The formation of F2-isoprostane or F4-neuroprostane acts as a specific marker for lipid peroxidation. Their presence shows that arachidonic residues (F2-isoprostane) and docosahexaenoic chains (F4-neuroprostane) are subject to excessive oxidation in animals deprived of ascorbic acid. Turning this point around, the presence of sufficient ascorbate protects highly unsaturated chains such as 22:6 (DHA) against lipid peroxidation. Note that in mice, where these data were collected, the presence of F4-neuroprostane serves as a generalized marker of oxidative stress because DHA is enriched in membranes of virtually all cells of this animal. However, in humans, where most of the DHA in the body is selectively targeted to the brain and nervous system, the presence of F4-neuroprostane is primarily a marker for lipid peroxidation of specialized neuronal membranes and is consistent with the concept of lipid whiskers of DHA phospholipids largely occurring in neurons (Chapter 8).

In summary, the essential roles of vitamins C and E established in nutritional studies have been confirmed using knockout mutants in mice. Blocking the major transporter of vitamin C (ascorbic acid) entering the brain and thus preventing this antioxidant from being concentrated in neurons is a lethal mutation and can be explained in part by a model involving DHA peroxidation.

Biochemical cross talk among at least three distinct classes of antioxidants operating in neurons—vitamins C and E along with glutathione—is likely essential for long-term brain function. We have said relatively little about the important roles played by glutathione as an antioxidant. Harrison et al. (2010a) have found that total glutathione levels in the cortex, cerebellum, and liver are higher in mutants of mice unable to concentrate ascorbate in neurons. These researchers point out that high glutathione levels might compensate for a lack of ascorbate as an antioxidant.

Finally, we propose that specialized DHA-enriched membranes of neurons are especially vulnerable to oxidative damage with aging and thus require specialized antioxidant-based defenses that must function efficiently for a lifetime. The point is underscored that a major role of antioxidant protection in the brain is to save energy. There is increasing data that suggest that the robust antioxidant systems of the human brain described above can weaken during aging, perhaps being upregulated by the FOXO system discussed in Chapter 17.

REFERENCES

Harrison, F. E., and J. M. May. 2009. Vitamin C function in the brain: Vital role of the ascorbate transporter SVCT2. *Free Radic. Biol. Med.* 46:719–730.

Harrison, F. E., S. M. Dawes, M. E. Meredith et al. 2010a. Low vitamin C and increased oxidative stress and cell death in mice that lack the sodium-dependent vitamin C transporter SVCT2. *Free Radic. Biol. Med.* 49:821–829.

Harrison, F. E., M. E. Meredith, S. M. Dawes et al. 2010b. Low ascorbic acid and increased oxidative stress in gulo(-/-) mice during development. *Brain Res.* 1349:143–152.

Hill, K. E., T. J. Montine, A. K. Motley et al. 2003. Combined deficiency of vitamins E and C causes paralysis and death in guinea pigs. *Am. J. Clin. Nutr.* 77:1484–1488.

Murphy, M. P., A. Holmgren, N. G. Larsson et al. 2011. Unraveling the biological roles of reactive oxygen species. *Cell Metab.* 13:361–366.

Sotiriou, S., S. Gispert, J. Cheng et al. 2002. Ascorbic-acid transporter Slc23a1 is essential for vitamin C transport into the brain and for perinatal survival. *Nat. Med.* 8:514–517.

10 Oxygen-Dependent Damage (Lipid Peroxidation) of DHA Membranes of Neurons Is Inevitable and Requires Novel Mechanisms for Long-Term Protection

In a strange quirk of nature, specialized membrane disks essential for sensing light in rod cells of the eye age more rapidly than any other class of membranes in the human body. The light-sensing protein rhodopsin is the predominant protein embedded in these membranes, and this gave rise to their name, *rhodopsin membrane disks*. Historically, rhodopsin membrane disks have been workhorses in the field of DHA membranes and have provided much of the background chemical information needed to understand mechanisms of lipid peroxidation of DHA membranes during aging and neurodegeneration. Some selected highlights from research on lipid peroxidation of rhodopsin disks are as follows:

- High DHA content fosters rapid turnover. The highest level of DHA reported in rhodopsin membrane disks (found in certain rodents) is 53 percent of total fatty acids.
- Photo-oxidation accelerates membrane damage.
- Rhodopsin membrane disks contain high levels of di-DHA molecular species of phospholipids. With 12 double bonds per molecule, these membrane building blocks are among the most sensitive to oxidation in nature.
- Truncated DHA phospholipids as by-products of lipid peroxidation have been identified and synthesized (see Chapter 8).
- Rhodopsin and other membrane proteins are readily oxidized and inactivated in "aging" disks.

- Too much DHA in rhodopsin disks of transgenic mice is harmful to vision (see Chapter 2).
- Oxidation products of DHA such as found in rhodopsin disks score positive in the Ames mutagen test (see Chapter 13).
- Research on rhodopsin disks played a seminal role in the development of the lipid whisker model of aging (oxidatively damaged) membrane surfaces (see Chapter 8).

Rhodopsin disks were also used as research tools for calibrating and defining the importance of extreme motion in membrane biochemistry (Chapter 5) including vision and sensory perception. These data show that DHA chains are responsible for rapid lateral movement of membrane proteins. How extreme membrane motion facilitated by DHA is harnessed for extreme flight is discussed in Chapters 11 and 12. We have selected several topics from the above list for further discussion because of their important implications for neuron health and decline.

10.1 RAPID TURNOVER OF RHODOPSIN DISKS

Rhodopsin membrane disks in rod cells of the eye are targets of photo-oxidation during waking periods (i.e., when the eye is open) and are subject to dark or auto-oxidation in the absence of light (reviewed by Giusto et al., 2000). Due to their high DHA content, it is predictable that these specialized bilayers are unstable and require continuous renewal. The life span of a newly minted membrane entering the bottom of the stack (composed of about 1000 disks) and exiting at the apex is about 10 days (Figure 10.1). Roughly 100 disks, or 10 percent of total disks, are removed as spent or senescing disks from the top of the stack daily, being replaced with an equal number of newly minted disks on the bottom. The cellular physiology of renewal of rhodopsin disks involves an elaborate process with spent disks being shed and taken up and processed by specialized scavenging cells called *pigmented epithelial cells*. See Chapter 8 for roles of lipid whiskers in signaling phagocytic cells. Within pigmented epithelial cells, fragments of senescing disks become surrounded by cell membranes to form inclusions called *phagosomes*. The phagosomes are subsequently degraded within the pigmented epithelial cells with undamaged DHA being salvaged and used for new disk synthesis. Phagocytosis of spent disks is accelerated as expected when the disks in the intact eye are exposed to excessive light, with prolonged intense light causing permanent damage and in extreme cases, blindness. Interestingly, the life span of disks in rod cells of frogs' eyes is significantly increased, which is consistent with cool temperatures in their environment acting to lower rates of damage to DHA membranes. As described in Chapter 2, overexpression of DHA synthesis in transgenic mice has the same effect as too much light, accelerating the aging of disks with collateral damage spreading and resulting in the death of rod cells.

The experiment summarized in Figure 10.1 shows how scientists were able to "tag" newly minted membrane disks entering the bottom of the stack with a radioactive amino acid, which is incorporated primarily into rhodopsin protein, shown as dots. In this experiment an intense band of radioactivity is first seen in newly minted disks entering the bottom of the stack of membrane disks. In this case it is

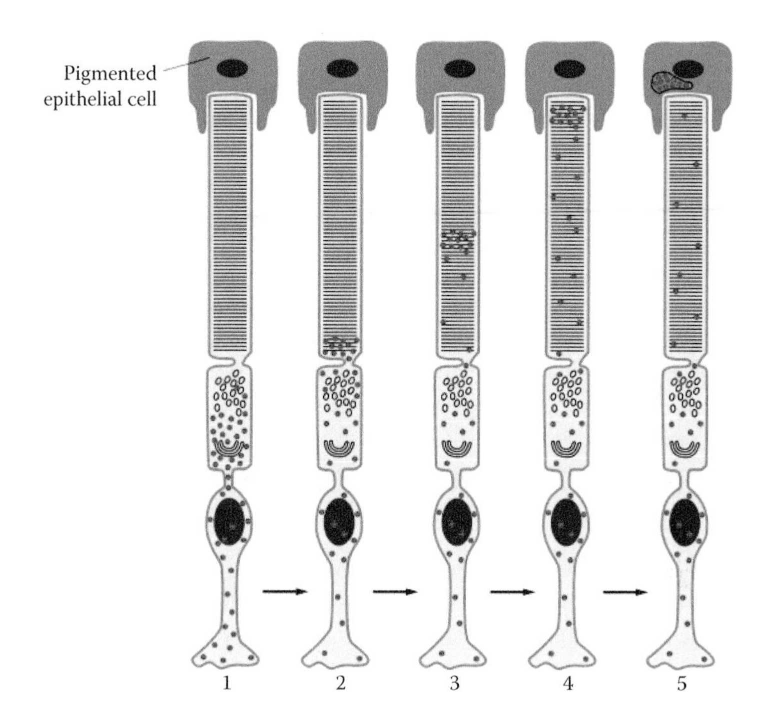

FIGURE 10.1 Rhodopsin disk membranes enriched with DHA are oxidatively unstable. Radioactive proteins (dots) are used as reporters to show that rhodopsin disk membranes in rod cells of the eye "age" rapidly. Note that newly minted membrane disks harboring newly synthesized rhodopsin proteins enter the bottom of the stack and emerge at the top in a time period of about 10 days. Light damage caused by photo-oxidation is thought to rapidly destroy high levels of DHA along with rhodopsin in rhodopsin disks requiring constant recycling of these specialized membranes. (From Alberts, B. et al., *Molecular Biology of the Cell*, Copyright 1994. Reproduced by permission of Taylor & Francis Group, LLC, a division of Informa plc.)

not DHA that is labeled but newly synthesized rhodopsin protein inserted into newly synthesized disks and acting as a marker to follow the progression of disks over time. During a period lasting about 10 days, the band of radioactivity moves from the bottom to the top of the stack where spent or "old" disks are removed. Loss of radioactive rhodopsin shows that these membranes are being continually recycled at a dramatically fast pace. Note that these data simultaneously time the life span of both rhodopsin and the DHA-enriched membranes that embed this protein.

Thus, conditions of warm body temperatures, light and high fluxes of O_2, combined with the high DHA composition of rod cell disks make these membranes ideal "substrates" for chemical attack by oxygen. The retina has evolved several defense mechanisms that protect against oxidative damage. Conventional defenses include high levels of protective enzymes (found in photoreceptor outer segments) that intercept and destroy reactive oxygen species before they attack rhodopsin membranes. Antioxidants (see Chapter 9) are a main line of defense against auto-oxidation in the retina (Robison, Kuwabara, and Bieri, 1982; Valk and Hornstra, 2000). Vitamin E appears to be located strategically in rhodopsin disks and protects

DHA from oxidation. Researchers have shown that incorporation of vitamin E into DHA-enriched membrane vesicles significantly reduces damage to DHA. A variety of other antioxidants also help protect our vision. Even though the retina has evolved a battery of defenses usually efficient enough in most cells to prevent oxidative damage, the presence of high levels of DHA, especially in the presence of visible light, appears capable of overwhelming these defensive systems in rod cells within a span of about 10 days. This necessitates additional protection in the form of active remodeling and highly sophisticated membrane renewal mechanisms. With aging or through environmental insults, defenses in the eye might begin to deteriorate, resulting in the death of rod cells and serious damage to vision. For example, at high light intensities or elevated enrichment with DHA, the risks associated with DHA begin to outweigh the benefits, helping to trigger a pathological state.

10.2 RHODOPSIN AND OTHER PROTEINS ARE TARGETS OF OXIDATIVE DAMAGE

Stadtman (2006) developed a unified concept of protein oxidation that he applied to explain why rhodopsin disks age so rapidly. According to this unified mechanism, reactive oxygen species (ROS) generated from any source can damage and destroy the functions of critical proteins such as rhodopsin. Stadtman recognized that DHA in rhodopsin disks has the potential of a dual effect on protein integrity—generating both oxidative radicals (ROS) that can directly attack critical amino acid residues such as methionine in proteins, and a source of alkylating molecules whose potent chemistry also directly targets proteins. Emphasis here is on the interplay between DHA oxidation and protein damage.

As discussed above, rhodopsin disks are widely used substrates for studies of oxidative damage to membranes because of high rates of damage and the presence of high levels of rhodopsin protein, representing about 90 percent of total protein content of these membranes. Modern analytical and chemical synthesis techniques have allowed biochemists to use rhodopsin membranes as tools for developing a picture of interplay between damage to proteins and DHA (Hollyfield, Perez, and Salomon, 2010; Lu et al., 2009; Warburton et al., 2005). Some of the highlights are that photo-oxidation is a powerful force in destroying DHA chains present in almost half of the phospholipids forming the surfaces of rhodopsin disks. Detailed chemical analysis coupled with synthetic chemistry has led to the identification of essentially all of the degradation products predicted by the chemical theory of lipid peroxidation, including potent alkylating fragments that can readily react with and damage proteins such as rhodopsin. In addition, damaged DHA is also a source of ROS, in this case being generated within the membrane. Normally that portion of a membrane protein embedded in the membrane is believed to be partially protected against insults such as ROS being generated in the aqueous cytoplasm. However, powerful Fenton chemistry triggered by DHA oxidation and occurring deep in the membrane can target membrane proteins by oxidizing methionine residues and causing other damage. In addition, toxic by-products of DHA oxidation momentarily trapped within the membrane where they are generated are believed to be in a position to react directly with rhodopsin or other proteins. Thus, the current picture is that oxidative damage to DHA and proteins populating rhodopsin disks are interlinked chemical processes, together driving the remarkably fast aging of these specialized membranes essential for vision.

10.3 SPERM TAIL MEMBRANES AS SURROGATES FOR AXONS

Axons and tails of sperm cells share a number of properties including similar morphology, excitatory electrophysiology involving Na^+/K^+ circulation, mechanical stretching, and high enrichment with DHA and similar molecular species of DHA phospholipids. The general importance of DHA as a building block for healthy sperm is well established, but with a surprise: the apparent strategic localization or targeting of DHA (Connor et al., 1997). One of the most unusual chemical features of sperm membranes is the differential enrichment of the tail membrane versus head membrane—with 19.6 percent DHA in the tail compared to 1.1 percent in the head based on analysis of monkey sperm. These values are expected to apply to human sperm. Arachidonic acid (20:4) is also preferentially targeted to the tail of monkey sperm—6.4 versus 1.6 percent. The overall differences in total unsaturated fatty acids are 34.1 percent in the tail versus 12.1 percent in the head membrane.

Studies on spermatogenesis show that sperm production is an imperfect process under the best of conditions, subject to numerous forms of environmental stress. In normal semen, about 20 million sperm are present per cubic centimeter, of which about 5 million are defective. The sensitivity of sperm biogenesis to environmental stresses helps explain why sperm counts may plummet following sessions in saunas and hot tubs or after wearing tight-fitting cycling shorts, each of which causes heat buildup around the scrotum. The external location of the scrotum is thought to be a cooling mechanism against heat stress leading to higher levels of viable sperm. The fact that sperm tail membranes are highly enriched with DHA plays into this scenario, raising the likelihood that protection of DHA might be important in male fertility.

Chemical studies on animal and human sperm established that DHA-enriched phospholipids are highly susceptible to oxidation, and this process is associated with progressive and irreversible loss of structural integrity, motility, viability, and metabolic activity of the sperm cells (Aitken and Clarkson, 1987; Alvarez and Storey, 1995). Exogenously applied lipid peroxides, the first product of lipid peroxidation, are powerfully spermicidal against washed human sperm treated with as little as 30 nM of lipid peroxide per milliliter. Treated sperm become irreversibly immobile within a few minutes. Lipid peroxides are major products of chemical oxidation of DHA and polyunsaturated oils in membranes. Fatty acid peroxides are not free radicals but are readily converted to lipid peroxyl radical in the presence of trace amounts of ferric iron (Fe^{3+}). Lipid peroxyl radical is a free radical and can extract an electron from a double bond of DHA or polyunsaturated fatty acids, yielding lipid peroxide and a lipid free radical. Thus, the chemical steps of oxidation of DHA in cellular membranes can be divided into three stages, as follows:

- Initiation by Fe^{3+} or another free radical generates a fatty acid free radical.
- Propagation of lipid oxidation occurs as a chain reaction.
- Termination occurs when available substrates are depleted or an antioxidant such as vitamin E annihilates lipid free radicals participating in the reaction.

It is important to note that O_2 does not directly interact with double bonds of fatty acids but does so extremely rapidly once initiation occurs. The term *lipid peroxidation* is used to define this series of reactions because of the important role played by the lipid peroxyl radical. Hydrogen peroxide (H_2O_2) is classified as a ROS because in the presence of Fe^{3+} the simplest form of a peroxyl radical ($H_2O_2 \rightarrow HOO\cdot$) is formed. H_2O_2 is universally produced in respiring cells and is dangerous to DHA membranes because of its ease of conversion to the free radical form, which causes propagation of lipid oxidation to occur (see Chapters 14 and 17 for further details). Classic chemical oxidation protocols using iron or ascorbate catalysis of oxidation of sperm phospholipids show that DHA is preferentially targeted and rapidly lost from the membrane in contrast to the 16:0, which remains relatively constant as expected (Alvarez and Storey, 1995).

Also, $18{:}1_{n9}$, $18{:}2_{n6}$, and cardiolipin, the latter diagnostic of sperm mitochondrial phospholipids, are relatively stable toward this lipid oxidation protocol. However, $20{:}4_{n6}$ present at levels of approximately 10 percent of DHA is subject to oxidation, though at somewhat lower rates than DHA. The stability of cardiolipin is consistent with low levels of DHA being incorporated into ram sperm mitochondrial membranes in contrast to tail membranes.

The targeting of DHA to tail membranes keeps this potentially damaging source of ROS away from the stored DNA (see Chapter 13). DHA also seems to be excluded or targeted away from membranes of multiple mitochondria that fill a space lying just underneath the DNA packaged in the headspace. According to this model both space and time work to prevent damage to germ-line DNA caused by the accumulation of toxic ROS radicals (Fraga et al., 1991), which are readily derived from DHA.

Avoidance of O_2 (Gray et al., 2004), which readily drives lipid peroxidation of DHA, has been shown to be a simple but effective mechanism to protect DHA-enriched cells against oxidative stress. The avoidance approach maximizes benefits of DHA and minimizes oxidative risk. However, it comes as a surprise that O_2 avoidance by virtue of low oxygen levels in regions of the female reproductive tract might be a mechanism for protecting sperm and its DNA against DHA-mediated oxidative stress. The apparent lack of most conventional ROS defenses in sperm is also consistent with a diminished need for such antioxidation mechanisms in the low O_2 world of the female reproductive tract. Note that the entire reproductive tract is not devoid of O_2 because the egg in the oviduct requires respiration for survival. A review of the benefits and risks of reactive oxygen species in human sperm cells has recently been published and provides a comprehensive overview of this field (Koppers et al., 2008).

To summarize this section, sperm tails have proven to be important research tools acting as surrogates for axons (see lucid review by Storey, 2008, on the history of sperm metabolism and lipid peroxidation of sperm membranes). The following conclusions from studies of DHA-enriched tail membranes of sperm cells likely apply to axons (also to synapses and synaptic vesicles):

- DHA is the most readily peroxidized fatty acid in sperm cell membranes.
- The rate of lipid peroxidation increases in proportion to the number of double bonds in the fatty acid chain.
- The lipid peroxidation rate is proportional to O_2 levels up to 95 percent O_2.
- Membrane peroxidation proceeds by established chemical rules for lipids.
- Loss of sperm motility occurs in proportion to oxidative membrane damage.
- Oxidative damage causes loss of membrane permeability barriers to ions, sugars, and other metabolites.
- A scrotal temperature in humans of 30°C compared to body temperatures of 37°C is estimated to lower membrane oxidation sixfold.

10.4 SLOW TURNOVER OF DHA IN THE BRAIN WAS NOT EXPECTED

Rhodopsin membrane disks in rod cells of the eye provide the clearest picture yet of events likely occurring in aging, DHA-enriched membranes of neurons. However, there is one major difference between rhodopsin disks and neuron membranes. Brain cells are never exposed to light. Thus, the dark world of neurons works to their great advantage with respect to protecting DHA membranes. Nevertheless, neurons are destabilized because they live in perhaps the most oxygenated region of the human body. This high oxygenation is required to support the frantic respiratory activity characteristic of neurons. Therefore, the DHA membranes of neurons, especially the vast axon network of nonmyelinated connections, including synapses, are constantly exposed to a flux of oxygen being delivered across an expansive surface area; this exposure to oxygen leads to membrane damage. Such a highly oxygenated condition essential to sustain the energy needs of neurons predisposes neuronal membranes to chemical attack by oxygen and its reactive derivatives. The following conditions in the brain favor oxygen-mediated damage to neuronal membranes:

- Continuous presence of oxygen
- Warm temperatures (i.e., body temperature)
- Stray metals such as copper or iron in the free state (This can greatly speed up damage.)
- Six chemical double bonds present in each DHA chain that make this fatty acid the most readily attacked by oxygen in nature

The chemical stability of neuron membranes in an oxygenated environment is dependent to a large degree on the number of double bonds in DHA. The chemical principles governing lipid peroxidation of DHA are now understood in detail and are readily applied to neuronal membranes. For example, omega-3-enriched membranes are subject to chemical oxidation in which molecular oxygen through its ROS byproducts attacks the double bonds of this fatty acid. The oxidation of unsaturated fatty acids is a spontaneous chemical process, and as such will occur wherever the proper conditions exist, from a bottle of cooking oil to an axon membrane. Polyunsaturated fatty acids are far more susceptible to oxidation than monounsaturated fatty acids, with linolenic acid ($18:2_{n6}$) common in vegetable oils such as corn or canola being oxidized at a rate more than 10-fold greater than the monounsaturated oil oleic acid (18:1n9), also prevalent in the same oils. Generally, oleic acid is used as a standard against which rates of oxidation of all members of the hierarchy of fatty acids can be judged. The chemical rule that applies is that for every double bond past two per fatty acid, the rate of oxidation is approximately doubled; thus, the rate of oxidation of DHA is calculated to be ~480 times the rate of oxidation of oleic acid, with a single double bond (see Figure 8.1). The theoretical values quoted are for DHA peroxidation such as occurs in neurons.

An adult brain weighs about 3 pounds (cauliflower size), and its dry weight is composed of about 50 percent fat. Of this an estimated 5 to 10 grams of DHA is present. However, in order to estimate the amount of DHA needed by the brain over an average lifetime (i.e., 80 years), it is necessary to know the rate at which DHA turns over in the brain. Recent studies (Umhau et al., 2009) show that DHA in the human brain is stable, taking greater than five years for complete replacement of all DHA in neurons. However, after only 49 days, 5 percent of total DHA is turned over, consistent with the idea that at least a certain fraction of DHA is somewhat less stable in neurons. Thus, over a lifetime complete turnover would take place at least 16 times for a total of about 80 to 160 grams of DHA. For comparison, DHA turnover in a rat brain is roughly three times faster with a 30 percent reduction in 105 days. Just one generation of omega-3 deprivation in rats led to behavioral changes (DeMar et al., 2006). Also note that the values for DHA turnover discussed above are averaged across all DHA-enriched membranes and do not distinguish among classes. This leaves room for the possibility that one class of DHA-enriched membranes might turn over faster than another. For example, there are data discussed later that neurons composing "gray matter" may be more sensitive to oxidation compared to "white matter."

From a chemical perspective the reported half-life for DHA measured in years is an amazing result given the oxygenated nature of the brain environment. Is this a clue that novel, yet to be discovered, mechanisms for protecting axons/synapses from oxidative damage have evolved in the human brain? We suspect that the answer to this question is yes and set about to conceptualize possible new protective mechanisms based on a comparative biological approach.

10.5 SIFTING THROUGH THE LIPOFUSCIN "GARBAGE PILE" FOR CLUES TO NEURON AGING

The "garbage pile" for neurons is called *neural lipofuscin*. As the name implies, lipofuscin consists of a high amount of indigestible lipid and protein products apparently derived from oxidatively damaged unsaturated fatty acids, especially mitochondrial membranes. Neural lipofuscin is recognized as a marker for aging neurons, and its proposed bioactivity ranges from being beneficial to harmful (see Gray and Woulfe, 2005, for a review).

The chemical composition of neural lipofuscin is controversial because of a great variability from granule to granule, cell to cell, and region to region. Recently, Ng and colleagues (2008) carried out an extensive analysis of retinal lipofuscin. The presence of oxidative protein modifications generated from reactive lipid fragments such as derived from DHA is consistent with cross talk between unsaturation and membrane proteins that together help drive membrane aging. In turn, bioactive components of lipofuscin seem to contribute to pathogenesis (e.g., retinal degenerative disease such as macular degeneration).

It is important to recognize that the composition and rates of accumulation of lipofuscin in different cells such as between retinal epithelial cells and neurons, likely depend to a large extent on the classes of membranes being degraded. For example, rhodopsin disks, which are highly enriched with DHA, seem to be a predominant source of lipofuscin in the retina. In contrast, turnover of neural mitochondrial membranes might be a major driver of accumulation of lipofuscin in neurons. It would be of great interest to know if DHA content is a major determinate of rates of lipofuscin production. So far it seems reasonable to assume that lipofuscin is a sign of aging of both membranes and cells. But depending on specific biochemical needs, different cell types or even specialized neurons located in different regions of the brain might each contribute to different patterns of lipofuscin composition.

10.6 SEQUESTRATION OF METALS KNOWN TO PROMOTE DHA OXIDATION IS VITAL FOR NEURON HEALTH

Chemists use traces of copper or iron as catalysts to dramatically increase oxidation rates of DHA and other unsaturated fatty acids. We next describe how increasing membrane unsaturation in a living yeast cell used as a surrogate for neurons simultaneously increases sensitivity to copper. Yeast serves as a viable model system for demonstrating the linkage among membrane unsaturation, metal homeostasis, and lipid peroxidation in a living cell.

For an historical background, orchardists routinely use concentrated copper spray to kill certain pathogenic fungi threatening their crops. For example, peach and nectarine crops in California are often devastated by a fungus that causes leaf-curl, a pathogen that eventually destroys much of the new leaf structure, essentially robbing the plant of energy. A widely used treatment against this fungus involves spraying the dormant trees with soaking doses of a strong solution of free copper, usually as copper sulfate. Trees and the ground underneath turn blue from the amount of copper in the spray. Horticulturists believe that copper kills the leaf-curl fungus by attacking its polyunsaturated membranes. In essence, copper is thought to accelerate oxidative membrane damage eventually killing the fungus. To test this idea we have studied the effects of copper toxicity using transgenic yeast cells, expressing a fungal desaturase. Note that the membranes of yeast cells (*Saccharomyces cerevisiae*) normally contain only monounsaturated fatty acids. The result is that transgenic cells with polyunsaturated membranes are far more sensitive to the lethal effects of copper. This is not an original finding and confirms work by earlier researchers where hypersensitivity to copper was achieved by feeding polyunsaturated fatty acids to a mutant dependent on unsaturated fatty acids for growth (Avery, Howlett, and Radice, 1996). Because yeast can be grown anaerobically, we also experimented with the close linkage between oxygen and copper toxicity with the clear result that oxygen is required for toxicity. Years of spraying copper on tree crops has led to the selection of a mutant strain of a common bacterial plant pathogen of these crops that has become tolerant or resistant to copper spray. Resistant bacteria have been found to protect themselves by effluxing toxic levels of copper from the cytoplasm and depositing it in a chemically inert form on the outside of the cell.

Thus, yeast has proven to be an important research tool for detailed studies of copper toxicity. It should also prove to be a versatile model for researching the linkage between DHA membranes and dementia. Recall that mutants of common brewer's yeast dependent on unsaturated fatty acids can be forced to build their membranes using fatty acids representing each step of the unsaturation ladder from one to six double bonds. Whereas initial experiments have already been reported showing DHA supports growth of yeast, we are not aware that copper toxicity has been tested in these strains whose membranes would be expected to be hypersensitive to lipid peroxidation.

We next discuss the nature of copper homeostasis and toxicity in humans with implications for neurodegeneration. In humans copper is an essential micronutrient for a healthy brain. Copper works in partnership with at least 13 enzymes that drive a crucial array of biochemical reactions that underpin brain health and development. For example, this battery of enzymes includes conventional detoxifying enzymes that destroy toxic oxygen radicals before they attack essential cellular molecules such as DHA in the membrane or even our chromosomes. These beneficial biochemical roles define copper as an essential micronutrient in our diet. Copper is available from a wide variety of foods, and copper deficiency, though rare, has major effects on health. Copper toxicity is more widely reported than copper deficiency and includes a rare familial form called *Wilson's disease* (Ala et al., 2007). The defective gene has been identified and is involved in copper transport and detoxification of excess copper from the body (Bull et al., 1993). Copper toxicity associated with Wilson's disease is readily treated when detected early. Therapy is lifelong. The main point here is that copper toxicity is fatal if left untreated, and the brain is one of the major targets. About 60 percent of Wilson's disease patients display neurological disorders including tremors, muscle spasms, speech problems, and an unsteady walk. In some cases symptoms of brain damage are diagnosed before damage to the liver, the second major target organ.

Because copper is both essential and highly toxic (much like our model for DHA), its circulation in the body and to the brain is highly regulated and involves many checks and balances. For example, the brain quickly converts free copper, which is believed to be the toxic form, to a sequestered storage or bound state. When the liver as the major storage organ for essential minerals such as copper, iron, and manganese becomes diseased or overburdened, this family of potentially toxic minerals ends up in the brain, which serves as a secondary, but dangerous, storage depot. With so many sophisticated steps needed for protecting neurons against the toxicity of free copper, it is easy to see how imbalances in one of these stages might occur, especially with aging. The linkage between copper, iron, manganese, and other metals and degeneration is the subject of ongoing research (Strausak et al., 2001).

10.7 HYPOTHESIS NUMBER 1: MYELIN FORMS A PROTECTIVE SHIELD AGAINST OXIDATIVE DAMAGE TO AXONS

The retina of the eye contains some 125 million rod cells and about 6 million cones whose DHA-enriched membrane disks function as sensors of light. In this highly oxygenated and often illuminated environment, DHA oxidation is rapid, with complete turnover of rhodopsin disks in rod cells occurring in about ten days. As described above these membranes are recognized as being among the most oxidatively unstable in nature. On the other hand, convincing data also described above show that DHA-enriched membranes of neurons turn over much more slowly. At first glance, the apparent oxidative stability of DHA-enriched membranes of neurons might be explained simply because of the absence of light. However, we suggest this is only part of the story and that other yet undiscovered protective mechanisms have evolved to protect neuron membranes.

The first novel mechanism proposed here involves myelin sheaths (see Figure 1.1b) acting as protective barriers against oxidation of axon membranes. Although there is little direct evidence for this specific protective effect of myelin, we suggest that it deserves consideration. It is well known that many axons in the brain and nervous system are wrapped layer after layer with a fatty substance called *myelin*. Myelin is composed of about 80 percent lipid, including 15 percent cholesterol, along with about 20 percent protein. Roughly 100,000 miles of axonal membranes in the adult brain are coated with myelin. Myelin plays several essential roles as insulation and is important for maximizing speed and energy efficiency in the brain and nervous system. The importance of myelin is shown in patients with demyelinating diseases. The most common is multiple sclerosis (MS) caused when a person's immune system becomes overactive and attacks his own myelin. MS sufferers eventually develop severe sensory and motor defects including paralysis and blindness. Nerves in MS patients cannot efficiently conduct neural impulses.

The importance of myelin in brain function can also be seen in the unique pattern of myelination during development of the brain (Bock et al., 2009; Diau et al., 2005). Myelination begins in the nerve fibers of the spinal cord after about five months of gestation. Myelination of brain axons and dendrites starts in the ninth prenatal month. Myelination is a slow process advancing through several stages in which the myelin wrapping thickens progressively and is biochemically transformed into a mature composition. The paces of myelination of different parts of the brain are markedly different. These genetically programmed patterns of myelination reflect the development of critically important functions in different regions of the brain. In essence, myelination modulates the speed at which many essential brain functions progress. Axonal fibers in older brain regions controlling basic functions tend to get myelinated first, well before fibers that control more sophisticated mental abilities.

The conventional view of adult myelin concerns its role as insulation that speeds up electrical impulses resulting in a leap in neuron efficiency. This is undoubtedly a cardinal function, but myelination might also serve a secondary role as insulation against membrane damage caused by oxygen. According to this model a great deal of total axon circuitry lies beneath myelin sheaths and is at least partially protected against oxidative membrane damage, compared to nonmyelinated neuron membranes. Membranes at synapses and synaptic vesicles are not myelinated and, thus, are perhaps exposed to higher levels of oxidative damage. Recall that the initial cost of building a vast myelin network composed of cholesterol-rich membranes is extremely high, and we hypothesize that one way to recoup this expenditure is through multiple uses—electrical insulation and protective insulation against oxygen. Myelinated axons still have sufficient surface area for oxygen diffusion to occur. This insures that myelinated axons are able to obtain sufficient O_2 from their surroundings to carry out critical internal cellular functions such as mitochondrial respiration. Thus, we are asserting that myelination might help create a localized low-oxygen world surrounding the surface area of axons. We envision this mechanism of defense as a barrier against saturating levels of oxygen coming in direct contact with DHA-enriched membranes of axons. One purpose of this kind of O_2 avoidance might be to increase the effective "life span" of axon membranes before resurfacing is necessary. While seemingly a patchwork way of protecting its DHA membranes, given the tight energy budget of neurons, every little bit of energy efficiency is important for maintaining a healthy brain, especially during aging.

Recent studies on DHA circulation and turnover in humans provide some clues about myelination as a potential protective layer against oxidation (Umhau et al., 2009). When radioactive DHA was injected into human subjects, more DHA was taken up by gray matter than by white matter. Recall that white matter is more heavily myelinated. One explanation for this result is that DHA is less protected and thus turns over more rapidly in gray matter compared to white matter (see Umhau et al., 2009, for additional explanations).

10.8 HYPOTHESIS NUMBER 2: TAURINE AS AN ANTIOXIDANT

The role of taurine as an osmoregulant important in stretch growth of neurons is introduced in Chapter 7 with emphasis here on a second possible important role in protection against oxidative damage. During the past decade there have been major advances in the cloning and molecular biology of taurine transporters that provide clues about bioactivity of taurine (El-Sherbeny et al., 2004; Heller-Stilb et al., 2002; Ni et al., 2008; Paskowitz, LaVail, and Duncan, 2006; Rockel et al., 2007; Tabassum et al., 2007). Briefly, taurine pumps such as found in neurons are integral membrane proteins fueled by Na^+ gradients. Using knockout mutants of mice, it has been shown that mice missing taurine transporters display retinal pathologies. Photoreceptor degeneration of the rod-cone type along with apoptosis of these critical cells for vision is characteristic of taurine transporter-minus mutants. Molecular studies with mice are consistent with research on cats, which show that dietary deficiency of taurine leads to progressive retinal degeneration. The pattern of retinal degeneration in taurine-deficient mice resembles that seen in some forms of human retinitis pigmentosa (Rockel et al., 2007). According to these authors it remains to be determined whether retinal dystrophy in mice is directly caused by a lack of taurine or whether taurine deficiency leads to a disturbance in the osmotic balance specific to

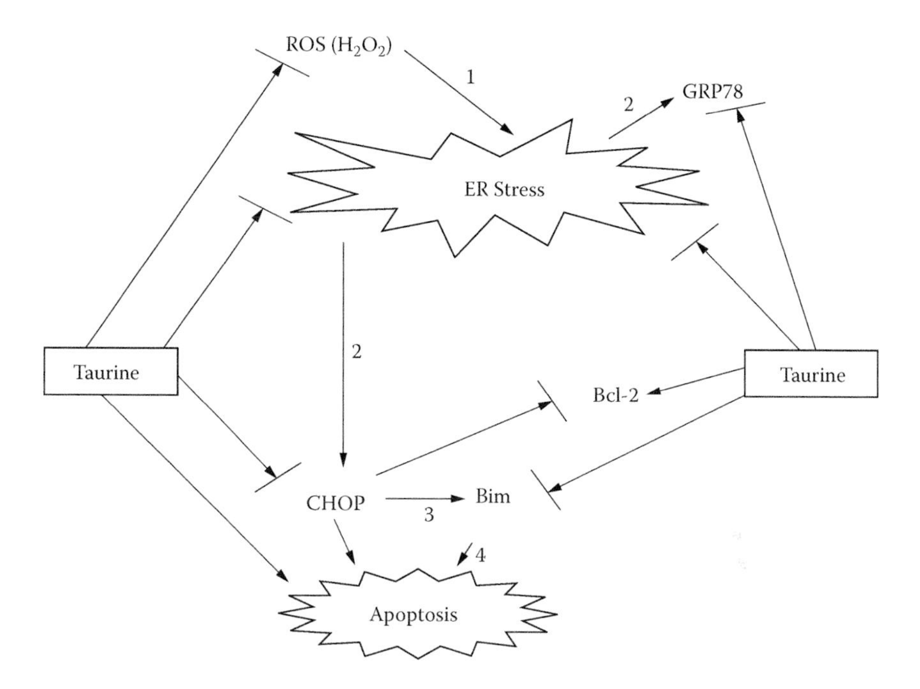

FIGURE 10.2 Model for taurine as an antioxidant. Taurine, an osmoregulatory molecule, might also protect cells against oxidative stress. Taurine is proposed to protect the endoplasmic reticulum against oxidative damage. CHOP and GR878 shown being upregulated are markers of increased oxidative stress. Taurine is also proposed to protect cells by downregulation of Bim and upregulation of Bcl-2. (Copyright 2010 Prentice and Wu: Licensee BioMed Central Ltd.)

photoreceptor cells, or both. This same question also applies to the biochemical roles of taurine in brain function (see Chapter 7 for a discussion of the potential role of taurine in neuron growth and osmoregulation).

Defining a specific biochemical role of taurine as an oxidative protector (Green et al., 1991) is complicated by its likely contribution of multiple critical functions that are difficult to separate. Dual roles that come to mind are as conventional osmotic balancing molecules versus oxidative protectors. We have already mentioned how taurine might directly lodge itself in neuron membranes and plug energy leaks while at the same time act as an important cytoplasmic osmoregulant in neurons (see Chapter 7). Because taurine stands out as being especially important in the brain and retina, this brings DHA into the picture. The point is that we believe that DHA predisposes neurons to excessive oxidative stress. Indirect evidence links taurine with protection against oxidative stress (Heller-Stilb et al., 2002; Ni et al., 2008). Our current working model is that taurine and perhaps other organic osmoregulatory compounds such as glycine betaine play multiple biochemical roles including conventional osmolytes, bulking agents to plug membrane defects (see Chapter 7), and as oxidative protective molecules (Figure 10.2) (Pan et al., 2010). Our reasoning is as follows:

- DHA-enriched cells such as rods, cones, sperm, and neurons require powerful protection systems against oxidative stress.
- Relatively high cellular levels of taurine already present and acting as osmoprotectants might be the secret to the antioxidant powers of this molecule.
- Taurine has chemical properties (e.g., charge-neutral chemical nature accounting for its lipophilic properties) allowing it to penetrate into and perhaps protect DHA-enriched membranes.

At first glance, taurine does not fit the classic view of an antioxidant. Taurine seems to have some antioxidant activity in the human brain but the effect might be indirect (Aruoma et al., 1988). A key finding is that taurine somehow reduces levels of reactive oxygen species though many mechanisms can be postulated to account for this effect. One idea is that taurine plays multiple antioxidant roles including direct scavenging of reactive oxygen species and an indirect role triggering upregulation of conventional enzymatic defenses against oxidative stress. The latter mechanism has recently been described by Yu and Kim (2009). As described above, axons in the brain and nervous system can be divided into two broad classes—myelinated and demyelinated. If it is assumed that myelin sheaths act to shield axons at least partially against oxidative damage to DHA-enriched membranes, then it follows that demyelinated axons would be expected to be more sensitive to lipid peroxidation. One mechanism to counter any additional threat of lipid peroxidation of demyelinated axons is to specifically increase oxidative defenses in this class of axon. Theoretically, taurine might play such a protective role against oxidative stress.

10.9 HYPOTHESIS NUMBER 3: MELATONIN AS AN ANTIOXIDANT

It has been proposed that melatonin, which acts as an antioxidant in chemical studies, plays a similar role in living animals, especially in the brain and central nervous system (Reiter, Manchester, and Tan, 2010; Tan et al., 2010). However, data derived from studies of knockout mutations of melatonin synthesis in mice show that melatonin is not an essential antioxidant (Kasahara et al., 2010). Furthermore, selection of mice with knockout mutations in melatonin genes appears to be a beneficial but accidental step in domestication of mice suitable for commercial breeding colonies. Thus, definitive experiments on melatonin as an antioxidant remain to be done.

In summary, neurons and all cells and organisms that produce DHA must simultaneously develop novel mechanisms of protecting their DHA membranes against oxidative damage. In the human brain the net effect is to enhance the benefits of DHA balanced against risks. Neurons require a lifetime of protection. A failure to protect DHA membranes in neurons is a recipe for energy uncoupling and oxidative stress, which leads to neurodegeneration. Because oxidative damage to DHA membranes is a chemical reaction and is dependent on oxygen, a common mechanism of protection seen in nature involves simply avoiding oxygen. One class of human cells, sperm, has evolved to travel in the low oxygen world of the female reproductive tract, a niche where risks of lipid peroxidation of DHA are minimal. Thus, the DHA-enriched tail membrane of sperm is protected during the journey to fertilize the egg. The brain has evolved robust protection mechanisms against oxidative damage of DHA membranes. We discuss three novel mechanisms that might operate in neurons. One is based on myelination acting as a protective sheath; a second involves taurine acting directly or indirectly as an antioxidant; and in a third mechanism melatonin might act as an antioxidant. In all three cases definitive experiments remain to be performed. In Chapter 17 we discuss a powerful genetic mutation that might protect neuron membranes against oxidative damage as well as the role of microglia in both protecting and endangering neurons against oxidative stress.

REFERENCES

Aitken, R. J., and J. S. Clarkson. 1987. Cellular basis of defective sperm function and its association with the genesis of reactive oxygen species by human sperm. *J. Reprod. Fertil.* 81:459–469.

Ala, A., A. P. Walker, K. Ashkan et al. 2007. Wilson's disease. *Lancet* 369:397–408.

Alvarez, J. G., and B. T. Storey. 1995. Differential incorporation of fatty acids into and peroxidative loss of fatty acids from the phospholipids of human spermatozoa. *Mol. Reprod. Dev.* 42:334–346.

Aruoma, O. I., B. Halliwell, B. M. Hoey et al. 1988. The antioxidant action of taurine, hypotaurine and their metabolic precursors. *Biochem. J.* 256:251–255.

Avery, S. V., N. G. Howlett, and S. Radice. 1996. Copper toxicity towards *Saccharomyces cerevisiae*: Dependence on plasma membrane fatty acid composition. *Appl. Environ. Microbiol.* 62:3960–3966.

Bock, N. A., A. Kocharyan, J. V. Liu et al. 2009. Visualizing the entire cortical myelination pattern in marmosets with magnetic resonance imaging. *J. Neurosci. Methods* 185:15–22.

Bull, P. C., G. R. Thomas, J. M. Rommens et al. 1993. The Wilson disease gene is a putative copper transporting P-type ATPase similar to the Menkes gene. *Nat. Genet.* 5:327–337.

Connor, W. E., R. G. Weleber, C. DeFrancesco et al. 1997. Sperm abnormalities in retinitis pigmentosa. *Invest. Ophthalmol. Vis. Sci.* 38:2619–2628.

DeMar, J. C. Jr., K. Ma, J. M. Bell et al. 2006. One generation of *n*-3 polyunsaturated fatty acid deprivation increases depression and aggression test scores in rats. *J. Lipid Res.* 47:172–180.

Diau, G. Y., A. T. Hsieh, E. A. Sarkadi-Nagy et al. 2005. The influence of long chain polyunsaturate supplementation on docosahexaenoic acid and arachidonic acid in baboon neonate central nervous system. *BMC Med.* 3:11.

El-Sherbeny, A., H. Naggar, S. Miyauchi et al. 2004. Osmoregulation of taurine transporter function and expression in retinal pigment epithelial, ganglion, and Müller cells. *Invest. Ophthalmol. Vis. Sci.* 45:694–701.

Fraga, C. G., P. A. Motchnik, M. K. Shigenaga et al. 1991. Ascorbic acid protects against endogenous oxidative DNA damage in human sperm. *Proc. Natl. Acad. Sci. USA* 88:11003–11006.

Giusto, N. M., S. J. Pasquaré, G. A. Salvador et al. 2000. Lipid metabolism in vertebrate retinal rod outer segments. *Prog. Lipid Res.* 39:315–391.

Gray, D. A., and J. Woulfe. 2005. Lipofuscin and aging: A matter of toxic waste. *Sci. Aging Knowledge Environ.* 2005(5):re1.

Gray, J. M., D. S. Karow, H. Lu et al. 2004. Oxygen sensation and social feeding mediated by a *C. elegans* guanylate cyclase homologue. *Nature* 430:317–322.

Green, T. R., J. H. Fellman, and A. L. Eicheretal. 1991. Antioxidant role and subcellular location of hypotaurine and taurine in human neutrophils. *Biochim. Biophys. Acta* 1073:91–97.

Heller-Stilb, B., C. van Roeyen, K. Rascher et al. 2002. Disruption of the taurine transporter gene (*taut*) leads to retinal degeneration in mice. *FASEB J.* 16:231–233.

Hollyfield, J. G., V. L. Perez, and R. G. Salomon. 2010. A hapten generated from an oxidation fragment of docosahexaenoic acid is sufficient to initiate age-related macular degeneration. *Mol. Neurobiol.* 41:290–298.

Kasahara, T., K. Abe, and K. Mekadaetal. 2010. Genetic variation of melatonin productivity in laboratory mice under domestication. *Proc. Natl. Acad. Sci. USA* 107:6412–6417.

Koppers, A. J., G. N. DeIuliis, J. M. Finnie et al. 2008. Significance of mitochondrial reactive oxygen species in the generation of oxidative stress in spermatozoa. *J. Clin. Endocrinol. Metab.* 93:3199–3207.

Lu, L., X. Gu, L. Hong et al. 2009. Synthesis and structural characterization of carboxyethylpyrrole-modified proteins: Mediators of age-related macular degeneration. *Bioorg. Med. Chem.* 17:7548–7561.

Ng, K. P., B. Gugiu, K. Renganathan et al. 2008. Retinal pigment epithelium lipofuscin proteomics. *Mol. Cell Proteomics* 7:1397–1405.

Ni, L., P. Guo, K. Reddig et al. 2008. Mutation of a TADR protein leads to rhodopsin and Gq-dependent retinal degeneration in Drosophila. *J. Neurosci.* 28:13478–13487.

Pan, C., G. S. Giraldo, H. Prentice et al. 2010. Taurine protection of PC12 cells against endoplasmic reticulum stress induced by oxidative stress. *J. Biomed. Sci.* 17(Suppl 1):S17 (doi:10.1186/1423-0127-17-S1-S17).

Paskowitz, D. M., M. M. LaVail, and J. L. Duncan. 2006. Light and inherited retinal degeneration. *Br. J. Ophthalmol.* 90:1060–1066.

Reiter, R. J., L. C. Manchester, and D. X. Tan. 2010. Neurotoxins: Free radical mechanisms and melatonin protection. *Curr. Neuropharmacol.* 8:194–210.

Robison, W. G., T. Kuwabara, and J. G. Bieri. 1982. The roles of vitamin E and unsaturated fatty acids in the visual process. *Retina* 2:263–281.

Rockel, N., C. Esser, and S. Grether-Becketal. 2007. The osmolyte taurine protects against ultraviolet B radiation-induced immunosuppression. *J. Immunol.* 179:3604–3612.

Stadtman, E. R. 2006. Protein oxidation and aging. *Free Radic. Res.* 40:1250–1258.

Storey, B. T. 2008. Mammalian sperm metabolism: Oxygen and sugar, friend and foe. *Int. J. Dev. Biol.* 52:427–437.

Strausak, D., J. F. Mercer, H. H. Dieter et al. 2001. Copper in disorders with neurological symptoms: Alzheimer's, Menkes, and Wilson diseases. *Brain Res. Bull.* 55:175–185.

Tabassum, H., S. Parvez, H. Rehman et al. 2007. Nephrotoxicity and its prevention by taurine in tamoxifen induced oxidative stress in mice. *Hum. Exp. Toxicol.* 26:509–518.

Tan, D. X., L. C. Manchester, E. Sanchez-Barcelo et al. 2010. Significance of high levels of endogenous melatonin in mammalian cerebrospinal fluid and in the central nervous system. *Curr. Neuropharmacol.* 8:162–167.

Umhau, J. C., W. Zhou, R. E. Carson et al. 2009. Imaging incorporation of circulating docosahexaenoic acid into the human brain using positron emission tomography. *J. Lipid Res.* 50:1259–1268.

Valk, E. E., and G. Hornstra. 2000. Relationship between vitamin E requirement and polyunsaturated fatty acid intake in man: A review. *Int. J. Vitam. Nutr. Res.* 70:31–42.

Warburton, S., K. Southwick, R. M. Hardman et al. 2005. Examining the proteins of functional retinal lipofuscin using proteomic analysis as a guide for understanding its origin. *Mol. Vis.* 11:1122–1134.

Yu, J., and A. K. Kim. 2009. Effect of taurine on antioxidant enzyme system in B16F10 melanoma cells. *Adv. Exp. Med. Biol.* 643:491–499.

Section IV

Revised Membrane Pacemaker Theory of Aging and Age-Dependent Diseases

Major scientific breakthroughs often occur when old, established concepts or theories are abandoned in the face of new contrarian evidence. Such is the case for the field of aging and age-dependent diseases. The theory that reactive oxygen species (ROS) directly cause aging has a large following among scientists in this field. What if ROS is not the direct cause of aging? Then a new theory must arise to take its place. In the field of aging, this is a breathtaking possibility that is now taking hold. A revised mitochondrial theory of aging has now been successfully tested in mice and is consistent with similar data on human aging. These data show that mitochondrial DNA polymerase, which produces new copies of mitochondrial DNA each time mitochondria divide, acts as an energy-based pacemaker of aging. That is, the biochemical process of replicating DNA is not perfect and results in a linear accumulation of mtDNA mutations over time. According to the revised theory, the increasing numbers of mtDNA mutations gradually lower energy production of trillions of mitochondria that work as powerhouses in cells throughout the body. Decreasing energy supply acts like an invisible thread connecting the ultimate fate of all human cells. With aging, a threshold of reduced energy level is breached that triggers the natural and well-established process of programmed cellular death or apoptosis. Thus, aging can be regarded as a grand "energy stress disease" in which energy is depleted to the point that cellular life is gradually snuffed out. Is this revised mitochondrial theory of aging applicable to premature aging in neurons? Recall that neurons belong to a rare class of human cells that seldom divide, in essence holding the brain hostage to any premature death of large numbers of neurons by any cause. Note that Alzheimer's disease can be defined as a cellular or neuron disease characterized by the death of too many neurons. Thus, using the revised mtDNA theory of aging as a stepping-stone it is possible to delve more deeply into the molecular basis of the roles of DHA as a pro-aging molecule in the brain. We propose that DHA membranes working as gatekeepers responsible for a significant amount of the energy supply of neurons, behave as a second energy-based pacemaker for neurodegeneration, which is age-dependent. According to this hypothesis, DHA is at the heart of neurodegeneration because it mediates energy uncoupling unique to neurons, thus elevating energy stress in these specialized cells. These considerations have led to the concept that dual energy pacemakers—mitochondria and membranes—determine the life span of neurons and cells in general. A chapter on cancer is included because this disease, like neurodegeneration, is age-dependent and likely shares common threads with all age-dependent diseases and aging itself.

11 Revised Mitochondrial Theory of Aging and Brain Span

Energy is the basis of all cellular activity ranging from heavy lifting to intense thinking. An estimated total of 10 quadrillion (10,000,000,000,000,000) mitochondria act as power sources in human cells. These bacteria-like organelles, strategically located in cells where they are needed, burn fat and other energy-rich foodstuffs to produce the common energy currency of all cells called *ATP* (adenosine triphosphate). The process of converting food to useful energy by mitochondria is called *respiration* and consumes most of the oxygen carried by our circulating red blood cells. Mitochondria likely originated from bacteria or archaea but long ago lost their free-living lifestyle in favor of a permanent endosymbiotic relationship within animal and plant cells. These energy-transducing organelles, which still retain some properties of bacteria, are strictly dependent on their animal hosts for growth but have retained a miniature chromosome of their own called *mitochondrial DNA* (mtDNA). The mtDNA encodes several genes essential for energy production, and in a strange quirk of nature these genes hold life and death powers over the human organism—acting as a master pacemaker for aging. We begin this story by describing the general nature of energy production by mitochondria and end by defining how energy stress can arise and create aging and age-related diseases including Alzheimer's disease.

11.1 NATURE OF MITOCHONDRIA AS POWERHOUSES FOR THE CELL

The fundamental properties of mitochondria as powerhouses of the cell are similar across life forms. This allows us to introduce the subject of mitochondrial energy production using one of our favorite birds—the hummingbird (Figure 11.1a). Mitochondria (Figure 11.1b) are the power plants fueling the rapid rates of wing beats of this remarkable tiny bird. Among smaller bumblebee-size hummingbirds, wing beats of about 90 per second have been recorded in normal flight with bursts of up to 200 beats per second during mating acrobatics. Numerous adaptations are required to provide sufficient power for hummingbird flight (Suarez, 1998; Suarez et al., 1991). A robust heart and circulatory system, far larger than normal for a bird of this size, is an essential adaptation to carry enough oxygen and energy food for flight muscles. These breast muscles account for almost one third of total body weight. Virtually all of the oxygen used by animals is consumed by mitochondria. These organelles are packed nearly to the limits inside hummingbird flight muscle

(a)

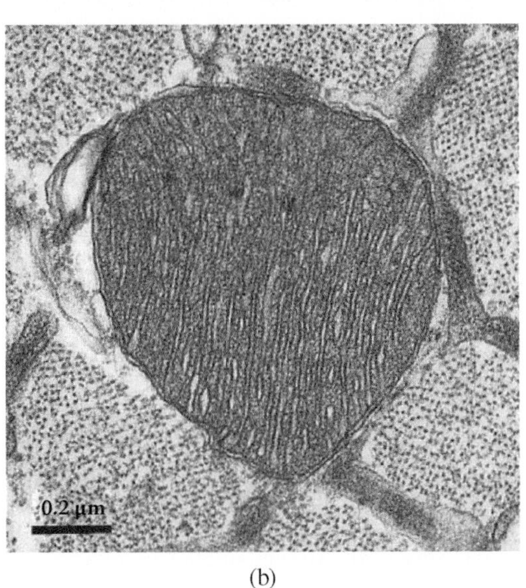

(b)

FIGURE 11.1 Mitochondria as powerhouses of the cell. (a) A hummingbird, shown hovering and gathering nectar, has wing beats as fast as 200 per second. (Photo courtesy of Claudia Bodmer.) (b) Single mitochondrion from hummingbird flight muscle cells showing numerous membrane cristae, which house components of the electron transport chain. (Reproduced with permission of *J. Exp. Biol.*, R. Suarez, 1998, Oxygen and the Upper Limits to Animal Design and Performance, *J. Exp. Biol.* 201: 1065–1072; Micrograph courtesy of Dr. Odile Mathieu-Costello, University of California, San Diego.) (c) Electron transport chain features a series of proton pumps. See text for details. (Courtesy of Gary Kaiser.)

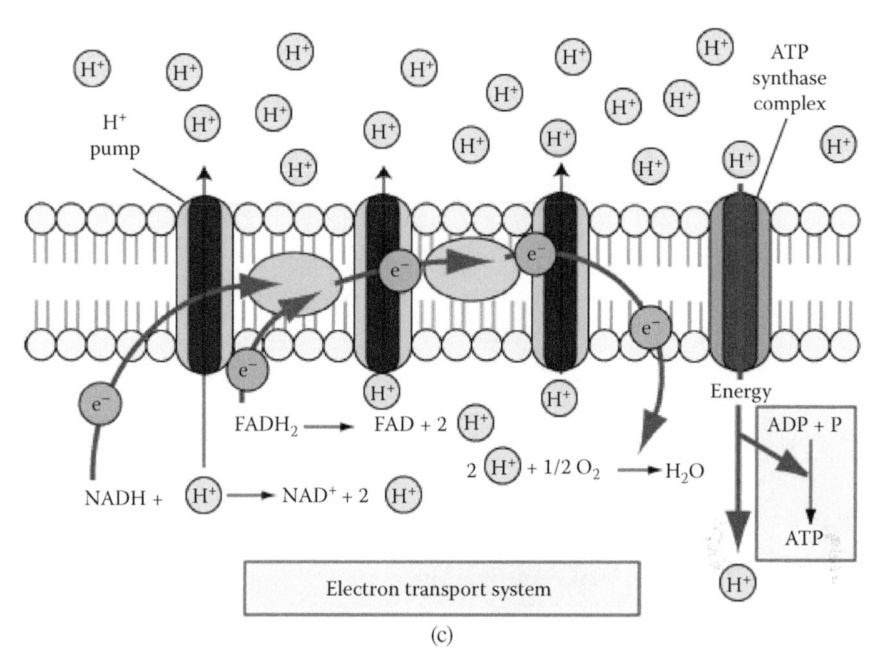

FIGURE 11.1 *(Continued)*

cells (30 to 35 percent of total cellular volume). An expert will notice the close packing of what appear as stripes in this electron micrograph taken of a slice or thin section across a single hummingbird mitochondrion (Figure 11.1b). These stripes are internalized membrane structures called *cristae*, which are packed to the limit with molecular machines that carry out respiration, as shown in Figure 11.1c. The chain of events shown in Figure 11.1c is called an *electron transport chain* and is one of the marvels of nature. The purpose of this chain is to bleed off energy from high-energy electrons derived by the breakdown of food and entering the chain in the form of the ubiquitous electron carrier NADH. NADH is the cardinal electron donor feeding into the chain and is derived from the breakdown of fats, sugars, and other energy-rich foodstuffs. Recall that glucose is the ultimate donor of high-energy electrons for mitochondria in neurons. Electrons donated by NADH enter the electron transport chain as high-energy electrons and then lose energy in discreet packets. Energy provided by the electrons is used directly to fuel powerful pumps that efflux protons (H^+) from inside to outside of the cell. The concept of inside versus outside of membranes is difficult to visualize in the electron micrograph but becomes clearer in Figure 11.1c. Note in Figure 11.1c that H^+ are concentrated on the outside membrane surface compared to the inside. Pumping protons to the outside against this proton gradient depletes the energy levels of electrons moving down the chain. Spent electrons, which have lost a good bit of their energy while passing through the respiratory chain, are deposited or donated to oxygen to form water as a by-product. (Note that whales die if they drink seawater and produce enough water as a by-product of respiration to supply their needs.) Energy stored in proton gradients (proton motive energy), like water stored behind a dam, is recovered by an amazing spinning turbine

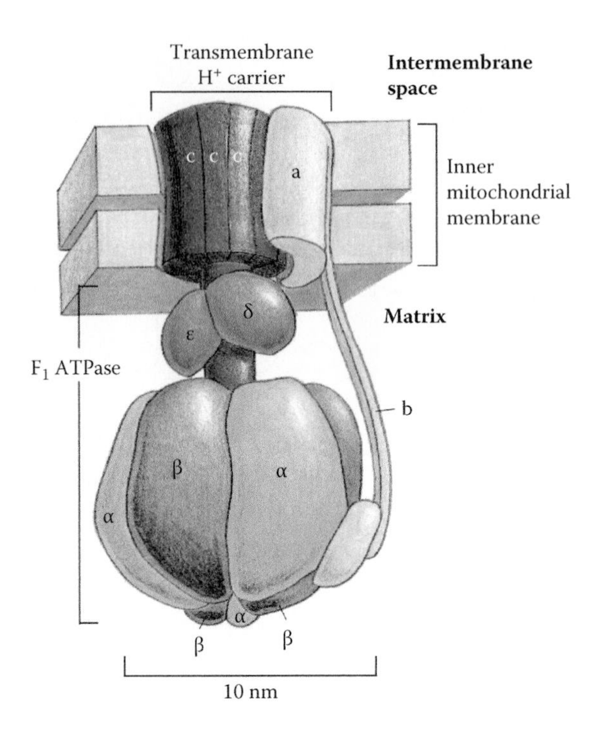

FIGURE 11.2 ATP synthase is a molecular machine that is embedded in the inner mitochondrial membrane and spins. Energy stored in proton (H^+) gradients, high outside and low inside, fuels this miniature rotary engine that transduces proton motive energy into pyrophosphate bond energy as ATP. (From Alberts, B. et al., *Molecular Biology of the Cell*, Copyright 2008. Reproduced by permission of Taylor & Francis Group, LLC, a division of Informa plc.)

called *ATP synthase*, which reverses the flow of protons (H^+) from outside to inside and in the process generates ATP (Figure 11.2). Protons (H^+) concentrated on the outside of the mitochondrial membrane enter ATP synthase and bind individually to each of 12 turbine blades causing the molecular turbine to rotate. The net effect is the conversion of proton power (outside) → mechanical power → chemical power in the form of ATP (inside). Proton circulation is completed by the powerful proton pumps of the electron transport chain, thus regenerating the proton gradient. ATP is used to power numerous cellular reactions such as the flight muscle action of hummingbirds or to maintain Na^+/K^+ levels in axons. Proton energy gradients are used directly to power numerous cellular processes ranging from driving flagellar motors of bacteria to energizing uptake of neurotransmitters in synaptic vesicles of neurons.

11.2 MITOCHONDRIAL DNA IS A CIRCULAR MINIATURE CHROMOSOME ENCODING 37 GENES ESSENTIAL FOR ENERGY PRODUCTION

Each mitochondrion, such as shown in Figure 11.1b, is estimated to contain two to ten mtDNA copies (Figure 11.3). In humans, 1000 to 10,000 separate copies of

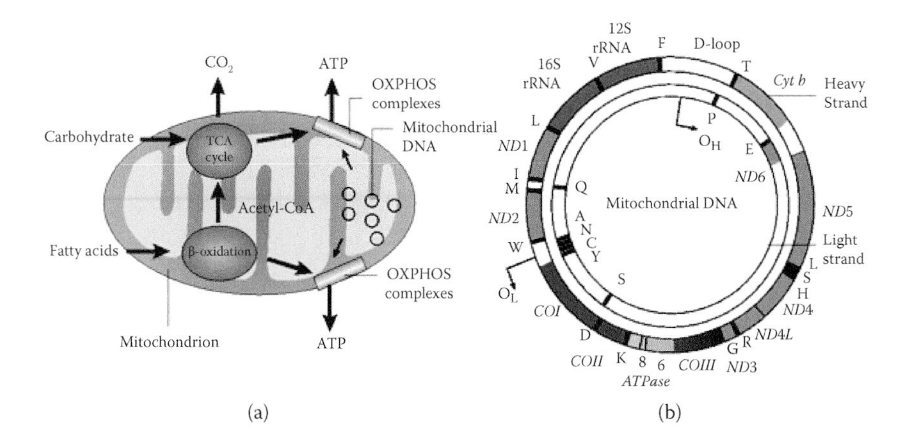

FIGURE 11.3 Human mitochondrial DNA (mtDNA) is a circular double helix encoding essential energy-producing genes. The genes that encode the subunits of complex I (ND1 through ND6 and ND4L) are shown in blue; the terminal complex, cytochrome *c* oxidase (COI through COIII) is shown in red; cytochrome *b* of complex III is shown in green; and the subunits of ATP synthase (ATPase 6 and 8) are shown in yellow. RNA genes are also listed (purple and black slashes). According to the revised mitochondrial theory of aging, mutations occurring during replication of mtDNA decrease energy production, causing aging. (Reprinted by permission from Macmillan Publishers Ltd: *Nat. Rev. Genet.*, R. W. Taylor, and D. M. Turnbull, Copyright 2005.) **(See color insert.)**

mtDNA are present per cell. Human mitochondrial DNA is composed of 16,569 base pairs or twice this many nucleotides forming the famous double-helix structure. The mtDNA forms a closed circular structure as shown in Figure 11.3. The two intertwined strands of mtDNA that make up the double helix are referred to as the *heavy strand* and the *light strand*. They each have distinct nucleotide compositions. The heavy strand encodes 28 genes compared to 9 genes for the light strand of the double helix, for a total of 37. Thirteen genes code for proteins (polypeptides), all of which play primary roles in energy production. For example, seven of these genes encode different subunits of complex I, the first proton pump receiving its electrons from NADH and starting the electron transport chain. We will see later that genes for complex I, also called *NADH dehydrogenase* and shown as the first member of the electron transport chain (Figure 11.1c), represent targets for mutations that serve as the pacemaker of aging. Mutationally modifying this powerful proton pump can affect energy homeostasis. Mutations in numerous other genes also modulate energy production and energy homeostasis. For example, two subunits of ATP synthase, the rotary enzyme necessary for production of ATP (Figure 11.2), are encoded by mtDNA, and energy production can be modulated by mutations in these genes. Genes encoding subunits of other essential components are also present along with 24 genes encoding for RNA products needed for assembling new proteins in mitochondria.

New copies of mtDNA are produced during mitochondrial division or fission by a replicating machine called the *DNA polymerase gamma enzyme complex*. This enzyme is composed of a larger catalytic DNA polymerizing enzyme encoded by the *POLG* gene and a smaller subunit coded for by the *POLG2* gene. Note that nuclear genes encode these DNA-replicating enzymes, which are sent or targeted from the nucleus into mitochondria. Placing DNA polymerase genes in the nucleus rather than them being encoded by mtDNA might have the effect of protecting these essential genes from high levels of mutations seen in mitochondria (Longley et al., 2005). The revised mitochondrial theory of aging featuring energy stress, not oxidative stress, as the cause of aging is based on studies with a now famous line of mice—mutator mice. These data are discussed next.

11.3 DATA DERIVED FROM MUTATOR MICE HAVE LED TO A NEW THEORY OF AGING IN WHICH ENERGY STRESS REPLACES OXIDATIVE STRESS AS THE PRIMARY CAUSE OF AGING

A great deal of correlative evidence linking mitochondrial dysfunction and aging has emerged over the last several decades (Trifunovic and Larsson, 2008). The creation of mtDNA mutator mice provides the clearest picture yet that fast-forwarding mtDNA mutations causes premature aging (Trifunovic et al., 2004). Thus, the first direct evidence is available showing that mutational loss of mitochondrial energy production is a major causal factor in mammalian aging. Mutator mice were genetically engineered with a defective catalytic subunit of mitochondrial DNA polymerase (POLGA), which speeds up random accumulation of mtDNA mutations. Mutations in mtDNA of mutator mice accumulate more rapidly compared to wild-type controls. Studies of mutator mice show that increased levels of mutations in mtDNA and subsequent energy stress can directly cause or fast forward numerous age-related symptoms in mice (Trifunovic and Larsson, 2008; Trifunovic et al., 2004; Tyynismaa and Suomalainen, 2009) as follows:

- Weight loss
- Progressive hearing loss
- Heart disease
- Osteoporosis
- Reduced subcutaneous fat
- Alopecia
- Kyphosis
- Anemia
- Sarcopenia
- Reduced fertility
- Decreased spontaneous activity

The mtDNA mutator mice appear completely normal at birth and early adolescence but subsequently display many features of premature aging now considered to be caused by energy stress. For example, hearing loss occurs at roughly double the rate in mutator mice as in controls (Niu et al., 2007; Someya et al., 2008). At the cellular level mtDNA mutations eventually cause irreplaceable cell losses through programmed cellular death. The biggest surprise from this work is that premature aging in mutator mice appears to occur without a major increase or spike in levels of reactive oxygen species or oxidative stress, as predicted according to the conventional reactive oxygen species (ROS) theory of aging. These data, along with other clues in the literature, have resulted in a fresh look at the fundamental role of reactive oxygen species in aging and age-dependent diseases.

Mutator mice have very high levels of single base mutations (point mutations) as well as high levels of deletions of linear stretches of mtDNA (linear deletion mutants). This is an important advance in the field of aging (Edgar et al., 2009). The current model for premature aging in mutator mice is that accumulation of point mutations rather than deletions of mtDNA lead to production of respiratory chain subunits with single amino acid substitutions along the peptide chain. These defects are believed to destabilize respiratory chain complexes, eventually slowing energy production below a critical threshold, which triggers widespread cellular death during aging.

Recently, mutations in another gene, the *cis d 2* gene in mice, with homologues in humans, have been shown to accelerate aging in mice as well as humans (Chen et al., 2009). In knockout mutants of mice disruption of CIS D 2, a redox-active, iron-sulfur protein targeted to mitochondria, leads to accelerated aging. Knockout mice display thinner bones and hair, corneal opacities and degeneration, decreased muscle mass, prominent eyes, and protruding ears, all of which are consistent with premature aging. Mitochondria isolated from the mutant mice show a defect in respiration, an indication that ATP production is slowed.

11.4 HUMANS WITH MUTATOR GENES DISPLAY SYMPTOMS OF PREMATURE NEURODEGENERATION BEFORE SYMPTOMS OF AGING

Obviously there are no genetically engineered mutator people walking around in which human DNA polymerase gamma is purposely altered to create a wave of mtDNA mutations such as seen in mutator mice. However, the replication fidelity of DNA polymerase can be altered by naturally occurring mutations found infrequently in the general population. Mutations in the catalytic subunit of DNA polymerase (POLG) are believed to create the same kind of mutator effect as seen in mice, increasing mutations in mtDNA and lowering cellular energy levels. Other genes supporting mtDNA replication and repair also exhibit a mutator effect when humans acquire these mutated genes. A wide spectrum of disease states in humans arise as the result of naturally occurring DNA polymerase gamma, mutations creating pathologies as follows:

- Parkinson's
- Breast cancer
- Ophthalmoplegia (PEO)
- Apert syndrome
- Neuropathy
- Dysarthria
- Sensory ataxic neuropathy, dysarthria, and ophthalmoparesis (SANDO)
- Male infertility

These and other human diseases and symptoms attributed to the mutator properties of mtDNA polymerase are reviewed by several authors (Bensch et al., 2009; Chan and Copeland, 2009; Hudson and Chinnery, 2006; Hudson et al., 2008; Invernizzi et al., 2008; Longley et al., 2005; Luoma et al., 2004; Singh et al., 2009; Taylor and Turnbull, 2005; Turnbull et al., 2010; Van Goethem et al., 2003). These new findings are revitalizing research in the fields of human aging, dementia, and cancer and reinforce the critical roles played by mitochondria and energy stress in aging and age-related diseases. However, there is little mention if any regarding what we suggest are critical roles played by membranes generally and DHA specifically in age-related diseases. This is not surprising because the DHA principle first appeared in print in 2009.

A recent comprehensive analysis of POLG-mediated human diseases shows that mutations of POLG often cause neurodegeneration (Wong et al., 2008). These data bring the fields of aging, neurodegenerative diseases, and DHA membrane research closer together. The key point is that many of these phenotypes presumably caused by energy stress have been traced to premature death of neurons in the brain or peripheral nervous system. Obviously membranes of neurons are unique compared to most human cells in being enriched with DHA. Do DHA membranes hold special power over aging in neurons? Does the presence of DHA in neurons bias the effects of human mutator genes toward neurological or neurodegenerative phenotypes?

11.5 SELECTIVE TARGETING OF DHA AWAY FROM MITOCHONDRIAL MEMBRANES AS A SECRET TO A LONG HUMAN BRAIN SPAN

In developing the central theme for this book (i.e., DHA-degenerative diseases), we sought some clue from the literature on DHA targeting in different cells and specialized membranes that might help tie this concept together. The question asked at the beginning of our search was, “Could the DHA principle in any way help explain in biochemical terms the extreme (30-fold) difference in longevity and consequently brain span between a mouse versus a human?” One needs only to assume that the long brain span of humans has some great selective advantage to our species to arrive at the starting point of a compelling concept. Perhaps the human brain in contrast to the mouse brain evolved a biochemical mechanism aimed at increasing brain span to keep up with life span, by somehow maximizing the benefits versus risks of DHA functioning in neuron membranes. This led to the proposition of selective DHA targeting developed in this section, in which specific membranes within neurons are enriched with DHA while other classes of membranes in the same cell are not. Based on considerations discussed in Chapter 12, we predicted that human neurons might target DHA away from their mitochondria as a mechanism to enhance longevity of these critical organelles in neurons, ultimately increasing brain span. The converse prediction is that due to their evolutionarily honed short life span, mice would gain little from selective DHA targeting and might even be harmed. DHA enrichment of membranes in mice mitochondria seems to be necessary to turbocharge energy production to keep up with the hyperactive metabolism of this tiny animal. For example, the DHA content of membranes of heart cells in mice is normally around 30 percent of total fatty acids, rising to 50 percent during dietary supplementation with fish oil (references in Chapter 2). Thus, the DHA level in the heart of a mouse is in the order of 30 times that of a human (Chapter 12) in contrast to neurons of mouse versus man in which such dramatic differences in DHA levels are not observed.

A selective targeting model for DHA is consistent with data that show that neuron membranes are divided into two distinct domains (Figure 11.4) identified by the presence of specific membrane proteins targeted to axonal versus headgroup or dendrite

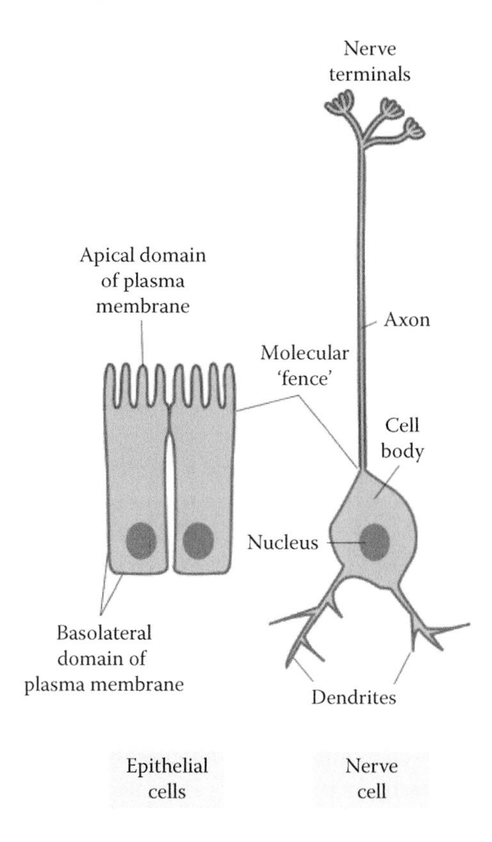

FIGURE 11.4 DHA in humans is believed to be selectively targeted to membranes of axons and synaptic vesicles and away from mitochondria. As shown, the plasma membrane surrounding neurons is divided into distinct domains—axon and nerve terminals versus cell body and dendrites. Membrane domains occur in other cells including epithelial cells shown for comparison as well as sperm (the latter is discussed in Chapter 8). Specific lipids including DHA and proteins are believed to be targeted to each domain through sophisticated patterns of vesicle trafficking. (Pfenninger, K. H., 2009, *Nature Reviews Neuroscience* 10:251–261.) In essence, membrane domains of nerve cells are separated by a molecular fence consisting of a meshwork of membrane proteins tightly associated with underlying actin cytoskeleton. This tight junction in neurons is called an *axonal hillock* and works by keeping membrane proteins from diffusing between the two distinct domains. This barrier might also localize specific lipid species including DHA from crossing boundaries. (From Alberts, B. et al., *Molecular Biology of the Cell*, Copyright 2008. Reproduced by permission of Taylor & Francis Group, LLC, a division of Informa plc.)

domains (Pfenninger, 2009). It is possible that each neuron membrane domain, as in the case of sperm (see Chapter 7), has a unique DHA composition with the axon and synaptic vesicles being enriched with DHA in contrast to the neuron head domain. Recall that synaptic vesicles that are enriched with DHA are derived from the *trans* Golgi apparatus and directly transported to the presynaptic region of axons. Newly synthesized lipids destined for mitochondrial membranes originate from the endoplasmic reticulum with their own mechanism of targeting distinct from that of synaptosomes. Hence, trafficking and targeting of new membrane lipids for building or repairing mitochondrial membranes versus axons and synaptosomal membranes are markedly different. Furthermore, mitochondria have evolved a novel mechanism for biosynthesis of molecular species of cardiolipin which is specific for these organelles (see Chapter 12). Thus, selective DHA targeting mechanisms seem to have been honed by Darwinian selection, opening the door for the evolution of distinct patterns of DHA targeting in neurons of mouse versus man. Perhaps the greatest advantage of the targeting of DHA into specific neuron membrane domains involves maximizing benefits while reducing oxidative risks associated with this highly unstable molecule. For example, we suggest that human sperm and human neurons share a need for targeting DHA to locations or domains where it is most advantageous. The converse might also be true.

Data on the comparative DHA composition of membranes of mouse versus man support a selective DHA targeting model. Specifically, data on the DHA molecular species of cardiolipin in the inner membrane of mitochondria seem to fulfill our prediction. Cardiolipin is the signature lipid in mitochondria, and its levels of DHA and unsaturation in general vary greatly among different animals or cells, as shown in Figure 11.5. As discussed in Chapter 12, we interpret high DHA levels in mitochondrial cardiolipin as an indication of energy stress in a cell, and vice versa. This idea is consistent with data on the evolution of extreme flight in hummingbirds where DHA levels in cardiolipin and other phospholipids are high and life spans are low. In humans with long life spans, DHA is rarely incorporated into cardiolipin and when it is DHA-cardiolipin is generally considered to be a pathological or suicidal molecule (see Chapter 2). Note from data summarized in Figure 11.5 that human mitochondria still retain a good bit of biochemical flexibility in terms of diversity in mitochondrial membrane unsaturation levels with the most common unsaturated molecular species in humans being $(18:2)_4$-CL (see Chapter 12). For example, $(18:2)_4$-CL is found in human heart muscle whose mitochondria are considered to be relatively hard working. As heart rates increase dramatically in tiny mammals, including miniature voles, to greater than 1000 beats per minute, DHA enrichment of mitochondrial cardiolipin is thought to increase proportionally (Chapter 12). Historically, comparative data on the direct relationship between DHA levels and heart rates first alerted scientists of a possible inverse relationship between DHA content of membranes and life span and led to the membrane pacemaker theory of aging discussed in detail in Chapter 12.

According to the membrane theory of aging, targeting DHA randomly into all membranes of human neurons would be predicted to put dramatic downward pressure on brain span. Instead, human neurons appear to target DHA away from cardiolipin (Kiebish, Han, and Seyfried, 2009; Kiebish et al., 2008; Kirkland et al., 2002; Söderberg et al., 1992; Yabuuchi and O'Brien, 1968), a finding that is consistent with

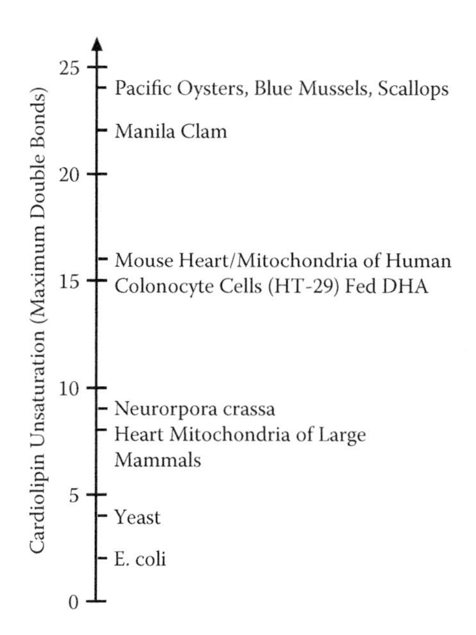

FIGURE 11.5 DHA-cardiolipin beneficial in some animals is considered to be pathological in humans. Note that maximum levels of unsaturated bonds in cardiolipin across life forms vary from a total of 2 double bonds in *Escherichia coli* to 24 in cold-adapted marine animals, including oysters and mussels. Thus, the most highly unsaturated species of cardiolipin contains four DHA chains, as discussed in Chapters 2 and 12. DHA-cardiolipin is believed to turbocharge energy production in mitochondria of fast muscles of certain animals including hummingbirds but acts as a cellular assassin in senescing human colonic cells (Chapter 2). The cardiolipin molecular species ($18{:}2_4$-CL) is considered the most unsaturated class of CL routinely present in human mitochondria. Specific targeting of DHA away from cardiolipin in human neurons is proposed to be a quantum leap forward in evolution to our long brain span. (From Valentine, R., and Valentine, D., *Omega-3 Fatty Acids and the DHA Principle*. Copyright 2009. Reproduced by permission of Taylor & Francis Group, LLC, a division of Informa plc.)

the absence of DHA-cardiolipin in most human mitochondria. This can be explained by the two-stage process used by mitochondria to synthesize cardiolipin. In the first stage cardiolipin synthetase located specifically in mitochondria joins two phospholipids together through their head group, generating a relatively saturated species of cardiolipin. The second stage involves a remodeling process that adds increasing levels of unsaturated chains to cardiolipin. The molecular species $(18{:}2)_4$-CL is believed to be the most common highly unsaturated molecular species of CL found in human neuron mitochondria. In contrast, mitochondria of mice routinely synthesize and incorporate DHA into cardiolipin of their mitochondrial membranes.

Mouse mitochondria have evolved along a different track from humans, incorporating DHA into cardiolipin in all of their mitochondria including those of neurons. Perhaps bioenergetic gain is more important in mice than neuron longevity. DHA-CL-enriched mitochondria in mice are expected to undergo more cycles of replication and acquire mtDNA mutations at a faster rate compared to human mitochondria

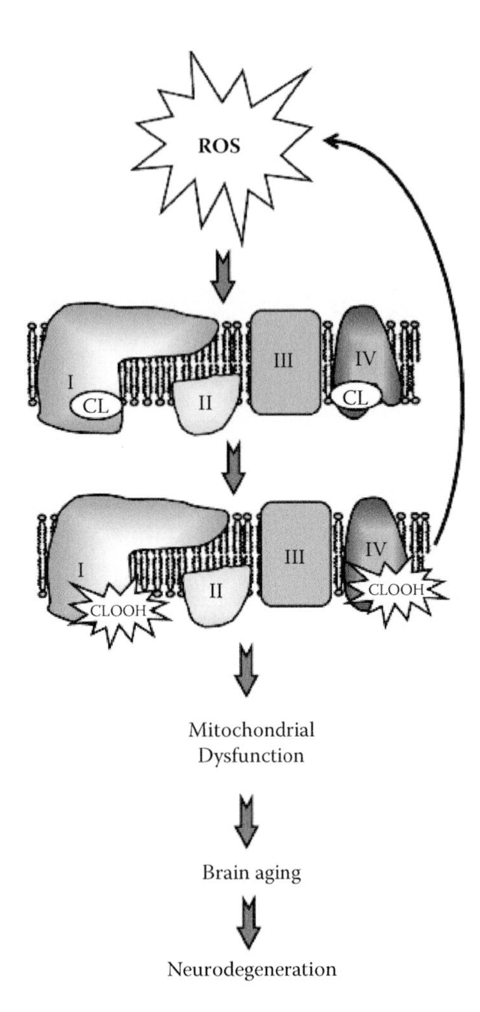

FIGURE 11.6 Possible role of cardiolipin and oxidative energy stress in brain aging. In this diagram reactive oxygen species (ROS) generation during brain aging causes CL peroxidation, uncoupling energy production by mitochondria. (Reprinted from *Neurochem. Int.*, 58, Paradies, G., G. Petrosillo, V. Paradies et al., Mitochondrial Dysfunction in Brain Aging: Role of Oxidative Stress and Cardiolipin, 447–457. Copyright 2011, with permission from Elsevier.)

with more saturated molecular species of CL. In humans, it is proposed that targeting DHA away from cardiolipin goes a long way in extending the effective life span of neurons and the brain. However, neurons still depend on DHA as a vital membrane building block for axons and synaptic vesicles but seem to maximize its benefits by avoiding incorporation into mitochondria. Selective targeting of DHA away from neuron mitochondria is regarded as a mechanism to ensure longevity of the human brain but requires that such targeting mechanisms remain functional for a lifetime. Should the mechanism governing selective DHA targeting in human neurons become enfeebled with aging, then the harmful side of DHA might be activated through the formation of DHA-cardiolipin in neuron mitochondria or DHA phospholipids in other membrane classes. As discussed in Chapter 16, DHA-CL peroxidation in mitochondria of neurons in small mammals including rats begins as early as 3 hours after traumatic brain injury. For a recent review of the literature linking cardiolipin and neurodegeneration see Paradies et al. (2011) (Figure 11.6). Also see the article by Gruber and colleagues (2011) for data on the role of CL peroxidation on aging in *Caenorhabditis elegans.*

In summary, new discoveries using mutator mice show that mutations accumulating in a linear fashion in mtDNA steadily decrease energy production in mitochondria, causing aging. A similar, naturally occurring mutator gene in humans causes neurodegeneration. We propose that random distribution of DHA in human neuron membranes would predispose neurons to energy stress leading to premature apoptosis of these critical brain cells. Selective targeting of DHA to specialized neuron membranes including axons and synaptic vesicles and away from mitochondria is proposed to be a critical mechanism helping explain the long brain span in humans. Thus, we consider selective DHA targeting to be a signature system for protecting neurons against aging. The role of membranes as pacemakers of aging is discussed in the next chapter.

REFERENCES

Bensch, K. G., J. L. Mott, S. W. Chang et al. 2009. Selective mtDNA mutation accumulation results in beta-cell apoptosis and diabetes development. *Am. J. Physiol. Endocrinol. Metab.* 296:E672–E680.

Chan, S. S., and W. C. Copeland. 2009. DNA polymerase gamma and mitochondrial disease: Understanding the consequence of POLG mutations. *Biochim. Biophys. Acta* 1787:312–319.

Chen, Y. F., C. H. Kao, and Y. T. Chenetal. 2009. Cisd2 deficiency drives premature aging and causes mitochondria-mediated defects in mice. *Genes Dev.* 23:1183–1194.

Edgar, D., I. Shabalina, Y. Camara et al. 2009. Random point mutations with major effects on protein-coding genes are the driving force behind premature aging in mtDNA mutator mice. *Cell Metab.*10:131–138.

Gruber, J., L. F. Ng, S. Fong et al. 2011. Mitochondrial changes in ageing *Caenorhabditis elegans*—What do we learn from superoxide dismutase knockouts? *PLoS One* 6:e19444.

Hudson, G., and P. F. Chinnery. 2006. Mitochondrial DNA polymerase-gamma and human disease. *Hum. Mol. Genet.* 15 Spec. No 2:R244–R252.

Hudson, G., P. Amati-Bonneau, E. L. Blakely et al. 2008. Mutation of OPA1 causes dominant optic atrophy with external ophthalmoplegia, ataxia, deafness and multiple mitochondrial DNA deletions: A novel disorder of mtDNA maintenance. *Brain* 131:329–337.

Invernizzi, F., S. Varanese, A. Thomas et al. 2008. Two novel POLG1 mutations in a patient with progressive external ophthalmoplegia, levodopa-responsive pseudo-orthostatic tremor and parkinsonism. *Neuromuscul. Disord.* 18:460–464.

Kiebish, M. A., X. Han, H. Cheng et al. 2008. Cardiolipin and electron transport chain abnormalities in mouse brain tumor mitochondria: Lipidomic evidence supporting the Warburg theory of cancer. *J. Lipid Res.* 49:2545–2556.

Kiebish, M. A., X. Han, and T. N. Seyfried. 2009. Examination of the brain mitochondrial lipidome using shotgun lipidomics. *Methods Mol. Biol.* 579:3–18.

Kirkland, R. A., R. M. Adibhatla, J. F. Hatcher et al. 2002. Loss of cardiolipin and mitochondria during programmed neuronal death: Evidence of a role for lipid peroxidation and autophagy. *Neuroscience* 115:587–602.

Longley, M. J., M. A. Graziewicz, R. J. Bienstock et al. 2005. Consequences of mutations in human DNA polymerase gamma. *Gene* 354:125–131.

Luoma, P., A. Melberg, J. O. Rinne et al. 2004. Parkinsonism, premature menopause, and mitochondrial DNA polymerase gamma mutations: Clinical and molecular genetic study. *Lancet* 364:875–882.

Niu, X., A. Trifunovic, N. G. Larsson et al. 2007. Somatic mtDNA mutations cause progressive hearing loss in the mouse. *Exp. Cell Res.* 313:3924–3934.

Paradies, G., G. Petrosillo, V. Paradies et al. 2011. Mitochondrial dysfunction in brain aging: Role of oxidative stress and cardiolipin. *Neurochem. Int.* 58:447–457.

Pfenninger, K. H. 2009. Plasma membrane expansion: A neuron's Hurculean task. *Nature Reviews Neuroscience* 10:251–261 (doi: 10.1038/nrn2593).

Singh, K. K., V. Ayyasamy, K. M. Owens et al. 2009. Mutations in mitochondrial DNA polymerase-gamma promote breast tumorigenesis. *J. Hum. Genet.* 54:516–524.

Söderberg, M., C. Edlund, I. Alafuzoff et al. 1992. Lipid composition in different regions of the brain in Alzheimer's disease/senile dementia of Alzheimer's type. *J. Neurochem.* 59:1646–1653.

Someya, S., T. Yamasoba, G. C. Kujoth et al. 2008. The role of mtDNA mutations in the pathogenesis of age-related hearing loss in mice carrying a mutator DNA polymerase gamma. *Neurobiol. Aging* 29:1080–1092.

Suarez, R. K. 1998. Oxygen and the upper limits to animal design and performance. *J. Exp. Biol.* 201:1065–1072.

Suarez, R. K., J. R. Lighton, G. S. Brown et al. 1991. Mitochondrial respiration in hummingbird flight muscles. *Proc. Natl. Acad. Sci. USA* 88:4870–4873.

Taylor, R. W., and D. M. Turnbull. 2005. Mitochondrial DNA mutations in human disease. *Nat. Rev. Genet.* 6:389–402.

Trifunovic, A., and N. G. Larsson. 2008. Mitochondrial dysfunction as a cause of ageing. *J. Intern. Med.* 263:167–178.

Trifunovic, A., A. Wredenberg, M. Falkenberg et al. 2004. Premature ageing in mice expressing defective mitochondrial DNA polymerase. *Nature* 429:417–423.

Turnbull, H. E., N. Z. Lax, D. Diodato et al. 2010. The mitochondrial brain: From mitochondrial genome to neurodegeneration. *Biochim. Biophys. Acta* 1802:111–121.

Tyynismaa, H., and A. Suomalainen. 2009. Mouse models of mitochondrial DNA defects and their relevance for human disease. *EMBO Rep.* 10:137–143.

Valentine, R. C., and D. L. Valentine. 2009. *Omega-3 Fatty Acids and the DHA Principle*. Boca Raton, FL: Taylor & Francis.

Van Goethem, G., J. J. Martin, B. Dermaut et al. 2003. Recessive POLG mutations presenting with sensory and ataxic neuropathy in compound heterozygote patients with progressive external ophthalmoplegia. *Neuromuscul. Disord.* 13:133–142.

Wong, L. J., R. K. Naviaux, and N. Brunetti-Pierrietal. 2008. Molecular and clinical genetics of mitochondrial diseases due to POLG mutations. *Hum. Mutat.* 29:E150–E172.

Yabuuchi, H., and J. S. O'Brien.1968. Brain cardiolipin: Isolation and fatty acid positions. *J. Neurochem.* 15:1383–1390.

12 Development of Dual Energy Pacemaker Theory of Aging

Role of the Membrane

Gudbjarnason and colleagues after measuring the DHA content of heart muscles of various mammals proposed a provocative concept for predicting life span (Gudbjarnason, 1989; Gudbjarnason et al., 1978). Their DHA theory of aging states that the longest-lived mammals will have the lowest DHA content in contrast to mammals with a short life span exhibiting greater DHA enrichment of their membranes. Their data on DHA content of heart muscle are consistent with an inverse relationship between DHA levels and life span. This simple relationship allowed Gudbjarnason to correctly place mammals on a ladder scaled according to life span. Gudbjarnason originally noted that heart rates of mammals are directly related to levels of DHA in cardiac tissue. That is, the mammal whose heart beats the fastest has the highest enrichment of DHA in its heart muscle, and vice versa. Thus, large whales with the lowest heartbeat and lowest DHA content are predicted to be the longest-lived mammal, a point discussed in detail later in this chapter. Whereas Gudbjarnason is credited with opening up this field, Hulbert and Pamplona, respectively, are regarded as co-fathers and champions of the membrane pacemaker theory of aging (Couture and Hulbert, 1995; Hulbert, 2003, 2010; Pamplona, Barja, and Portero-Otin, 2002; Pamplona et al., 1998) discussed in detail next.

The main purpose of this chapter is to revise the membrane pacemaker theory in light of advances in the molecular biology of aging. The finding that mtDNA mutations are generated directly by errors of DNA replication rather than by reactive oxygen species (ROS) (Chapter 11) sets the stage for the discussion. We suggest that the membrane is a primary target of oxidative damage both directly and indirectly generating energy stress, causing aging. Even healthy membranes including mitochondrial, plasma, and miniaturized synaptosomal membranes of neurons waste energy in the form of spontaneous leakage or uncoupling of proton, sodium, and potassium gradients. According to this concept the membrane is an important gatekeeper of the cellular energy supply, estimated to be responsible for about half of the energy budget of neurons (see Chapter 6). Thus, an integrated picture of energy deficiency or stress as a driver of the aging process must take into account not only energy production via mitochondria but energy wasted or saved at the membrane level according to the following rule of bioenergetics:

Net cellular energy available to a neuron equals total energy produced minus the energy lost by membrane-based energy uncoupling.

The key point here and as discussed elsewhere in this book is that energy wasted at the membrane level is a significant factor capable of tipping the balance in neurons from energy efficiency to energy deficiency or energy stress. We assert that energy uncoupling mediated by membrane fatty acids can be divided into two primary parts—conformationally generated energy wastage and futile energy cycling caused by oxidative damage.

12.1 ENERGY UNCOUPLING AND THE IMPORTANCE OF THE MEMBRANE AS CRITICAL GATEKEEPER FOR CONSERVING OR WASTING ENERGY

Almost 50 years ago Peter Mitchell won a Nobel Prize for the then revolutionary concept that energy contained in gradients of protons (proton motive energy or chemiosmotic energy) is the primary energy form in cells (see Figure 11.1c). Note in Figure 11.1c that high levels of H^+ on the outside of the membrane represent a form of stored energy that is accessed by ATP synthase to produce ATP. Also note that the respiratory chain housed in a mitochondrial membrane is responsible for maintaining high proton levels outside in contrast to lower levels inside. For many years membranes were taken for granted as an apparently robust physical barrier or permeability barrier against spontaneous leakage of protons downhill, from high to low. Spontaneous leakage wastes or uncouples energy supply because this chemical process robs ATP synthase of proton motive fuel needed to drive its miniature turbine blades in forming ATP. Such energy uncoupling is sometimes referred to as futile proton cycling because protons that leak into the cell must be pumped out again in order to maintain acid balance and ensure a strong proton electrochemical gradient. Protons also enter the cell during ATP production and other cellular processes, including uptake of metabolites. When proton energy uncoupling occurs across membranes, as it inevitably does, the wasted energy appears as heat.

In the 1980s a controversy arose when it was reported that chemically defined membrane vesicles, also called *liposomes* (Figure 12.1) used as surrogates for membranes operating in living cells, spontaneously leak protons (Nichols and Deamer, 1980). The absence of proteins in these vesicles ruled out dysfunctional membrane proteins as being responsible for proton leakage and spotlighted membrane lipids as major contributors of permeability properties. These data were confirmed using membrane vesicles derived from bacteria where membrane lipids were also found to govern proton permeability. These data led membrane biochemists to propose that lipid-based membranes surrounding bacteria, mitochondria, and indeed all life forms display varying degrees of energy uncoupling (see Valentine and Valentine, 2009, for a review of lipid-mediated energy uncoupling; also see Chapter 8). We propose that the magnitude of energy uncoupling at the level of membrane fatty acid structure-function is great enough to contribute to aging and neurodegeneration. Pioneering research on the contribution of membrane fatty acids to futile energy cycling across

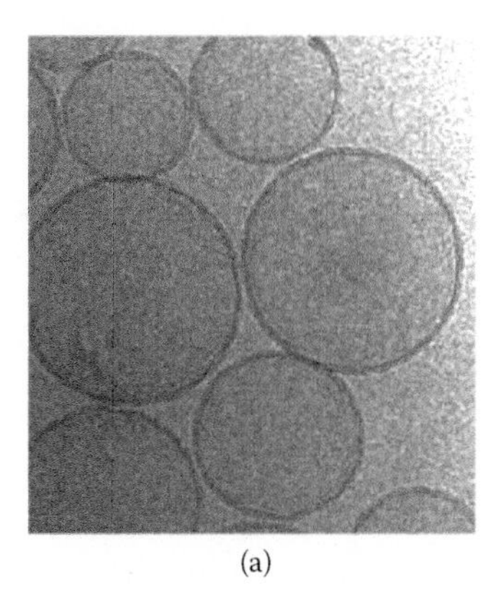

(a)

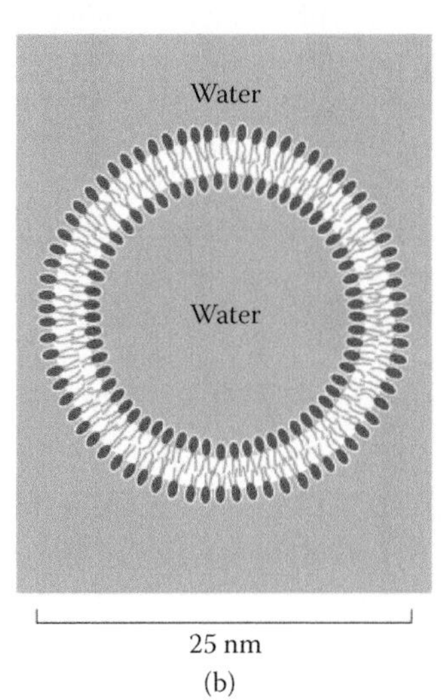

(b)

FIGURE 12.1 Membrane vesicles have been used to show the importance of fatty acid structure-function in proton-sodium bioenergetics. The original finding that membrane fatty acid shape and conformations are key determinants of cation permeability properties of membranes was controversial. It is now thought that futile proton and cation cycling across the lipid portion of the membrane wastes a great deal of energy in human cells and likely contributes to aging and neurodegeneration. (a) Electron micrograph of membrane vesicles (liposomes), which are formed spontaneously from phospholipids. Contrast the absence of membrane proteins in these vesicles with the crowded conditions of synaptic vesicles (Chapter 6). (Reprinted from *Meth. Enzymol.*, 391, Frederick, P., and H. Hulbert, Cryoelectron Microscopy of Liposomes, 431–448, Copyright 2005, with permission from Elsevier.) (b) Bilayer structure of membrane vesicle. (From Alberts, B. et al., *Molecular Biology of the Cell*, Copyright 2008. Reproduced by permission of Taylor & Francis Group, LLC, a division of Informa plc.)

membranes featured saturated fatty acid chains essentially inert toward peroxidation. Thus, it is unlikely that cation leakage in liposomes used as research tools for these studies was caused by oxidative damage to membrane architecture. That is to say that membrane or fatty acid conformational dynamics can account for cation leakage in these experiments. However, the picture of cation uncoupling across membranes becomes more complicated when membranes are blended with DHA or polyunsaturated phospholipids. In this case covalent modification of fatty acid chains by oxidation can change permeability properties significantly.

Energy-transducing membranes including plasma and mitochondrial membranes (Porter and Brand, 1993) are now considered to be inherently leaky for protons and sodium or converted to a leaky state by peroxidation. This means that membrane surface area becomes a fundamentally important parameter in gauging the magnitude of energy lost by futile cation cycling. For example, the surface area of the inner membrane of mitochondria is roughly fivefold greater than that of the smooth outer membrane. Also in cells with numerous mitochondria, the combined surface area of inner mitochondrial membranes will likely greatly surpass that of the plasma membrane. This leads to the idea that energy losses across mitochondrial membranes might be much greater compared to the plasma membrane. However, neurons appear to be an exception because of their vast membrane surface area represented by axons, dendrites, synapses, and synaptic vesicles. Thus, an extraordinary amount of energy wasted across their plasma membranes may be the price paid for a fast neural communication network essential for human survival. We have chosen a series of case histories to illustrate the importance of membrane unsaturation in life spans of a hierarchy of animals ranging from flies to whales.

12.2 INSECTS DRAMATICALLY INCREASE LIFE SPAN BY DIETARY MANIPULATION AND OPTIMAL TUNING OF POLYUNSATURATED FATTY ACID SYNTHESIS

Most terrestrial insects including flies, bees, and butterflies contain no detectable DHA and often only traces of EPA in their membranes (see Chapter 4); thus the original DHA pacemaker theory of aging does not apply to these insects. Instead, the more generalized membrane polyunsaturation theory of aging (Hulbert et al., 2007) seems more likely to apply to terrestrial insects including the fruit fly, *Drosophila*, which is used as an important model organism for research on aging. Membrane lipid dynamics and molecular biology of membrane fatty acids have not been studied in great detail in *Drosophila*, but a picture can be pulled together from studies of other insects. Bees, for example, appear to extend the life span of their queen by delivering more saturated fatty acids in the form of "royal jelly" (Haddad, Kelbert, and Hulbert, 2007). In contrast, foraging bees, which have the shortest life span in the hive, upregulate levels of polyunsaturated chains presumably as an adaptation for flight. Thus, in bees gross membrane composition in terms of levels of saturated fatty acids, monounsaturated fatty acids, and polyunsaturated fatty acids, the latter mainly 18:2 and 18:3, can vary dramatically. For example, 18:2n6 levels in workers are 10 times higher than levels in the queen, and levels of 18:3n6 within different classes of workers also show a 10-fold variation. These data suggest that foraging bees upregulate membrane unsaturation levels for flight in contrast to the queen, who is fed more saturated chains by her nurses, a diet believed to prolong her reproductive life for supplying eggs to sustain the colony. A queen ant fed royal jelly for her entire life can live up to 20 years (versus 1 year for a queen bee), 20 times longer than her worker ants.

Given that bees and ants appear to use dietary manipulation to regulate by 10- to 20-fold the life span of the queen versus a foraging worker, we wondered if other insects including *Drosophila* might modulate their life spans dependent on other dietary or environmental signals. The following environments and conditions have been reported in the literature to extend life span in *Drosophila*:

- Cool temperatures
- Dietary caloric restriction
- Mild stresses including heat shock, cold stress
- Hypergravity
- Low levels of radiation
- Reduced flight activity

The strongest environmental stimulus that affects life span in *Drosophila* is environmental temperature, with flies cultured at 18°C living twice as long as those grown at 25°C. Temperature also modulates heart rate from 271 beats per minute at 22°C compared to 511 beats per minute at 38°C. Shifts in gross membrane fatty acid composition as seen in honeybees and believed to accompany such environmentally mediated extension of life span have not been documented for *Drosophila*. However, studies of membrane biochemistry in *Drosophila* have led to a remarkable working model that links environmental stimulus, membrane fatty acid unsaturation, and life span. This model is based on data from the pioneering research done by Schlame and colleagues on the molecular biology of cardiolipin (CL) synthesis in *Drosophila* (Malhotra et al., 2009; Schlame et al., 2005; Xu et al., 2009). These researchers found that the prominent species of CL in *Drosophila* is $(18{:}2)_4$-CL (Schlame et al., 2005). They pointed out that even though 18:2 is a minor fatty acid in *Drosophila* membrane phospholipids, a mechanism triggered by environmental signals has evolved to recruit the relatively low level of 18:2 in *Drosophila* membranes making it available to generate a CL molecular species with up to four 18:2 chains. The specificity for CL molecular species does not reside in *Drosophila* taffazin, an enzyme that catalyses synthesis of $(18{:}2)_4$-CL. Rather unknown environmental signals are proposed to modulate the synthesis of molecular species including $(18{:}2)_4$-CL (Malhotra et al., 2009). These data suggest that the environment somehow directs the remodeling of CL in *Drosophila* by a mechanism that mobilizes existing 18:2 chains in contrast to newly synthesized chains. The net effect is roughly a doubling of the number of double bonds in CL from about four to eight without the need to increase the overall level of polyunsaturated chains, which become targets of oxidative damage.

Dependent on the following assumptions, these data are consistent with the membrane pacemaker theory of aging applied to *Drosophila*:

- Temperature is one of the signals modulating unsaturation levels and molecular species of cardiolipin.
- Cool temperature (17°C) increases levels of cardiolipin polyunsaturation (e.g., $(18:2)_4$-CL) but does not appreciably alter gross levels of unsaturated fatty acids compared to growth at 25°C.
- Cool temperatures are proposed to decrease rates of membrane peroxidation, helping to double longevity, according to the membrane pacemaker model.

A linkage between temperature and aging in hummingbirds is discussed shortly, and in Chapters 11 and 17 we apply CL pacemaker theory to help explain neurodegeneration in humans. The role of DHA-CL as an assassin of senescing colonic epithelial cells has already been discussed in Chapter 2, whereas the benefits of DHA-CL in enabling extreme flight of hummingbirds are discussed below. Also, growth of the nematode *Caenorhabditis elegans* at 15°C has been found to significantly increase life span (Huang, Xiong, and Kornfeld, 2004) in spite of the fact that EPA levels in membranes rise sharply at this temperature (Tanaka et al., 1999). According to one scenario this rise in membrane unsaturation might be counterbalanced by decreased lipid peroxidation due to cool temperatures along with decreased rates of respiration.

There is another aspect of aging in *Drosophila* that might be explained by the revised membrane pacemaker concept. The heart organ of *Drosophila*, which pumps hemolymph through the body, is considered to be perhaps the hardest-working organ with the hardest-working mitochondria in *Drosophila*. It is known that *Drosophila* is subject to serious "heart disease," raising the question of whether the heart might be a critical pacemaker for aging in this animal (Ocorr et al., 2007; Vogler, Bodmer, and Akasaka, 2009). We propose that the level of CL unsaturation reaches a peak in mitochondria of cardiac cells of *Drosophila*. The mitochondrial membranes of heart cells in *Drosophila* are expected to be enriched with $(18:2)_4$-CL or $(16:2)_4$-CL, perhaps generating unusually high levels of oxidative stress and causing energy stress and early apoptosis of heart cells compared to cells of most other organs.

Upregulation of the transcription factor dFOXO in *Drosophila* has been found to increase life span, obviously extending the functional life of the heart (Hwangbo et al., 2004). The role of FOXO in human aging is discussed in Chapter 17. There are similarities between the roles of polyunsaturation in the aging of the heart of *Drosophila* and the roles of DHA in aging of the human brain, the latter also discussed in more detail in Chapter 17.

12.3 FIRST LESSON FROM HUMMINGBIRDS: DHA TURBOCHARGES MITOCHONDRIA

The study of the bioenergetics of hummingbird flight introduced in Chapter 11 provides further lessons on energy stress and the importance of mitochondria in aging. We suggest that the adaptations cited in Chapter 11 necessary for hummingbird flight—robust heart and circulation system, huge breast muscle, massive numbers of mitochondria per cell, and double the level of cristae within individual mitochondria—do not fully explain remarkable wing beats of up to 200 per second. We propose that at least three additional adaptations in bioenergetics play important roles in enabling the extreme flight of hummingbirds as follows:

- Elevated temperatures of around 40.6°C to 41.7°C (compared to 37°C for humans) are a mechanism to speed up energy production. (Note that the temperature of breast muscle often plummets during the chill of the night.)
- DHA in mitochondrial membranes works to supercharge the electron transport chain, maximizing energy production for flight.
- DHA-CL in flight muscle mitochondria is the molecular species enabling extreme flight.

It is well known that rates of biochemical processes, including energy production by the electron transport chain, predictably rise with temperature. Hulbert and colleagues (Infante, Kirwan, and Brenna, 2001) have proposed that DHA increases rates of respiration in hummingbird mitochondria (Hulbert and Else, 1999). We have developed a molecular model to explain how DHA-cardiolipin supercharges electron transport and energy production as an adaptation for extreme flight (Valentine and Valentine, 2009). A corollary to the second point is that the absence of DHA in mitochondrial membranes, typical of most animals including humans, is a sign that these mitochondria are not subjected to a regime of extreme energy stress such as seen in hummingbirds. Thus, these less active mitochondria are expected to divide at a slower rate and consequently accumulate mtDNA mutations at a slower rate.

12.4 SECOND LESSON FROM HUMMINGBIRDS: THE HARDEST WORKING MITOCHONDRIA WEAR OUT FASTEST

According to the rate-of-living theory, the hardest working mitochondria wear out more quickly and consequently undergo, over a lifetime, more rounds or cycles of division including more replication cycles of their mtDNA (for an historical perspective on the rate-of-living theory see Speakman, 2005; Speakman et al., 2002). According to the revised mtDNA theory of aging (Chapter 11), each round of replication exposes DNA to natural errors or mutations that occur and accumulate as new strands of DNA are being laid down. Though the chance of a mutation is small at each round of DNA replication, over a lifetime this probability increases greatly. Also keep in mind that vast numbers of mitochondria undergo division. For example, the total number of mitochondria operating in neurons of the human brain is estimated to be at least 100 trillion. Mitochondria in hummingbird flight muscle have evolved to accommodate extreme flight, which in turn depends on extreme bioenergetics along with specialized membranes to support hyper-respiration of these mitochondria. We suggest that hummingbirds evolved their unique mitochondria to keep up with energy demands for powering extreme flight. Thus, during flight these mitochondria are considered to operate in an *energy stressed* state. We predict that mitochondria for extreme flight and any mitochondria forced to operate near maximum capacity will wear out more quickly and require more cycles of division compared to mitochondria operating well below maximum wattage. The point is that each division of mitochondria increases the probability of the accumulation of mutations in mtDNA—mutations that eventually diminish cellular energy supplies toward a critical minimum threshold below which cells undergo apoptosis. This concept is essentially a molecular restatement of the rate-of-living theory, incorporating recent advances in the molecular biology of aging.

12.5 THIRD LESSON FROM HUMMINGBIRDS: "ENERGY-STRESSED" MITOCHONDRIA ACCUMULATE MUTATIONS FASTER

The relatively high rate of mtDNA mutations found in hummingbirds (Bleiweiss, 1998) is consistent with the revised mtDNA theory of aging. As background, hummingbirds must feed voraciously and frequently simply to stay alive. Thus, any serious reduction of energy efficiency of flight muscle mitochondria caused by aging might quickly end the life of these birds. In the world of ecology it is difficult to find a better example of the consequences of age-dependent decline in mitochondrial energy supply and its corresponding effects on the life span of an animal.

Unfortunately, the very adaptation that powers their amazing wing beats is predicted to greatly shorten their life span (Hulbert et al., 2007). Life span among hummingbirds has a lower range of four to five years and an upper range of about twelve. A medium-sized hummingbird (broad-billed) banded and released in the wild holds the current longevity record of 12 years. Broad-billed hummingbirds have adapted to high elevations, and their short migratory patterns take them to similar elevations and temperature patterns throughout their relatively short North–South seasonal migration route.

Hummingbirds adapted to life at high elevations display decreased rates of mutations in their mtDNA (Bleiweiss, 1998; this paper lists references of earlier work on the effect of the environment on mutation frequencies). These data suggest that the high elevation environment directly influences rates of mtDNA mutations, in this case prolonging life span. One scenario to explain the contribution of elevation to rates of mtDNA mutations is that at high elevation, nighttime temperatures almost always plummet. This would allow high-elevation hummingbirds to dependably harness extreme flight to acquire nectar during mild daytime temperatures while spending a large part of their life in a state of deep torpor brought on by consistently low nighttime temperatures. Note that hummingbirds require about 20 minutes to awaken from the torpid state. Body temperature often drops dramatically from 40.6°C to 15.6°C. Also note that the torpid state caused by rapidly dropping temperatures characteristic of the Sonoran ecosystem also drastically lowers respiration rates and energy consumption. Thus, the many hours each day spent in deep nocturnal torpor are expected to lower rates of oxidation of DHA, saving energy and in turn helping lower spontaneous rates of mtDNA mutations. These data suggest that a lifetime of low nighttime temperatures inducing a deep, torpid state might significantly increase the life span of high-elevation hummingbirds. In essence, this example raises the possibility that high-elevation hummingbirds have adapted to use low temperature as an "avoidance" mechanism to protect their DHA membranes and increase life span. As mentioned above *Drosophila* seem to share properties with hummingbirds in that cool temperatures of 17°C compared to 25°C can double their life span.

12.6 LESSONS FROM SMALL MAMMALS: LESS DHA, LONGER LIFE SPAN

Hulbert and colleagues have proposed that relatively high DHA levels in membranes of mice predict the short life span of three to four years for this small rodent (Hulbert, 2005; Hulbert and Else, 1999; Hulbert, Faulks, and Buffenstein, 2006). Note that the DHA pacemaker theory of aging also correctly predicts the long life span of humans based on the fact that DHA levels in human membranes are much lower compared to mice. Independent research by Pamplona and associates (Magwere et al., 2006; Pamplona, Barja, and Portero-Otin, 2002; Pamplona et al., 1998) led to a modification of the DHA pacemaker theory. Pamplona's laboratory found that the levels of membrane polyunsaturated fatty acids in mammals ranging from mouse to man also reliably predict life spans. Again an inverse relationship between membrane polyunsaturation and longevity is recorded. Because DHA is present in small amounts in most classes of human cells as well as cells of other larger mammals, the polyunsaturation pacemaker is considered as a more generalized mechanism to explain the membrane's contribution to aging (Hulbert et al., 2007).

Two tiny mammals shown in Figure 12.2—naked mole rats versus mice—are at the center of a critical test of the membrane pacemaker theory of aging, with implications for humans (Hulbert et al., 2006). Naked mole rats are mouse-sized mammals that exhibit exceptional longevity. This small rodent (~35 g) has a captive maximum life span of >28 years, which is the longest known for any rodent species and is eight to nine times greater than the typical three- to four-year life span of similar-sized mice. Naked mole rats not only exhibit extraordinary life spans, but these subterranean Northeast African rodents also show vitality and continue to breed well into their third decade of life. Due to their amazing longevity, naked mole rats are receiving increasing attention as a mammalian model to evaluate mechanisms that impact longevity and the aging process in humans.

We next look more closely at data generated during the mouse/naked mole rat study (Hulbert et al., 2006). A comprehensive analysis of the DHA content of a variety of tissues of naked mole rats is reported and contrasted with simultaneous studies of mice (Figure 12.2). The startling finding is that naked mole rats have only about one ninth the level of DHA compared to mice. For example, phospholipids from skeletal muscle, heart, kidney, and liver of naked mole rats contained an average 2.2 percent DHA (range 0.6 to 6.5 percent) versus 19.3 percent (range 11.7 to 26.2 percent) in mice. Even brain DHA content, which is constant among most mammalian species, is about 25 percent lower in naked mole rats compared to mice. This drop in brain DHA might be linked to the fact that naked mole rats are virtually blind, having evolved a fold of skin that prevents direct sunlight from reaching the eye. These subterranean animals retain only the ability to detect day versus night.

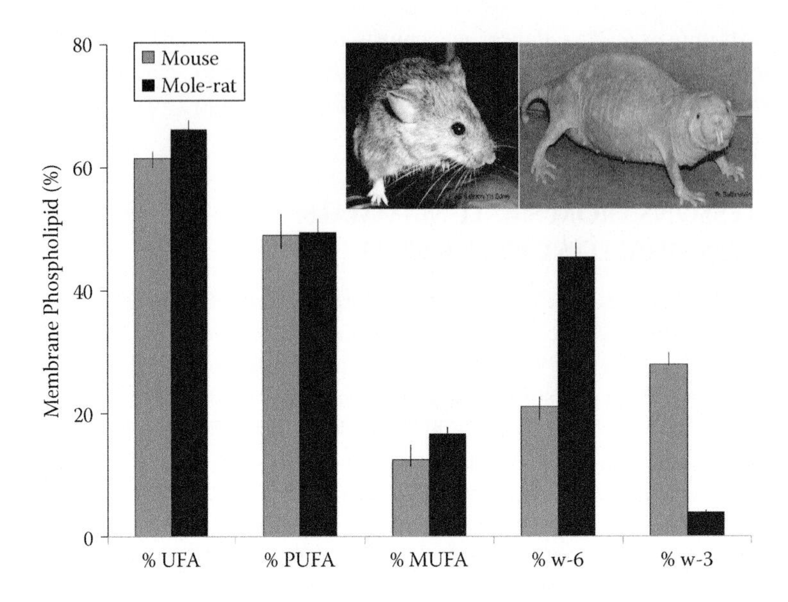

FIGURE 12.2 Membrane pacemaker theory of aging has been tested in naked mole rats versus mice. The extraordinarily long life span of the naked mole rat, a mouse-sized mammal, is explained by a dramatic decrease in DHA content, slowing aging compared to mice. That is, naked mole rats live eight to nine times longer than mice and have only about one ninth the level of DHA in their membranes. Even the level of DHA in the brain is significantly lower in the naked mole rat. According to the membrane pacemaker theory, high membrane DHA content and high overall membrane unsaturation levels of mice are consistent with their relatively short life span. (Courtesy of Yael Edrey and Rochelle Buffenstein.)

The percentage of total unsaturated chains in membranes between naked mole rats and mice is similar, which means that a major variable is DHA. Once again, the most interesting result is that naked mole rats have only about 10 percent as much DHA and live roughly nine times longer than mice. Hulbert and colleagues (2007) developed a model to explain the membrane's role as the pacemaker for aging. Their proposed mechanism is a modification of the conventional oxidative stress theory of aging (Figure 12.3). According to the conventional theory of aging, reactive oxygen species generated during electron transport in mitochondria are proposed to directly damage DNA, membranes, and proteins causing aging. As shown in Figure 12.3, DHA-polyunsaturated membranes are proposed to contribute to oxidative stress by increasing the size of the pool of reactive oxygen species, causing mutations in mtDNA and damaging other cellular constituents, ultimately decreasing life span. We propose a revision of the membrane pacemaker theory of Hulbert et al. (2007). Our model links oxidative stress to energy stress to aging. We believe that energy stress caused by oxidative stress acts synergistically with mutations of mtDNA to promote aging. Thus, in our model we place more emphasis on DHA and polyunsaturation as triggers for membrane damage, resulting in energy uncoupling powerful enough to help drive the aging process.

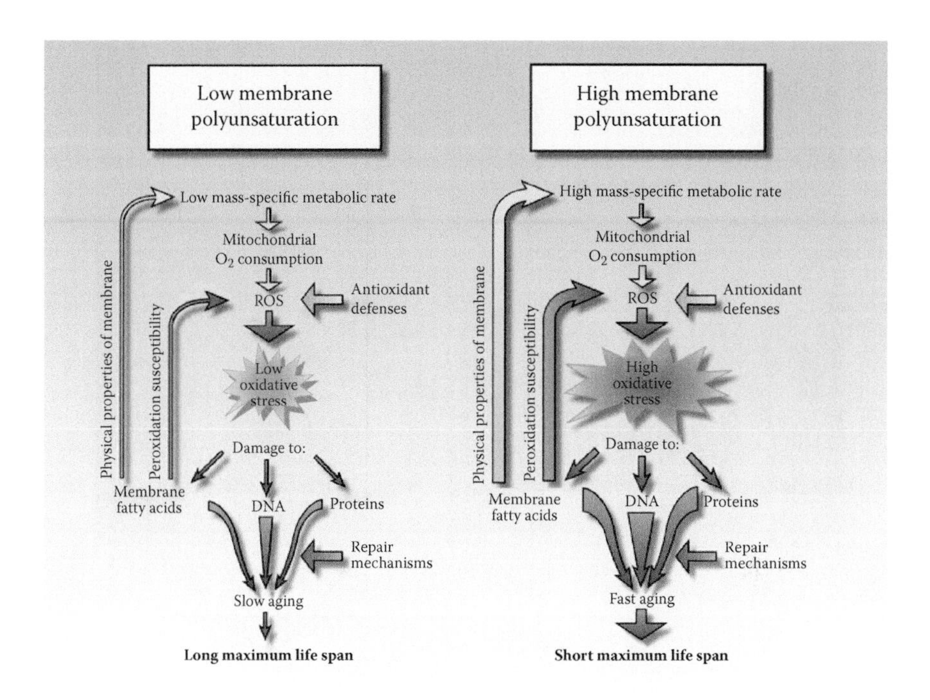

FIGURE 12.3 Peroxidation of unsaturated membranes is proposed as a mechanism to explain the membrane pacemaker theory of aging. The original membrane pacemaker model proposed by Hulbert and Pamplona states that increased levels of unsaturated chains in membranes, especially DHA or other polyunsaturated molecules, elevate rates of membrane peroxidation, causing aging. In this model peroxidation of polyunsaturated phospholipids boosts the pool of reactive oxygen species (ROS) to directly drive mutations in mitochondrial DNA. This mechanism is a derivative of the famous mitochondrial theory of aging featuring ROS as mutagen-generating mutations in mtDNA. This concept now requires some modification as discussed in the text. (Reproduced by permission of the American Physiological Society, Hulbert et al., 2007, *Physiol. Rev.*, 87: 1175–1213.) **(See color insert.)**

12.7 MEMBRANE PACEMAKER MODEL IS CONSISTENT WITH THE BOWHEAD WHALE AS LONGEST LIVED MAMMAL

Hulbert and colleagues and Pamplona and coworkers, respectively, have gone far beyond humans, mice, naked mole rats, and bees and have applied their membrane pacemaker theory to predict life spans in an increasing number of animals (Buttemer, Battam, and Hulbert, 2008; Hulbert, Beard, and Grigg, 2008). These researchers find repeatedly that the membrane pacemaker theory correctly predicts the life span of mammals both large and small. Gudbjarnason (Gudbjarnason, 1989; Gudbjarnason et al., 1978) suggested that whales would live longer than humans (Figure 12.4a). The small hummingbird shown in Figure 12.4b for comparison has a life span of about four to five years. We next discuss the claim that a giant baleen whale living in Alaska

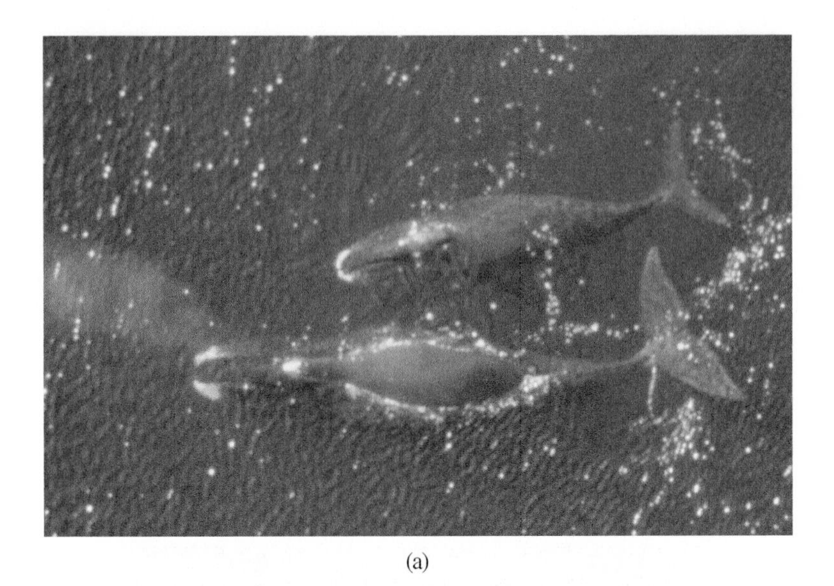

(a)

(b)

FIGURE 12.4 Large baleen whales as the longest living mammals? (a) Hydrogenation of DHA and other highly unsaturated fatty acids (HUFAs) by ruminant bacteria in the whale's forestomach along with selective targeting of DHA to specialized membranes may contribute to longevity of the whale. Bowhead whales might live up to two centuries. (Photo courtesy of the National Oceanic and Atmospheric Administration/Department of Commerce.) (b) The life span of a small hummingbird is about four to five years in contrast to more than a century for bowhead whales. Note that membranes of hummingbirds, unlike those of bowhead whales, are highly enriched with DHA in agreement with membrane unsaturation being a pacemaker of aging. (Photo courtesy of Claudia Bodmer.)

is the oldest living mammal. The bowhead is an Arctic right whale with a large, bow-shaped head that composes up to 40 percent of its body. In total, bulk bowheads are second only to blue whales. The bowhead has made numerous adaptations for life in Arctic waters, where it resides year around. The mouth of the bowhead, which gives rise to its name, is the largest and strongest of any mammal. The bowhead feeds by continuously swimming forward with its enormous baleen-covered mouth open. The baleen filter system allows grazing on copepods or zooplankton, which are the smallest yet most numerous animals in the Arctic food web. The mouth has a large upturning lip on the lower jaw that reinforces baleen plates in its lower jaw and prevents buckling due to the strong water pressure exerted on the jaw as it feeds.

The bowhead is the only baleen whale that spends its entire life in extremely cold Arctic waters, with the Alaskan population spending the winter months in the southwestern Bering Sea. The group migrates northward in the spring, following openings in the pack ice, into the Chukchi and Beaufort seas, hunting zooplankton. Like other right whales, bowheads swim slowly, and due to their approximate 1 foot thickness of blubber surrounding their bodies, they are extremely cold tolerant, carry a large reserve of stored energy preventing starvation when not feeding and have great buoyancy. During the whaling era this buoyancy was a curse, as diving to escape whalers is difficult, and once harpooned, bowheads remained on the surface. For whalers this was the "right whale."

Analyses of bowhead whale gastrointestinal content (Hazard and Lowry, 1984) show that at least in the Alaskan Beaufort Sea, zooplankton such as copepods and euphausiids are the principal prey. In samples of gastrointestinal contents from 15 whales these planktonic organisms made up 73 to 99 percent of the stomach contents by volume. These data are consistent with the unique mouth adaption of bowheads, which allows them to filter large amounts of zooplankton-rich waters to sustain their large bodies and energy reserves. Records show that other prey such as benthic organisms are also taken. Zooplankton, however, seem to be the main food source for bowheads, especially in the spring. The often-observed absence of any food in the stomachs of bowheads suggests that these whales depend on their enormous supply of blubber for energy when food is scarce, especially in winter months. Bowheads spend their entire lives in Arctic waters where much of the year is dark and not favorable for growth of their favorite prey. Thus, these whales can be considered to have adapted to a state of natural dietary restriction with intense feeding in spring and summer followed by extended periods when food is scarce. Their giant storage of blubber makes this pattern possible and might favor a long life span.

Jeffrey Bada of the Scripps Institute of Oceanography is quoted as saying in *National Geographic News* (July 13, 2006, news article) that about 5 percent of bowhead whales are over 100 years of age and in some cases 160 to 180 years old (Roach, 2006). He points out that these are truly aged animals, perhaps the most aged on earth. These estimates are based on chemical analysis involving levels of amino acid isomerization seen in proteins in the nucleus of the eye lens (George et al., 1999). Lens proteins when first synthesized during early growth and development contain only *l*-amino acids. These proteins once made remain permanently. This allows time for the chemical process of amino acid racemization ($l \rightarrow d$ amino acids) to occur, allowing Bada to use the ratio of l to d amino acids as a time clock for determining life span. These chemical studies, originally questioned by some researchers as not being biologically definitive, have since been reinforced by the work of wildlife biologist Craig George of the Alaska Department of Wildlife Management. In 1990 George examined several whales taken during an annual hunt by Inupiat Eskimos who have eaten whales as a major food source for centuries and are still permitted to harvest bowheads (George et al., 1999). George found stone harpoons imbedded in whales, a method of hunting replaced by metal harpoons around 1860 to 1870. This finding raised the possibility that bowheads may reach 100 years of age. Similar findings of antique ivory spear points in living whales in 1993, 1995, 1999, and 2007 triggered chemical studies by Bada as described above, searching for confirmation of bowhead ages. In May 2007, a bowhead caught off the Alaskan coast was discovered with the head of an explosive harpoon embedded deep under an old wound in its neck. Examination dated this arrow-shaped projectile manufactured in New Bedford, Massachusetts, to around 1890, a period when New Bedford was a whaling center. This suggests that this whale survived a hunt a century ago, placing its age at about 115 to 120 years. This value can be compared with the five whales dated by Bada's isomerization technique to be 91, 135, 159, 172, and the oldest at 211 years old. The oldest might have been swimming off the Alaskan coast during the presidency of Thomas Jefferson. Japanese scientists have dated blue whales up to 110 years of age and a fin whale at 114 using a different method not applicable to bowheads.

George suggested that the bowhead's long life span might be due to its environment—cold water without abundant food available for long periods—forcing it to maintain a great body mass, an effective system for fat storage, an efficient insulation mechanism to keep warm, and often a starvation diet. The stress of living in Arctic waters may nurture the whale's pattern of slow growth and long life. Does the membrane pacemaker theory explain the apparent long life span of bowheads?

Zooplankton common in the diet of bowheads are considered to be among the richest sources of DHA/EPA, comprising about 35 percent of total fatty acid in the diet (see data of Napp et al., 2006, discussed in Valentine and Valentine, 2009). In contrast, DHA/EPA content of blubber is reduced about 80 percent (Reynolds, Wetzel, and O'Hara, 2006). The reverse is true of monounsaturates whose content rises from about 5 percent in the diet to about 30 percent in blubber. These data can be explained if the bowhead and its copepod prey team up to dramatically, metabolically alter unsaturation levels of dietary fatty acids (Budge et al., 2008; Christie, 1981). Data on dietary habits of bowheads suggest that the whale's digestive system in conjunction with lipid dynamics and storage of lipids in its prey metabolically idealize the composition of fatty acids in a manner consistent with the membrane pacemaker theory of aging. Much of the responsible biochemistry has yet to be resolved. But one explanation is that whales, which evolved from ruminants (nearest living relatives are hippopotami), have retained powerful hydrogenation reactions carried out by gut symbionts that convert DHA and polyunsaturated fatty acids in the forestomach to more saturated forms. Hydrogenation of unsaturated fatty acids in the whale stomach prior to entering the bloodstream might be a win-win-win situation, first providing unsaturated fatty acids in bulk as electron acceptors supporting higher rates of anaerobic respiration and robust growth of microbial symbionts, increasing caloric content of hydrogenated storage lipids (i.e., blubber) and perhaps ultimately increasing longevity. Note that bacterial biomass is a source of food for ruminants. The meat of a bowhead is lean and has a fatty acid composition consistent with both premetabolism of highly unsaturated fatty acids (HUFAs) and the operation of mechanisms targeting DHA to specific cells and membranes and away from most other cells, as in humans. We interpret the data of Gudbjarnason et al. (Gudbjarnason, 1989; Gudbjarnason et al., 1978) on the fatty acid composition of whale heart tissue as evidence that DHA-enriched, molecular species of cardiolipin are virtually absent from cardiac mitochondria and likely most cells in the whale. The hard shells of zooplankton consumed in mass by bowheads might also favor this whale's longevity. Chitin, the polymeric material forming the shell of zooplankton, is composed of a repeating unit of glucosamine, an energy-rich substrate released by symbionts in the rumen. Gut microbes are known to hydrolyze chitin and release large amounts of glucose, a substrate that provides additional reducing power necessary to hydrogenate and restructure dietary fatty acids in the whale's forestomach, perhaps a biochemical process favoring longevity. Thus, it is proposed that dietary HUFAs including DHA and EPA act as terminal electron acceptors enhancing the growth of ruminant bacteria, which convert these chains to more saturated species, thus contributing multiple benefits to whales, including longevity.

In summary, we suggest that energy stress created by both conformational dynamics of polyunsaturated chains and oxidative damage to membranes are missing links in understanding aging and age-dependent diseases. The membrane is proposed as a second but interlocking pacemaker along with the mtDNA mutational pacemaker, both operating on an energy standard. This concept is a revision of the membrane pacemaker theory proposed by Hulbert and Pamplona and is based on established principles of bioenergetics and membrane biochemistry. Dietary and environmental modulation of life span can be explained by the revised membrane pacemaker theory of aging. Dual energy pacemakers are proposed to determine the rate of aging. The dual pacemaker concept of aging raises a number of questions requiring further research. Some of the following questions will be addressed in later chapters:

- What regulatory cascades govern up- and down-modulation of membrane unsaturation?
- How important is cardiolipin unsaturation as a pacemaker of aging?
- Is the dual pacemaker theory consistent with a linear process of energy stress during aging?
- Is the membrane the most important target of oxidative stress in neurons?
- Is neurodegeneration a special case of aging?
- How important is membrane peroxidation mediated by hyperactive macrophages, for neurodegeneration or for aging?

REFERENCES

Bleiweiss, R. 1998. Slow rate of molecular evolution in high elevation hummingbirds. *Proc. Natl. Acad. Sci. USA* 95:612–616.

Budge, S. M., A. M. Springer, S. J. Iverson et al. 2008. Blubber fatty acid composition of bowhead whales, *Balaena mysticetus*: Implications for diet assessment and ecosystem monitoring. *J. Exp. Mar. Biol. Ecol.* 359:40–46.

Buttemer, W. A., H. Battam, and A. J. Hulbert. 2008. Fowl play and the price of petrel: Long-living *Procellariiformes* have peroxidation-resistant membrane composition compared with short-living *Galliformes. Biol. Lett.* 4:351–354.

Christie, W. W. 1981. *Lipid Metabolism in Ruminant Animals.* New York: Pergamon Press.

Couture, P., and A. J. Hulbert. 1995. Membrane fatty acid composition of tissues is related to body mass of mammals. *J. Membr. Biol.* 148:27–39.

George, J. C., J. Bada, J. Zeh, L. Scott et al. 1999. Age and growth estimates of bowhead whales (*Balaena mysticetus*) via aspartic acid racemization. *Can. J. Zool.* 77:571–580.

Gudbjarnason, S. 1989. Dynamics of n-3 and n-6 fatty acids in phospholipids of heart muscle. *J. Intern. Med. Suppl.* 731:117–128.

Gudbjarnason, S., B. Doell, G. Oskardottir et al. 1978. Modification of cardiac phospholipids and catecholamine stress tolerance. In: deDuve, C., and O. Hayaishi, editors. *Tocopherol, Oxygen and Biomembranes.* Amsterdam: Elsevier, pp. 297–310.

Haddad, L. S., L. Kelbert, and A. J. Hulbert. 2007. Extended longevity of queen honey bees compared to workers is associated with peroxidation-resistant membranes. *Exp. Gerontol.* 42:601–609.

Hazard, K. F., and L. F. Lowry. 1984. Benthic prey in a bowhead whale from the northern Bering Sea. *Arctic* 37:166–168.

Huang, C., C. Xiong, and K. Kornfeld. 2004. Measurements of age-related changes of physiological processes that predict lifespan of *Caenorhabditis elegans. Proc. Natl. Acad. Sci. USA* 101:8084–8089.

Hulbert, A. J. 2003. Life, death and membrane bilayers. *J. Exp. Biol.* 206:2303–2311.

Hulbert, A. J. 2005. On the importance of fatty acid composition of membranes for aging. *J. Theor. Biol.* 234:277–288.

Hulbert, A. J. 2010. Metabolism and longevity: Is there a role for membrane fatty acids? *Integr. Comp. Biol.* (doi:10.1093/icb/icq007).

Hulbert, A. J., L. A. Beard, and G. C. Grigg. 2008. The exceptional longevity of an egg-laying mammal, the short-beaked echidna (*Tachyglossus aculeatus*) is associated with peroxidation-resistant membrane composition. *Exp. Gerontol.* 43:729–733.

Hulbert, A. J., and P. L. Else. 1999. Membranes as possible pacemakers of metabolism. *J. Theor. Biol.* 199:257–274.

Hulbert, A. J., S. C. Faulks, and R. Buffenstein. 2006. Oxidation-resistant membrane phospholipids can explain longevity differences among the longest-lived rodents and similarly-sized mice. *J. Gerontol. A Biol. Sci. Med. Sci.* 61:1009–1018.

Hulbert, A. J., R. Pamplona, R. Buffenstein et al. 2007. Life and death: Metabolic rate, membrane composition, and life span of animals. *Physiol. Rev.* 87:1175–1213.

Hwangbo, D. S., B. Gershman, M. P. Tu et al. 2004. Drosophila dFOXO controls lifespan and regulates insulin signalling in brain and fat body. *Nature* 429:562–566.

Infante, J. P., R. C. Kirwan, and J. T. Brenna. 2001. High levels of docosahexaenoic acid (22:6n-3)-containing phospholipids in high-frequency contraction muscles of hummingbirds and rattlesnakes. *Comp. Biochem. Physiol. B Biochem. Mol. Biol.* 130:291–298.

Magwere, T., R. Pamplona, S. Miwa et al. 2006. Flight activity, mortality rates, and lipoxidative damage in Drosophila. *J. Gerontol. A Biol. Sci. Med. Sci.* 61:136–145.

Malhotra, A., Y. Xu, and M. Renetal. 2009. Formation of molecular species of mitochondrial cardiolipin. 1. A novel transacylation mechanism to shuttle fatty acids between *sn*-1 and *sn*-2 positions of multiple phospholipid species. *Biochim. Biophys. Acta* 1791:314–320.

Napp, J. M., L. E. Schaufler, G. L. Hunt, Jr. et al. 2006. Summer food web structure in the eastern Bering Sea: Fatty acid composition of plankton, fish, and seabirds around the Pribilof Islands. Presented at PICES 15th Annual Meeting, October 13–22, 2006, Yokohama, Japan.

Nichols, J., and D. Deamer. 1980. Net proton-hydroxyl permeability of large unilamellar liposomes measured by an acid-base titration technique. *Proc. Natl. Acad. Sci. USA* 70:2038–2042.

Ocorr, K., L. Perrin, H. Y. Lim et al. 2007. Genetic control of heart function and aging in Drosophila. *Trends Cardiovasc. Med.* 17:177–182.

Pamplona, R., G. Barja, and M. Portero-Otín. 2002. Membrane fatty acid unsaturation, protection against oxidative stress, and maximum life span: A homeoviscous-longevity adaptation? *Ann. NY Acad. Sci.* 959:475–490.

Pamplona, R., M. Portero-Otín, D. Riba et al. 1998. Mitochondrial membrane peroxidizability index is inversely related to maximum life span in mammals. *J. Lipid Res.* 39:1989–1994.

Porter, R. K., and M. D. Brand. 1993. Body mass dependence of H^+ leak in mitochondria and its relevance to metabolic rate. *Nature* 362:628–630.

Reynolds, III, J. E., D. L. Wetzel, and T. M. O'Hara. 2006. Human health implications of omega-3 and omega-6 fatty acids in blubber of the bowhead whale (*Balaena mysticetus*). *Arctic* 59:155–164.

Roach, J. 2006. Rare whales can live to nearly 200, eye tissue reveals. *National Geographic News* (http://news.nationalgeographic.com/news/2006/07/060713-whale-eyes.html).

Schlame, M., M. Ren, Y. Xu et al. 2005. Molecular symmetry in mitochondrial cardiolipins. *Chem. Phys. Lipids* 138:38–49.

Speakman, J. R. 2005. Body size, energy metabolism and life span. *J. Exp. Biol.* 208:1717–1730.

Speakman, J. R., C. Selman, J. S. McLaren et al. 2002. Living fast, dying when? The link between aging and energetics. *J. Nutr.* 132(6 Suppl 2):1583S–1597S.

Tanaka, T., S. Izuwa, K. Tanaka et al. 1999. Biosynthesis of 1,2-dieicosapentaenosyl-sn-glycero-3-phosphocholine in *Caenorhabditis elegans*. *Eur. J. Biochem.* 263:189–194.

Valentine, R. C., and D. L. Valentine. 2009. *Omega-3 Fatty Acids and the DHA Principle*. Boca Raton, FL: Taylor & Francis.

Vogler, G., R. Bodmer, and T. Akasaka. 2009. A *Drosophila* model for congenital heart disease. *Drug Discov. Today Dis. Models* 6:47–54.

Xu, Y., S. Zhang, and A. Malhotraetal. 2009. Characterization of tafazzin splice variants from humans and fruit flies. *J. Biol. Chem.* 284:29230–29239.

13 Membranes and Cancer

Most cancers, like neurodegeneration, are age dependent. This raises the possibility that membrane peroxidation could be involved in carcinogenesis. The purpose of this chapter is to show that high levels of DHA phospholipids, at least in certain human organs, correlate with and might be linked to tumor formation. We are not claiming that the native DHA molecule is a carcinogen; rather it is known that oxidation products of DHA are tumorigenic *in vitro*. Marnett and colleagues found that polyunsaturated fatty acids score positive in the Ames test (Marnett et al., 1985) and Fraga and colleagues (1991) showed that DNA of sperm seems to be a "hot spot" for mutagenesis. DHA has recently been shown to damage DNA (Kasai et al., 2005). It seems clear in these studies that DHA oxidation products, but not the intact DHA chain, are the mutagenic agents (Maekawa et al., 2006). Thus, the positive scoring of DHA in the Ames test opens up a new window toward identifying human organs in which DHA-linked tumors might appear. The results of this analysis are discussed below with an eye toward understanding the role of DHA as a common thread running through neurodegeneration, aging, and cancer.

13.1 RETINOBLASTOMA OCCURS IN A TISSUE WITH THE HIGHEST LEVELS OF DHA IN THE BODY

In our initial literature search for any linkages between DHA and cancer, we decided to first focus on any suspicious tumors in human organs with high DHA levels. At this point we ran headlong into the field of retinoblastoma, a good place to start because the retina has the highest levels of DHA in the body. We do not attempt to review this large field of literature here except to say that retinoblastoma is a childhood cancer and tumors develop from neural precursor cells in the immature retina (Classon and Harlow, 2002). About one child in 20,000 is afflicted. Unlike other forms of cancer including colon tumors, which require roughly five to six separate mutations (see Figure 13.1 for relationship between age, number of mutations, and cancer), an unusually small number of mutations early in life are responsible for converting neural precursor cells to a retinoblastoma in infants (Knudson, 1971). A few individuals with retinoblastoma have inherited a deletion of the retinoblastoma (*Rb*) gene, which is present in all somatic cells in the body. Children carrying one defective copy of the *Rb* gene are predisposed to cancer. A second mutation blocking the function of the remaining good copy results in retinoblastoma. The remarkable feature of familial retinoblastoma is that once one eye is affected the other eye is likely to follow shortly. Thus, the mutation frequency of immature neural cells in the retina seems to be extraordinarily high, and this tumor might be accounted for by a pair of mutations, one in each of the *Rb* genes. In the familial form a single mutation triggers the cancerous state. The *Rb* gene was the first class of genes called

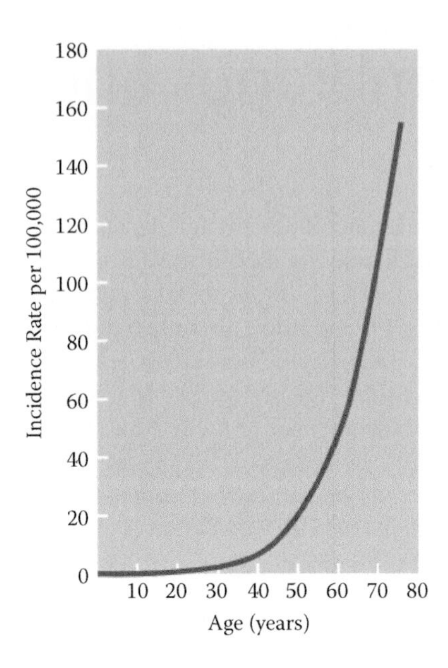

FIGURE 13.1 Cancer incidence as a function of age. The number of newly diagnosed cases of colon cancer in women in England and Wales in 1 year is plotted as a function of age at diagnosis and expressed relative to the total number of individuals in each age group. The incidence of cancer rises steeply as a function of age. If only a single mutation were required to trigger the cancer and this mutation had an equal chance of occurring at any time, the incidence would be independent of age. (Data from C. Muir et al., *Cancer Incidence in Five Continents*, Vol. V, Lyons: International Agency for Research on Cancer, 1987; From Alberts, B. et al., *Molecular Biology of the Cell*, Copyright 2008. Reproduced by permission of Taylor & Francis Group, LLC, a division of Informa plc.)

tumor suppressor genes to be identified, and many others have followed, adding an important chapter to the cancer field. A genetic defect targeted to a tumor suppressor gene increases the probability of cancer, but the data on the high rates recorded for retinoblastoma suggest a specialized mechanism is at work. There is no generally accepted mechanism to account for retinoblastoma, though various ideas have surfaced, including a genetic hot spot in the *Rb* gene and its relatively large target size for mutations. We propose that heavy trafficking of DHA needed for building the membranes for neurons and rods and cones subjects these cells to relatively high levels of oxidative stress directly or indirectly causing damage to chromosomal DNA. According to this scenario the window for DNA damage is open during childhood but closes later. The vulnerable period coincides with the most dynamic stage of neuron development and sculpting of the infant brain (see Chapter 3), a period when DHA circulation in retinal cells is proposed to be especially high.

Thus, the *Rb* gene can be considered as a genetic probe for mutations of the developing rods and cones of the retina with the first tumors in the familial disease appearing in as little as nine months after birth followed by multiple tumors forming in the diseased eye some months later. Tumors in the second eye follow shortly.

Familial retinoblastoma is considered a bilateral tumor because both eyes become diseased. However, a trilateral presentation is also recorded though more rare. The pineal gland, a rice grain–sized organ producing vital hormones such as melatonin, in the middle of the brain can also form tumors among heterozygous carriers of one defective *Rb* gene. Neurons composing the pineal gland are also enriched with DHA (Sarda et al., 1991; Zhang et al., 1998). Data from these papers show relatively high levels of DHA in membranes of the pineal gland of rats, levels dramatically increased by dietary sources of DHA. On the other hand, severe restriction of dietary DHA causes DHA in membranes to plummet, being replaced by more saturated chains including 18:2, which is elevated 12-fold in total membrane fatty acids. Diets rich in fish oil raise the levels of DHA significantly above a "normal" diet. Zhang et al. (1998) show that the biochemistry of the pineal gland, including production of melatonin, is significantly modulated in lockstep with changes of DHA levels.

Whereas these data do not establish a causal relationship between DHA and retinal or pineal cancer, they suggest a linkage. Our literature survey in search of other evidence for the roles of DHA in cancer next led to the field of brain tumors in children.

13.2 HEAVY TRAFFICKING OF DHA WITHIN THE DEVELOPING BRAIN MIGHT POTENTIATE BRAIN TUMORS IN CHILDREN

Brain tumors account for 24 percent of total cancers for children in contrast to 1.4 percent for adults. This is a dramatic difference that might be explained on the basis of active growth of brain cells during development giving rise to more cancer-causing mutations. However, the DHA principle might also be applicable, explaining the spike in incidence of brain tumors in children.

Whereas a great effort is now underway to determine external or environmental causes of childhood cancer, internal causes are still poorly understood. The class of glial cells called *microglia* is especially interesting (Allen and Barres, 2009; Azevedo et al., 2009; Pelvig et al., 2008). These cells include phagocytic cells involved in handling senescing DHA membrane fragments and synaptosomes generated during massive remodeling of the brain occurring in childhood. Microglia are capable of division, which might momentarily expose their nuclear DNA to both errors of replication as well as insults from toxic chemicals in the environment. During the handling of large amounts of DHA derived from axon pruning during brain development, by-products toxic to microglia are likely to be generated. Microglia roam throughout the brain and multiply when the brain is damaged or diseased. These phagocytic cells are expected to engorge themselves with large amounts of DHA during neuron remodeling in infants. DHA exposed to relatively harsh oxidative conditions expected inside microglia might generate toxic by-products capable of damaging the nuclear DNA of these cells (Kasai et al., 2005; Lee and Blair, 2000; Maekawa et al., 2006; Marnett et al., 1985). During brain development rapid rates of microglia division occur followed by apoptosis. Thus, the presence of DHA-mediated mutagenic products during microglial cell division would be the worst-case scenario because of the extreme vulnerability of nuclear DNA at this time. Another scenario involves DHA or its oxidation products acting as energy uncouplers at the membrane level, enabling the tumor process through energy deficiency. Thus, microglia damaged by DHA-derived by-products may give rise to brain tumors preferentially in children.

13.3 DHA IS A MAJOR BUILDING BLOCK FOR SPERM MEMBRANES: IS THERE AN INCREASED RISK OF CANCER IN THE TESTES?

DHA is needed as a building block for spermatogenesis (Chapter 2) and is present in relatively large amounts in the testes. In this section we explore whether the presence of this oxidatively unstable molecule might elevate cancer risk in the testes. Membranes of free-living human sperm cells are enriched with DHA and are readily attacked by molecular oxygen (Aitken and Clarkson, 1987; Alvarez and Storey, 1995), creating potentially carcinogenic products such as reactive oxygen species (ROS) as well as energy uncouplers (see Chapter 8). Chemical studies on human sperm established that DHA-enriched phospholipids are highly susceptible to ROS oxidation. These studies show that under laboratory conditions potentially toxic by-products are readily formed during oxidation of DHA membranes of sperm. DHA is localized in the tail membrane, which distances any procarcinogenic properties of DHA oxidation products from germ-line DNA stored in the head. Note that the head membrane is composed of more saturated fatty acids, also expected to prevent damage to DNA by products of lipid peroxidation.

There are at least three aspects of spermatogenesis including the presence of DHA that might put germ-line DNA at risk as follows:

- Much greater probability of mutations caused by errors of DNA replication during numerous cell divisions occurring during spermatogenesis.
- Increased exposure of sperm DNA to mutagens caused by replacement of a thick coating of histones generally covering DNA with protamines, the latter thought to be less protective.
- Large amount of DHA in the testes environment as a potential source of energy-uncoupling or mutagenic molecules.

Altogether these risks might lead to significantly higher rates of nuclear mutations targeting DNA of sperm.

In recent years, powerful DNA sequencing technology has been applied toward understanding the high levels of spontaneous mutations known to occur in sperm DNA (Qin et al., 2007). Qin and colleagues found that common sporadic Apert syndrome mutation in a human fibroblast growth factor receptor gene is 100 to 1000 times higher than expected. Positive selection in the testes seems to account for this dramatically increased rate. Hot spots for mutations in sperm DNA seem to have been eliminated as the primary cause of Apert mutations, but there remains the question of what caused the original mutation. We suggest that mutations in sperm nuclear DNA might be linked to the unique environment of the testes caused by the presence of DHA. It is of interest that the same gene governing Apert syndrome causing dwarfing also causes cancer when mutated at different sites (reviewed by Qin et al., 2007).

Clearly, heavy trafficking of DHA is necessary for spermatogenesis. Thus, germline DNA being packaged in sperm heads is exposed to any toxic or mutagenic effects of DHA. DHA deserves attention as a potential carcinogenic agent against nuclear DNA in this specialized environment.

13.4 CORRELATION BETWEEN HIGH LEVELS OF DHA IN BLOOD AND AGGRESSIVE PROSTATE CANCER

Prostate cancer is second only to lung cancer in tumor-dependent death of males. For example, one in six men will be diagnosed with prostate cancer sometime in their lifetime. Prostate cancer originates in various regions of the prostate gland and can be generally classified as being slow growing in a majority of cases, in contrast to a more aggressive stage that spreads to other organs. There is considerable controversy concerning the roles of diet and dietary lipids on the levels of prostate cancer. However, data from the nationwide Prostate Cancer Prevention Trial have led to a surprising finding on the role of DHA (Brasky et al., 2011). A subset of 3461 men ages 55 to 84 taking part in the much larger trial were divided into two groups—1803 controls versus 1658 participants, the latter chosen on the basis of relatively high blood levels of DHA. The group with high DHA levels included consumers of fish as well as fish oil, enriched supplements, or capsules. Prostate biopsy was used to confirm the presence or absence of prostate cancer in all study participants. The main conclusion is that men with the highest blood DHA levels were 2.5 times more likely to develop high-grade, aggressive prostate cancer than those with the lowest DHA levels. In contrast, levels of milder prostate cancer representing a majority of cases showed little difference between the controls versus those falling into the high blood levels of DHA. Brasky and colleagues (2011) point out that the beneficial effects of eating fish likely outweigh any harm related to prostate cancer risk. Data from this study show the complexity of human nutrition and its impact on cancer risks and help explain opposing conclusions from different trials. These authors advise caution in making assumptions and propose the association between blood levels of DHA and increased prostate cancer requires further research.

According to one scenario, high blood DHA levels are reflected in higher enrichment of DHA in membranes of prostate tissue, leading to a state of chronic inflammation. This idea is based on previous data showing that commonly used nonsteroidal anti-inflammatory drugs including aspirin can reduce the risk of prostate cancer. Now a dramatic lowering of rates of prostate cancer has been reported in men who take an acetaminophen tablet daily for at least five years (Jacobs et al., 2011). Overall, prostate cancer rates are reduced by 38 percent compared to controls, whereas the risk of the most aggressive form of prostate cancer is cut by 51 percent. This large epidemiologic study, called the Cancer Prevention Study II Nutrition Cohort, included 78,485 men. Participants answered questions about food consumption and drug use in 1992 and every two years thereafter. During this period 8092 cases of prostate cancer were reported, but those with a history of daily acetaminophen use for at least five years were less likely to have developed prostate tumors. Obviously further research is necessary to establish linkage among DHA, inflammation, and prostate cancer.

13.5 CHEMOTHERAPY PATIENTS ARE ADVISED NOT TO TAKE FISH OIL SUPPLEMENTS

During the course of chemotherapy tumor cells often develop a tolerance or resistance to the drugs. Many different resistance mechanisms have been uncovered, and one of the most recent involves the increased risk of resistance in patients taking supplements of fish oil (Roodhart et al., 2011). Omega-3 and omega-6 fatty acids enriched in fish oil have been identified as antichemotherapeutic agents in the treatment of certain cancers. These normally beneficial fatty acids are not directly harmful but are converted to oxidation products that are. Minute amounts of these by-products send signals to and activate stem cells to secrete chemicals that protect tumor cells against a range of chemotherapeutics. Thus, omega-3 and omega-6 fatty acids become harmful, acting as precursors for the formation of potent signaling molecules that alert and cause tumor cells to become drug resistant. These data reveal new targets to enhance the efficacy of chemotherapy in patients. One of the targets revealed in this study is the enzyme cyclooxygenase, which as discussed in Chapter 17 is subject to inhibition by common aspirin. Data presented by Roodhart and colleagues (2011) show that various specific inhibitors of cyclooxygenase, which produces the signaling molecules inducing resistance in tumor cells, block the harmful effects of fish oil. It is well known that omega-3 and omega-6 fatty acids in fish are converted by the cyclooxygenase pathway to yield a powerful class of hormones called *eicosanoids*, essential for human health and development. It is also well known that eicosanoids can be harmful, and the example discussed here seems to be another case in which too much of a good thing becomes harmful. We propose in Chapter 17 that overactive eicosanoid production in the brain may trigger neurodegeneration.

13.6 ASYMMETRICAL PHOSPHOLIPIDS AS POSSIBLE PROMOTERS OF BREAST CANCER

Aging is the single most powerful risk factor for breast cancer as it is for most cancers. This brings recent advances in aging research to bear on this important form of cancer. Attention here is on epithelial cells lining ducts or tubes that bring milk from the breast to the nipple. These cells are highlighted because about 80 percent of breast cancers start with the milk-producing epithelial cells. Our focus here is on what causes the earliest stages of breast cancer, and we approach the subject from the perspective of membranes and energy stress as possible enablers of the tumor state. We initially considered the idea that DHA delivered from its site of synthesis in the liver to duct cells might be involved in breast cancer, but a closer look led us in another direction.

As membrane ecologists we first explored any unusual properties of membranes of breast epithelial cells for clues about how breast cancer is triggered with aging. With

TABLE 13.1
Medium-Chain Fatty Acid Composition of Human Milk Fat

Fatty acid	8:0	10:0	12:0	14:0
Mole%	1.0 ± 0.7	2.4 ± 0.4	7.8 ± 2.3	11.5 ± 1.5

Source: Modified from Thompson, B. J., and S. Smith, 1985, Biosynthesis of Fatty Acids by Lactating Human Breast Epithelial Cells: An Evaluation of the Contribution to the Overall Composition of Human Milk Fat, *Pediatr. Res.*, 19: 139–143. With permission.

Note: Value derived from chromatographic analysis of milk samples from five selected donors.

energy or energy stress as the assumed cause of aging, we decided to evaluate a possible novel energy-uncoupling gene expressed uniquely in breast epithelial cells as a possible cause of breast cancer. A unique class of fatty acids (called medium-chain fatty acids) is made in bulk by breast epithelial cells and makes up about 20 percent of the total fatty acid chains in human milk (Thompson and Smith, 1985). The fatty acid composition of human milk fat showing the high content of medium chains is summarized in Table 13.1. An acyl chain–terminating enzyme called *thioesterase II* is expressed in breast epithelial cells and governs synthesis of medium-chain fatty acids in breast milk. Medium-chain fatty acids are shorter and more water soluble than their long-chain counterparts and thus are more readily digested as energy food, a selective advantage for the nutrition of human infants. Ruminants (for instance, cows) do not produce medium-chain fatty acids in their milk, whereas breast epithelial cells of mice and rats produce over twice the levels found in human milk. Breast epithelial cells of rats express more than 10 times the level of thioesterase II compared to humans. Low-fat diets are found to upregulate expression of thioesterase II by as much as sixfold with high-fat diets strongly downregulating activity. Thus, thioesterase II activity is believed to be regulated by a variety of signals.

Virtually all of the studies of the biological functions of medium-chain fatty acids produced by breast epithelial cells focus on the beneficial bioactivity of these chains as readily metabolized energy food in infants. However, we suggest that medium-chain fatty acids synthesized by breast duct cells might have dual chemical modes, similar to that proposed for DHA. The energy-uncoupling properties of medium-chain fatty acids when incorporated into mitochondrial membranes have been known for almost 50 years and were first studied in yeast. As background, when growing without air (e.g., in the wine vat) yeast synthesizes large amounts of medium-chain fatty acids, as do breast epithelial cells. These fatty acids are incorporated into membrane phospholipids including in the mitochondria. Medium-chain fatty acids, apparently working as membrane-based energy-uncouplers, are likely beneficial in wine making, allowing yeast to complete the breakdown of grape sugars creating a dry wine (Valentine and Valentine, 2009). A novel membrane phospholipid structure, in which a medium-chain fatty acid is paired with a longer chain, is proposed to cause energy uncoupling. The Nobelist Konrad Bloch and his student Howard Goldfine discovered this phospholipid structure almost 50 years ago (Goldfine and

Bloch, 1963), and subsequent researchers have focused on the membrane bioactivity of these asymmetrical phospholipids. The current model supported by convincing data is that *Saccharomyces cerevisiae* has evolved the medium-chain biosynthetic pathway as an alternative and essential mechanism for fluidizing its plasma membrane during periods when synthesis of long-chain monounsaturated fatty acids is shut down due to lack of oxygen. Note that yeast fatty acid desaturase, which adds one double bond to long-chain fatty acids, including 16:0 and 18:0, requires molecular oxygen (O_2) as a substrate. Thus, monounsaturated chains of long-chain length essential for aerobic growth are replaced in the membrane of anaerobic cells by medium-chain fatty acids, C8 through C14, the latter crucial for anaerobic growth. Relative abundances in terms of total fatty acids in membranes for C8 are usually low, about 1 percent of total fatty acids, whereas C10 (6 percent), C12 (23 percent), and C14 (14 percent) are more abundant in yeast growing without oxygen. It is of interest that breast epithelial cells produce a roughly similar pattern of chain lengths for transfer to the milk duct during breast-feeding. When yeast cells are grown aerobically, medium-chain fatty acids are harmful and quickly disappear from membranes. These data show that regulatory systems have evolved in yeast for timing the synthesis of medium-chain fatty acids to periods of growth when they are most beneficial, while quickly downregulating synthesis under respiratory growth when these chains act as energy uncouplers, as discussed next.

The energy-uncoupling activity of phospholipids containing medium-chain fatty acids in mitochondria of yeast was first recognized and studied in the 1970s. The morphology of mitochondria isolated from yeast tricked into enriching their mitochondria with medium-chain fatty acids was found to be normal compared to mitochondria of cells grown aerobically. Moreover, the respiratory or electron transport chain—measured as rates of O_2 consumed—was fully functional in mitochondria containing the following levels of medium-chain fatty acids: 2.3 percent C8, 15.3 percent C10, 13 percent C12, and 13.6 percent C14. However, biochemical studies show that proton bioenergetics in yeast mitochondria are completely uncoupled when mitochondrial membranes are enriched in medium-chain fatty acids, bringing ATP production by these organelles to a halt (Haslam et al., 1973). Potassium circulation is also disrupted. This defines a state of energy stress that might be accounted for by a dramatic thinning of the membranes (Figure 13.2). From this data it is concluded that phospholipids in which one fatty acid chain is much shorter than the other likely behave as potent proton energy uncouplers when present in mitochondrial membranes. These data are consistent with biophysical studies of the uncoupling of cation gradients caused by medium-chain fatty acids in phospholipids in chemically defined lipid vesicles (Paula et al., 1996). These authors showed that unusually thin membranes are also leaky for Na^+ and K^+, suggesting that futile cycling of H^+, Na^+, and K^+ will occur when membranes are thinned by any mechanism.

We are not aware of any biochemical data showing a direct causal relationship linking medium-chain fatty acids as energy uncouplers in breast epithelial cells and rates of breast cancer and present this scenario as a hypothetical model. There are several mechanisms that might account for the toxicity of medium-chain fatty acids in breast epithelial cells with a few listed as follows:

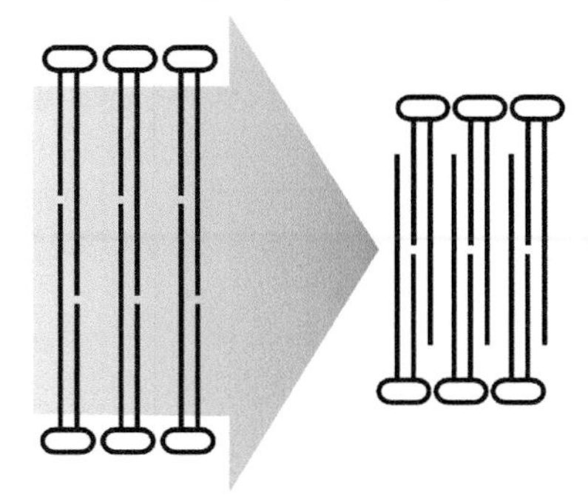

FIGURE 13.2 Medium-chain fatty acids as a possible cause of breast cancer. High levels of medium-chain fatty acids in human milk are proposed to become pro-tumorigenic when incorporated into membranes of breast epithelial cells where these fatty acids are synthesized. Data from yeast show that incorporation of medium chains causes excessive membrane thinning believed to cause energy uncoupling. Thus, breast epithelial cells are proposed to be unique among cells in the human body in being exposed to toxic effects of membrane thinning and energy stress caused by this class of fatty acid. (From Valentine, R., and Valentine, D., *Omega-3 Fatty Acids and the DHA Principle*. Copyright 2009. Reproduced by permission of Taylor & Francis Group, LLC, a division of Informa plc.)

- Direct energy uncoupling of mitochondrial proton bioenergetics by the free fatty acids creating energy stress due to futile proton cycling.
- Energy stress caused indirectly as the result of incorporation of medium chains into membrane phospholipids destroying permeability properties of mitochondrial membranes.
- Energy stress caused by futile cation cycling at the level of the plasma membrane.

Studies with yeast membranes enriched with medium-chain fatty acids are consistent with a water-wire mechanism, and we suggest that this mechanism might also occur in breast epithelial cells whenever medium-chain fatty acids are present and incorporated into membranes. There is a possibility that chain lengths C8 to C10 favor direct energy uncoupling by free fatty acids, whereas C12 to C14 chains are more likely incorporated into membranes and cause energy uncoupling due to defects in membrane permeability. It is clear that membrane phospholipids enriched with medium-chain fatty acids are powerful energy uncouplers.

According to our current working model, a normal epithelial cell is somehow triggered (e.g., hormone imbalance?) to express thioesterase II activity prematurely. In this scenario medium chains are produced and incorporated into membranes causing energy deficiency and perhaps the premature release of duct cells to the lumen region (Schafer et al., 2009). Unlike normal senescing epithelial cells, which die when detached, prematurely detached duct cells might retain some of their ability to reestablish contact with duct epithelial cells and avoid programmed cellular death (Figure 13.3). These defective cells are envisioned to continue to grow, filling the lumen of the duct in a manner reminiscent of polyps forming in the colon (see Chapter 2). Once growth arrest barriers are broken, continual growth increases the

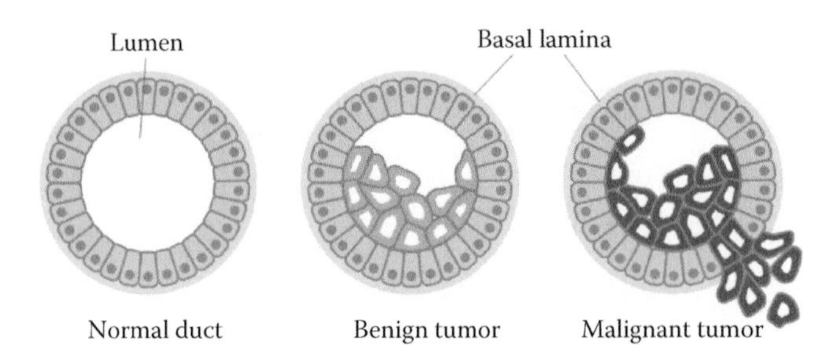

FIGURE 13.3 Development of breast cancer starting in a milk duct. Normal breast epithelial cells are attached to a basement membrane that provides them with survival signals. Senescing cells are continuously released and die, whereas tumorigenic cells survive detachment, grow, and fill the lumen of the milk duct. A benign glandular tumor (an adenoma) remains inside the basal lamina that marks the boundary of the milk duct, in this example, whereas a malignant glandular tumor (an adenocarcinoma) destroys duct wall integrity and spreads as shown. (From Alberts, B. et al., *Molecular Biology of the Cell*, Copyright 2008. Reproduced by permission of Taylor & Francis Group, LLC, a division of Informa plc.)

probability of mutations leading to breast cancer. In essence, we are proposing that medium-chain fatty acids might increase the risk of breast cancer. We have already discussed how incorporation of DHA from the diet into mitochondria of senescing colon epithelial cells acts as a deterrent against colon cancer (Chapter 2). We propose that medium-chain fatty acids have the opposite effect in milk duct epithelial cells, promoting rather than inhibiting cancer. According to this scenario gene silencing or therapeutic targeting of thioesterase II expression or activity might offer a new approach toward preventing breast cancer. Note that baby formula can be readily fortified with medium-chain fatty acids, making this class of easily digested oils available to infants even if an insufficient supply is produced in mother's milk. Pregnancy, which shuttles DHA to breast tissue and thus increases levels of DHA surrounding breast epithelial cells, might act to assassinate senescing, precancerous cells as it does with senescing colon cells. This model is consistent with the known benefit of pregnancy in lowering the risk of breast cancer.

13.7 DOES GENERALIZED CELLULAR ENERGY DEFICIENCY GOVERN RATES OF CANCER-CAUSING NUCLEAR MUTATIONS?

There are numerous clues in the extensive literature in the mutagenesis field that can be interpreted in the context of effects of energy stress on nuclear mutation frequency. Note that we define energy stress with broad strokes. Because momentum seems to be increasing in the energy-mutation field, a selected list of pioneering papers is compiled for further reading as follows: Brookfield, 2010; Drake, 2007; Fowler, Erickson, and Isbell, 1994; Gawel, Hamilton, and Schaaper, 2008; Kunkel, 2004; Loeb, Loeb, and Anderson, 2003; Mathews, 2006; Miller et al., 2002; Nooteboom et al., 2010; Rosenberg, 1997; Rosenberg and Hastings, 2003; Song et al., 2005; Stoler et al., 1999; Zinser and Kolter, 2004. Advances in DNA sequencing technology open up new windows of opportunity for tracing the evolutionary history of cells subjected to energy stress. Bacteria such as *Escherichia coli* have proven to be suitable subjects for such experimental evolution experiments. Data show that energy stress during the stationary phase or under energy starvation conditions appears to modulate mutation frequencies. These data support the view that energy stress helps govern DNA replication fidelity and might be a powerful evolutionary selective force to be reckoned with. However, lack of replication fidelity might also have its merits, such as giving evolution a jump start.

One of the big questions for future research concerns how DNA replication fidelity might turn with aging. Previously aging was thought to be mediated by ROS levels spiking late in life—a chain reaction of mtDNA damage caused by snowballing levels of ROS acting as a potent mutagen. Key data generated in the past few years challenge this view (see Chapter 11). According to the revised mitochondrial theory of aging, mtDNA mutations accumulate linearly with time and a sharp spike in ROS was not observed in mutator mice. MtDNA mutations are now thought to lower energy production in a steady fashion. These data cast doubt on the central dogma that states that ROS drives mutation rates of mtDNA. However, nuclear DNA is another story (Akimoto et al., 2005; Kiebish and Seyfried, 2005), and the central dogma that ROS causes mutations leading to cancer has yet to be overturned.

Pompei and coworkers generated a convincing set of data comparing cancer rates between mice and men (Pompei, Polkanov, and Wilson, 2001; Pompei and Wilson, 2001). These researchers collected and analyzed extensive data covering the age dependence of cancer in mice versus humans. The key point here concerns their conclusion that incidence of lethal cancers in mice is about 35 times higher than in humans—values for mice at 2.1 years of age versus 85 years for humans (i.e., 80 percent of the maximum lifetime for each animal). These researchers did not extend their analysis to include DHA levels versus cancer rates, but this trend is similar to DHA content versus the life span of mammals as summarized in Chapter 12. In essence, the data of Pompei and colleagues are consistent with a possible linkage between DHA levels and cancer rates in mouse versus man. These data might be explained by DHA acting directly as a mutagen or alternatively by DHA causing energy stress, or both.

In summary, DHA might well be a tumor-causing agent in certain DHA-enriched cells or organs, creating nuclear mutations through a conventional ROS → DNA mechanism. However, in another scenario DHA is envisioned to generate reactive oxygen species that damage membrane integrity, causing severe energy stress with potential effects on mutation rates. One of the distinguishing features between aging and age-dependent cancers concerns the roles of mtDNA mutations in aging versus nuclear mutations in cancer. Does energy stress govern the rate of accumulation of mutations of nuclear genes causing cancer? Is membrane polyunsaturation an important factor?

Finally, we included a chapter on cancer as a vehicle to explore common threads among aging, dementia, and cancer. In applying the DHA principle to cancer, several scenarios linking membranes to cancer rates have emerged that seem worthy of further consideration. Future advances in understanding cancer will likely find applications in the fields of aging and dementia, and vice versa.

REFERENCES

Aitken, R. J., and J. S. Clarkson. 1987. Cellular basis of defective sperm function and its association with the genesis of reactive oxygen species by human sperm. *J. Reprod. Fertil.* 81:459–469.

Akimoto, M., M. Niikura, M. Ichikawa et al. 2005. Nuclear DNA but not mtDNA controls tumor phenotypes in mouse cells. *Biochem. Biophys. Res. Commun.* 327:1028–1035.

Allen, N. J., and B. A. Barres. 2009. Neuroscience: Glia—More than just brain glue. *Nature* 457:675–677.

Alvarez, J. G., and B. T. Storey. 1995. Differential incorporation of fatty acids into and peroxidative loss of fatty acids from the phospholipids of human spermatozoa. *Mol. Reprod. Dev.* 42:334–346.

Azevedo, F. A., L. R. Carvalho, L. T. Grinberg et al. 2009. Equal numbers of neuronal and non-neuronal cells make the human brain an isometrically scaled-up primate brain. *J. Comp. Neurol.* 513:532–541.

Brasky, T. M., C. Till, E. White et al. 2011. Serum phospholipid fatty acids and prostate cancer risk: Results from the Prostate Cancer Prevention Trial. *Am. J. Epidemiol.* 23(12):1429–1439.

Brookfield, J. F. Y. 2010. Experimental evolution: The rate of adaptive evolution. *Curr. Biol.* 20:R23–R25.

Classon, M., and E. Harlow. 2002. The retinoblastoma tumour suppressor in development and cancer. *Nat. Rev. Cancer* 2:910–917.

Drake, J. W. 2007. Mutations in clusters and showers. *Proc. Natl. Acad. Sci. USA* 104:8203–8204.

Fowler, R. G., J. A. Erickson, and R. J. Isbell. 1994. Activity of the *Escherichia coli* mutT mutator allele in an anaerobic environment. *J. Bacteriol.* 176:7727–7729.

Fraga, C. G., P. A. Motchnik, M. K. Shigenaga et al. 1991. Ascorbic acid protects against endogenous oxidative DNA damage in human sperm. *Proc. Natl. Acad. Sci. USA* 88:11003–11006.

Gawel, D., M. D. Hamilton, and R. M. Schaaper. 2008. A novel mutator of *Escherichia coli* carrying a defect in the dgt gene, encoding a dGTP triphosphohydrolase. *J. Bacteriol.* 190:6931–6939.

Goldfine, H., and K. Bloch. 1963. Oxygen and biosynthetic reactions. In: Wright, B., editor. *Control Mechanisms in Respiration and Fermentation.* New York: Ronald Press, p. 81.

Haslam, J. M., T. W. Spithill, A. W. Linnane et al. 1973. Biogenesis of mitochondria. The effects of altered membrane lipid composition on cation transport by mitochondria of *Saccharomyces cerevisiae*. *Biochem. J.* 134:949–957.

Jacobs, E. J., C. C. Newton, V. L. Stevens et al. 2011. A large cohort study of long-term acetaminophen use and prostate cancer incidence. *Cancer Epidemiol. Biomarkers Prev.* 20(7):1322–1328.

Kasai, H., M. Maekawa, K. Kawai et al. 2005. 4-Oxo-2-hexenal, a mutagen formed by omega-3 fat peroxidation, causes DNA adduct formation in mouse organs. *Ind. Health* 43:699–701.

Kiebish, M. A., and T. N. Seyfried. 2005. Absence of pathogenic mitochondrial DNA mutations in mouse brain tumors. *BMC Cancer* 5:102.

Knudson, A. G. Jr. 1971. Mutation and cancer: Statistical study of retinoblastoma. *Proc. Natl. Acad. Sci. USA* 68:820–823.

Kunkel, T. A. 2004. DNA replication fidelity. *J. Biol. Chem.* 279:16895–16898.

Lee, S. H., and I. A. Blair. 2000. Characterization of 4-oxo-2-nonenal as a novel product of lipid peroxidation. *Chem. Res. Toxicol.* 13:698–702.

Loeb, L. A., K. R. Loeb, and J. P. Anderson. 2003. Multiple mutations and cancer. *Proc. Natl. Acad. Sci. USA* 100:776–781.

Maekawa, M., K. Kawai, Y. Takahashi et al. 2006. Identification of 4-oxo-2-hexenal and other direct mutagens formed in model lipid peroxidation reactions as dGuo adducts. *Chem. Res. Toxicol.* 19:130–138.

Marnett, L. J., H. K. Hurd, M. C. Hollstein et al. 1985. Naturally occurring carbonyl compounds are mutagens in Salmonella tester strain TA104. *Mutat. Res.* 148:25–34.

Mathews, C. K. 2006. DNA precursor metabolism and genomic stability. *FASEB J.* 20:1300–1314.

Miller, J. H., P. Funchain, W. Clendenin et al. 2002. *Escherichia coli* strains (ndk) lacking nucleoside diphosphate kinase are powerful mutators for base substitutions and frameshifts in mismatch-repair-deficient strains. *Genetics* 162:5–13.

Nooteboom, M., R. Johnson, R. W. Taylor et al. 2010. Age-associated mitochondrial DNA mutations lead to small but significant changes in cell proliferation and apoptosis in human colonic crypts. *Aging Cell* 9:96–99.

Paula, S., A. G. Volkov, A. N. Van Hoek et al. 1996. Permeation of protons, potassium ions, and small molecules through phospholipid bilayers as a function of membrane thickness. *Biophys. J.* 70:339–348.

Pelvig, D. P., H. Pakkenberg, A. K. Stark et al. 2008. Neocortical glial cell numbers in human brains. *Neurobiol. Aging* 29:1754–1762.

Pompei, F., M. Polkanov, and R. Wilson. 2001. Age distribution of cancer in mice: The incidence turnover at old age. *Toxicol. Ind. Health* 17:7–16.

Pompei, F., and R. Wilson. 2001. Age distribution of cancer: The incidence turnover at old age. *Hum. Ecol. Risk Assess.* 7:1619–1650.

Qin, J., P. Calabrese, I. Tiemann-Boege et al. 2007. The molecular anatomy of spontaneous germ line mutations in human testes. *PLoS Biol.* 5:e224.

Roodhart, J. M. L., L. G. M. Daenen, E. C. A. Stigter et al. 2011. Mesenchymal stem cells induce resistance to chemotherapy through the release of platinum-induced fatty acids. *Cancer Cell* 20(3):370–383 (doi: 10.1016/j.ccr.2011.08.010).

Rosenberg, S. M. 1997. Mutation for survival. *Curr. Opin. Genet. Dev.* 7:829–834.

Rosenberg, S. M., and P. J. Hastings. 2003. Microbiology and evolution: Modulating mutation rates in the wild. *Science* 300:1382–1383.

Sarda, N., A. Gharib, M. Croset et al. 1991. Fatty acid composition of the rat pineal gland. Dietary modifications. *Biochim. Biophys. Acta* 1081:75–78.

Schafer, Z. T., A. R. Grassian, L. Song et al. 2009. Antioxidant and oncogene rescue of metabolic defects caused by loss of matrix attachment. *Nature* 461:109–113.

Song, S., Z. F. Pursell, W. C. Copeland et al. 2005. DNA precursor asymmetries in mammalian tissue mitochondria and possible contribution to mutagenesis through reduced replication fidelity. *Proc. Natl. Acad. Sci. USA* 102:4990–4995.

Stoler, D. L., N. Chen, M. Basik et al. 1999. The onset and extent of genomic instability in sporadic colorectal tumor progression. *Proc. Natl. Acad. Sci. USA* 96:15121–15126.

Thompson, B. J., and S. Smith. 1985. Biosynthesis of fatty acids by lactating human breast epithelial cells: An evaluation of the contribution to the overall composition of human milk fat. *Pediatr. Res.* 19:139–143.

Valentine, R. C., and D. L. Valentine. 2009. *Omega-3 Fatty Acids and the DHA Principle*. Boca Raton, FL: Taylor & Francis.

Zhang, H., J. H. Hamilton, N. Salem Jr. et al. 1998. N-3 fatty acid deficiency in the rat pineal gland: Effects on phospholipid molecular species composition and endogenous levels of melatonin and lipoxygenase products. *J. Lipid Res.* 39:1397–1403.

Zinser, E. R., and R. Kolter. 2004. *Escherichia coli* evolution during stationary phase. *Res. Microbiol.* 155:328–336.

Section V

DHA Links Aging and Neurodegeneration

An epidemic of Alzheimer's disease is presently raging among the elderly of the world. The cause is known—it is aging. It remains a great challenge to understand the linkage between aging and neurodegeneration. We suggest that people are born with the seeds of neurodegeneration embedded in the signature cells of the brain—neurons. Furthermore, we suggest that DHA is closely linked to brain decline with aging.

The dynamic conformations of DHA chains working in membranes of neurons have hindered a precise explanation of their biochemical roles. This quandary has plagued membrane scientists for generations. In our previous book on this subject, we used a reductionist strategy involving bacteria to attempt to decipher a generalized membrane blending code applicable to all cells, including neurons. Data from ecological, physiological, biochemical, and recombinant DNA studies led to the concept of a tripartite fatty acid blending code in which membrane fatty acid chains contribute at least three essential roles to the cell: (1) The first role involves the conventional role of the membrane as a permeability barrier against the entrance and exit of metabolites. We add the caveat that fatty acids contribute to the exclusion of water from the membrane, a key defensive mechanism to conserve the cellular energy supply. In essence, blocking the entry of water into the interior of the membrane blocks water-dependent tunneling of protons that uncouple proton and cation energy gradients. (2) The second role involves a hierarchy of unsaturated fatty acids or equivalent structures that modulate membrane motional properties. Data from physical-chemical studies suggest that DHA sits at the top of the ladder of fatty acids in nature based on its extreme conformational dynamics in membranes. The dynamic shape shifting of DHA and other unsaturated fatty acids is harnessed to modulate motion within the membrane, motion that enables membrane components to collide and prevents lipid rafts from dominating the membrane surface. We suggest that extreme membrane motion contributed by DHA helps maximize the speed and repetition rates of neural impulses, a key milestone in the evolution of humans. (3) The third leg of the tripartite fatty acid blending code highlights the extreme oxidative instability of DHA in the presence of oxygen. Chemical oxidation targeting this highly unsaturated chain results in both defects in the membrane structure as well as the generation of a battery of toxic derivatives. The tripartite fatty acid blending code has led to the DHA principle, which states that DHA offers both great benefits as well as major risks, applied here to neurons. Our challenge in this section is to understand the impact of the delicate balancing act between the benefits and risks of DHA in neurodegenerative diseases. We have selected several specific neurodegenerative diseases to highlight the DHA principle as applied to neuron health and decline during aging.

14 Parkinson's Disease

There is an extensive history of experimental data linking neuron energy deficiency and Parkinson's disease. Before discussing recent advances a look back at some earlier experiments is revealing because these data suggest that energy stress, perhaps mediated by DHA and oxidative stress, causes premature death of parkinsonian neurons.

A few days after injecting a contaminated batch of a mood-altering drug he had brewed in a makeshift chemistry laboratory, a young drug user began to shake uncontrollably. The symptoms, though appearing far more quickly, mirrored those of classic age-dependent Parkinson's disease (PD). This led researchers to suspect that a powerful neurotoxin present as a contaminant was the cause. Chemical detective stories pursued almost simultaneously on both the West and East coasts led to the discovery of a potent, fast-acting neurotoxin called MPTP (1-methyl-4-phenyl-1,2,3,6-tetrahydropyridine). MPTP is converted to MPP^+ (1-methyl-4-phenylpyridinium) in the brain where it targets and kills parkinsonian neurons at rate orders of magnitude faster than age-dependent Parkinson's disease. From the onset of the discovery of MPTP in the early 1980s, it seemed clear that this neurotoxin must involve a novel mode of action to explain its roughly 1000-fold faster killing action of parkinsonian neurons. We suggest that the mechanism of action of MPTP can be explained in part by the presence of a membrane-mediated pacemaker operating in parkinsonian neurons that is fast-forwarded by this potent neurotoxin. We also propose that premature death of parkinsonian neurons in the age-dependent form of the disease can be explained by a mechanism involving DHA and the dual energy-pacemaker theory of aging.

14.1 PARKINSONIAN CHEMICALS LIKELY ACT BY ACCELERATING ENERGY-OXIDATIVE STRESSES VIA A CHAIN REACTION MECHANISM

During the era that young drug abusers were being diagnosed with symptoms of Parkinson's disease, farmers were using a molecule strikingly similar to MPTP for weed control (Figure 14.1a). This class of herbicides represented by paraquat is labeled *broad spectrum* against weeds and plants in general because it targets and destroys chloroplasts that carry out photosynthesis essential for plant growth. The sunlight-dependent effect is dramatic, causing a once verdant field of weeds or plants to be scorched brown starting about a day after application. The lethal activity of paraquat as an herbicide resides in its planar dual ring structure allowing delocalization of a single high-energy electron pirated from the photosynthetic apparatus.

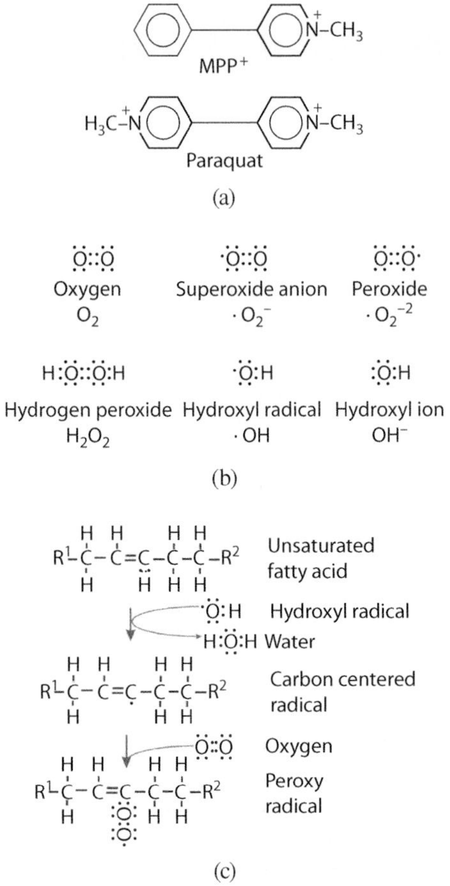

MPP+

Paraquat

(a)

Oxygen O_2

Superoxide anion $\cdot O_2^-$

Peroxide $\cdot O_2^{-2}$

Hydrogen peroxide H_2O_2

Hydroxyl radical $\cdot OH$

Hydroxyl ion OH^-

(b)

Unsaturated fatty acid

Hydroxyl radical

Water

Carbon centered radical

Oxygen

Peroxy radical

(c)

FIGURE 14.1 Similar structures of the weed-killer paraquat and the potent neurotoxin MPP+, both parkinsonian chemicals. (a) Note the remarkable similarity between these two high-energy electron carriers. (Reprinted from *Neuron*, 39, Dauer, W., and S. Przedborski, Parkinson's Disease: Mechanisms and Models, 889–909. Copyright 2003, with permission from Elsevier.) (b) A single high-energy electron carried by paraquat or MPP+ and transferred to O_2 generates a class of toxic oxygen derivatives called *reactive oxygen species* (ROS). Several ROS structures are shown. (c) The process of membrane fatty acid peroxidation is facilitated by ROS as shown in this diagram. A witch's brew of ROS including superoxide and hydroxyl radicals is generated from paraquat or MPP+ poisoning (see [b]). ROS can readily attack double bonds of DHA and other unsaturated membrane fatty acids causing lipid peroxidation. (Parts (b) and (c): Copyright 2010 BioTek Instruments, Inc., from Held, P., *An Introduction to Reactive Oxygen Species: Measurement of ROS in Cells.*)

Normally such high-energy electrons are carefully shepherded along electron transport chains where their electronic energy can be harnessed directly as a reducing power for carbon dioxide fixation or step-wise to produce ATP for plant growth. Electron transport chains operating in photosynthetic membranes of chloroplasts are remarkably similar to electron transport systems in mitochondria. Paraquat is a nonspecific, high-energy electron carrier. It short-circuits the photosynthetic electron transport chain robbing the plant of energy and producing a witch's brew of toxic or reactive oxygen species (ROS) (Figure 14.1b). The reduced form of paraquat donates one electron to oxygen generating the potent superoxide radical that can attack polyunsaturated membranes, as shown in Figure 14.1c. In one sense, spraying paraquat on a sunny day is like throwing a spark on a bone-dry pile of shavings, except paraquat radiates flameless chemicals scorching plant cells. Paraquat in the presence of light sparks an oxidative chain reaction that feeds on itself, breeding potent oxidative chemicals that degrade membranes and any other molecules in its vicinity. Thus, paraquat creates an oxidative chain reaction destroying the plant. It was not long before paraquat was suspected to be a neurotoxin and is now classified as a parkinsonian chemical. Paraquat also has the dubious distinction of being the chemical of choice for suicide in some countries. This farm chemical is cheap, readily available, and requires only about 10 mL of concentrate for a lethal dose. As a parkinsonian chemical, paraquat enters the brain and is converted by dark metabolism to its reduced state sparking a chain reaction targeting parkinsonian neurons.

14.2 MPTP: A POTENT NEUROTOXIN

Young drug users with Parkinson's symptoms had taken intravenously a contaminated derivative of the readily available drug meperidine (Demerol) (Dauer and Przedborski, 2003). This class of drugs has mood-altering effects. A neurotoxic contaminant called MPTP was identified as the toxic compound (Langston et al., 1983). Later biochemical studies showed that MPTP undergoes several transformations in the brain, finally yielding MPP^+ (Figure 14.2a). There is a remarkable similarity between symptoms of conventional Parkinson's disease and MPTP-induced brain disease. Specifically, both target a small region of the brain responsible for producing the neurotransmitter dopamine, which is a critical signaling molecule in neurons controlling muscle action. Death of about 60 to 70 percent of these neurons in the substantia nigra compacta region of the brain (also called parkinsonian/dopaminergic) causes symptoms. High-energy electrons generated as a normal part of metabolism in parkinsonian neurons are passed to MPP^+ that acts as a nonspecific electron carrier. Cycling of high-energy electrons through MPP^+ is believed to uncouple energy production and generates scorching levels of ROS including superoxide radicals. The rapid drop in energy caused by this neurotoxin seems to fast-forward the death of parkinsonian neurons, compressing the time frame of 50 or 60 years to merely days.

Thus, neurotoxic chemicals born in the drug world have proven to be valuable research tools in understanding how energy-oxidative stresses cause neuron death in PD. It is clear that once MPTP crosses the blood–brain barrier it is converted to a more toxic form (MPP^+); this toxin is concentrated in neurons using existing nutrient uptake pumps (Figure 14.2b). Once inside, MPP^+ can be taken up by mitochondria where it can catalyze the production of membrane-destroying oxidative radicals, quickly draining energy gradients perhaps via a water-wire mechanism (see Chapter 8). By attacking the mitochondria, these neurotoxins ultimately starve dopaminergic neurons of energy and trigger their death.

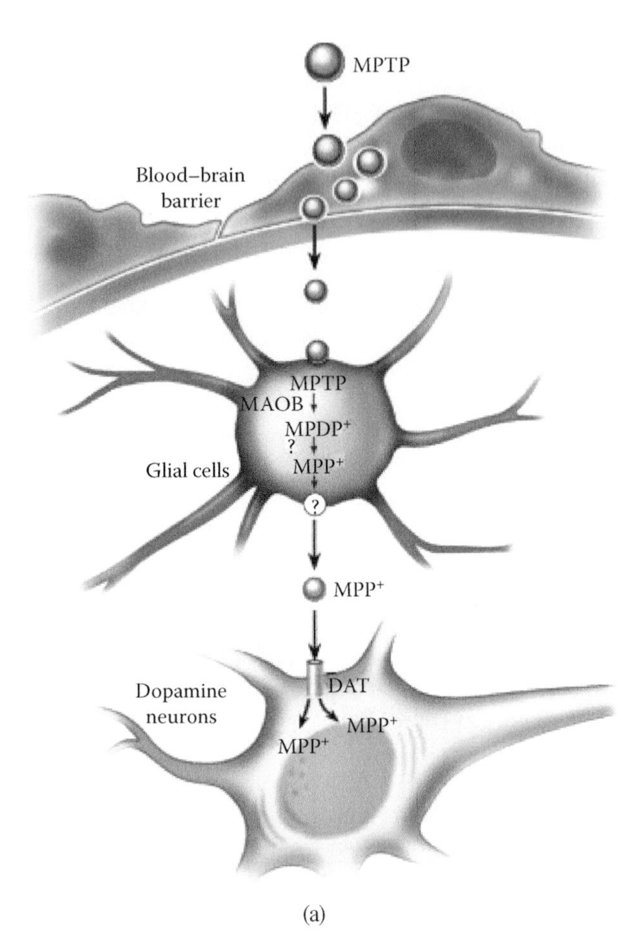

FIGURE 14.2 Mode of action of the potent neurotoxin MPTP (1-methyl-4-phenyl-1,2,3,6-tetrahydropyridine). (a) MPTP crossing the blood–brain barrier where it is converted by monoamine oxidase (MAO-B) to MPP^+ (1-methyl-4-phenylpyridinium) in glial cells. (b) MPP^+ might be concentrated in neurons via dopamine transporters. MPP^+ is proposed to catalyze the formation of superoxide radical in neurons, essentially fast-forwarding lipid peroxidation and destroying the permeability properties of mitochondria and plasma membranes. (Parts (a) and (b): Reprinted from *Neuron*, 39, Dauer, W., and S. Przedborski. Parkinson's Disease: Mechanisms and Models, 889–909. Copyright 2003, with permission from Elsevier.)

14.3 A CLOSER LOOK AT MPTP AS A MITOCHONDRIAL ENERGY POISON VERSUS AN OXIDATIVE THREAT

The discussion here focuses on how MPP^+, once inside a mitochondrion of a parkinsonian neuron, exerts its toxic activity. As shown in Figure 14.2b, MPP^+ inhibits mitochondrial energy production at the level of the first enzyme complex (complex 1) of the electron transport chain (Nicklas et al., 1987). Blocking complex 1 has the effect of decreasing ATP levels in parkinsonian neurons (Chan et al., 1991; Fabre

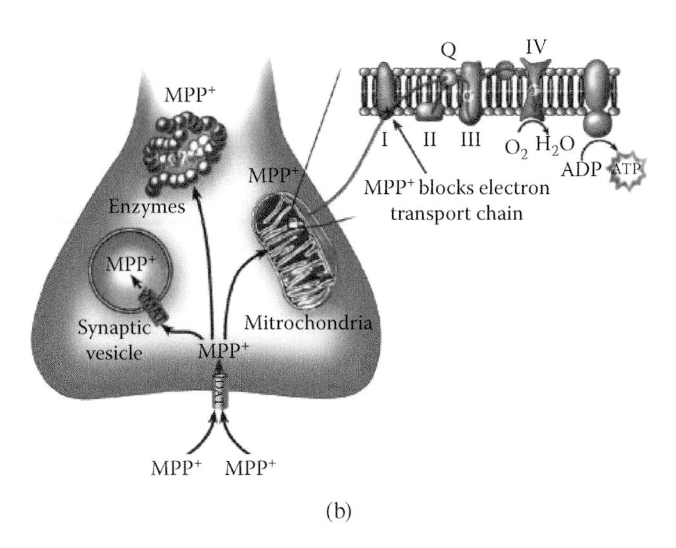

(b)

FIGURE 14.2 *(Continued)*

et al., 1999). Experiments on mitochondrial preparations isolated from the whole brain show that complex 1 must be inhibited by 70 percent before significant drops in ATP production occur (Davey and Clark, 1996). Mitochondria isolated from perhaps the hardest working region of the neuron, the synapse, were found to be far more sensitive to inhibition by MPP+ (Davey, Peuchen, and Clark, 1998). This raises the prospect that different classes of mitochondria localized in different regions of neurons may be more or less sensitive to MPP+ poisoning. Thus, different kinds of mitochondria may display different patterns of energy-oxidative stress tolerance with the hardest-working mitochondria being unusually sensitive to any further stresses.

In addition to inhibiting ATP production, MPP+ stimulates the production of ROS, especially superoxide (Hasegawa et al., 1997). In one study, transgenic mice overexpressing a gene coding for a well-known ROS detoxifying enzyme were protected against the killing of parkinsonian neurons by MPTP (Przedborski et al., 1992). This places ROS within the neuron but does not tell us how it participates in the killing cascade. However, in mice, DHA-enriched and tightly stacked mitochondrial membranes of parkinsonian neurons along with DHA-enriched membranes of axons are obvious targets for oxidative damage followed by energy uncoupling. Human parkinsonian neurons appear to target DHA away from their mitochondria (Chapter 11) in contrast to mice. Polyunsaturated chains, which replace DHA in mitochondrial membranes of human neurons, are the likely targets of oxidative damage by MPTP.

Following administration of MPTP, energy uncoupling and generation of ROS peak within hours, days before the death of parkinsonian neurons (Jackson-Lewis et al., 1995). Thus, MPTP does not instantaneously kill neurons but rather triggers a cascade that ultimately causes localized killing of most parkinsonian neurons (Vila et al., 2001).

Feeding studies using low doses of MPTP in mice along with studies of knockout mice targeting known genes in cellular death cascades reveal a complex molecular pathway involved in neuronal death. Mitochondrial mutations have been implicated in age-related Parkinson's, but we are not aware that mitochondrial DNA (mtDNA) mutations have been found responsible for the fast-track mechanisms involving MPTP. According to one model, energy stress caused by a variety of mechanisms is the mother of all stresses in Parkinson's disease. However, it remains difficult to separate the energy-uncoupling chemistry of MPTP from its role as an oxidative threat, again showing the close linkages between these two major stress cascades.

Since the discovery of MPTP as a neurotoxin, a battery of other chemicals and pathogenic agents have been identified that, broadly speaking, can be classified as being neuron-killing agents found in the environment. A short list of PD-causing agents with an emphasis on those toxins that might fast-track energy-oxidative stress in parkinsonian neurons is as follows:

- Contaminants in psychotic drugs (e.g., MPTP)
- Uptake or inhalation of the herbicide paraquat
- Exposure to the rat poison rotenone
- Carbon monoxide poisoning as in automobile exhaust
- Chronic use of hard water containing an excess of certain metals such as iron and manganese
- Brain trauma from contact sports such as football, rugby, and boxing (e.g., Mohammad Ali)
- Viral infection of the nervous system
- Pathogenic bacterial infections of the brain such as meningitis

Note that carbon monoxide behaves as a mitochondrial poison blocking energy production late in the respiratory chain in contrast to paraquat and MPTP, which are thought to act by uncoupling energy supply early in the chain. In light of recent data discussed below, it is now believed that a significant blockage of any stage of the electron transport chain of parkinsonian mitochondria, especially the first step at complex 1, might cause PD.

Recently, knowledge of the hallmark chemistry involving initiation and propagation of membrane peroxidation reactions has been applied in understanding and perhaps treating PD (Shchepinov et al., 2011). It has recently been reported that yeast cells enriched with polyunsaturated fatty acids (PUFAs) deuterated at *bis*-allylic sites are much more resistant to oxidative stress because of the isotope effect (Hill et al., 2011). These findings suggest a new approach to managing ROS-induced membrane damage in important neurodegenerative diseases such as Parkinson's disease (Shchepinov et al., 2011). These authors have tested the hypothesis that deuteration provides isotopic protection of PUFAs against oxidative damage and provides partial protection against nigrostriatal injury in a mouse model of Parkinson's disease. The neurotoxin MPTP is used in the mouse model to accelerate death of parkinsonian neurons. In the substantia nigra the number of nigral dopaminergic neurons following MPTP exposure in deuterium (D)-PUFA fed mice is 79.5 percent of the control versus 58.5 percent of the control in hydrogen (H)-PUFA fed mice. Biochemical studies of dopamine levels in brain tissue of D-PUFA versus H-PUFA fed mice show significant protection by D-PUFA of dopamine biosynthesis capacity. These researchers conclude that dietary D-PUFA partially protects against nigrostriatal damage from oxidative injury elicited by MPTP in mice. Thus, protection of PUFAs against peroxidation by introducing deuterium at the site of double bonds appears to be a universal mechanism explained by the isotope effect.

The final point in this section involves a word of caution regarding interpretation of data generated on the link between energy-oxidative stress and PD in mice versus humans. Specifically, the levels and localization of DHA membranes in parkinsonian neurons of mice versus humans are markedly different. DHA levels in mitochondrial membranes of parkinsonian neurons of mice are far more sensitive to peroxidation compared to mitochondria of humans. The take-home point is that the oxidative stress cascade operating in mice is significantly different from that in humans. From a broader perspective this means that the discovery of a linkage between oxidative stress and PD in mice does not guarantee that the precise mechanism operates in humans, and vice versa.

14.4 PARKINSONIAN NEURONS ARE LIKELY PREDISPOSED TO ENERGY STRESS MAKING THEM EXTREMELY SENSITIVE TO CHEMICALS IN THE ENVIRONMENT

Neurotoxins can destroy parkinsonian neurons in a matter of days to weeks. It is difficult to imagine how an mtDNA pacemaker (see Chapter 11) could be activated in a few days to kill parkinsonian neurons. A role for fast-forwarded mutation rates is not impossible but stretches the imagination. Another idea is that these toxins bypass the mtDNA energy-pacemaker and act directly and rapidly to generate catastrophic oxidative-energy stress, ultimately triggering cellular death. That is, neurotoxins such as MPTP might act by generating powerful enough stresses on their own to trigger an apoptotic cascade.

It has been proposed that parkinsonian neurons are predisposed to energy stress. A secret to the sensitivity of parkinsonian neurons might be found in a unique biochemical process operating in parkinsonian neurons. These neurons are unique in synthesizing the neurotransmitter dopamine using a two-stage biochemical process. The first step is carried out by an enzyme called *tyrosine hydroxylase*, which might be a source of dangerous ROS such as hydrogen peroxide (Pavon and Fitzpatrick, 2009). The operation of this enzyme in parkinsonian neurons is usually safe; but in the presence of paraquat, MPTP or carbon monoxide short-circuiting of electrons might occur, effectively creating another source for production of ROS, which can then attack DHA or polyunsaturated-enriched membranes and generate further energy stress. Several lines of chemical, biochemical, genetic, and pathological evidence are consistent with this model. For example, the solution of the structure of tyrosine hydroxylase using x-ray crystallography shows that iron atoms at active sites are normally safely buried in deep, narrow clefts in the enzyme surface (Goodwill et al., 1997). However, these molecular cavities in the enzyme are large enough to permit the entry of small electron carriers such as paraquat and MPP^+, perhaps short-circuiting the electron flow from tyrosine hydroxylase to form ROS. Carbon monoxide is also capable of short-circuiting this enzyme by binding directly to the iron forming the active site, turning the enzyme into a generator of ROS. According to this model parkinsonian neurons are predisposed by their own unique biochemical activity to produce an excess of membrane-destroying ROS, perhaps triggering a chain reaction of energy stress and cellular destruction. Dopamine, through its oxidation products, might also contribute toxicity for fast-forwarding the apoptotic cascade (Lotharius and O'Malley, 2000).

14.5 NEUROTOXIC DAMAGE TO PARKINSONIAN NEURONS MAY RECRUIT HYPERACTIVE MICROGLIA CAUSING MORE DAMAGE

It is well known that major defects in neurons can quickly recruit and activate microglia at the site of damage, in this case the parkinsonian region of the brain. The recruitment of hyperactivated microglia to this vulnerable region has been proposed to facilitate Parkinson's disease (McGeer, Yasojima, and McGeer, 2001). An oxidative burst of ROS by microglia following shortly after neurotoxic damage would be a worst-case scenario and is consistent with the rapidly unfolding time course of events following accidental ingestion of neurotoxic chemicals, including MPTP. A model to explain the linkage among DHA-hyperactive microglia-Alzheimer's disease is discussed in Chapter 17 and applies to Parkinson's disease.

14.6 MITOCHONDRIAL MUTATIONS AS A GENETIC PACEMAKER FOR AGE-DEPENDENT PARKINSON'S

Our discussion now shifts from fast-track mechanisms destroying parkinsonian neurons to the conventional age-dependent mechanism. Numerous mutations in genes causing mitochondrial dysfunction are implicated in age-dependent Parkinson's. These include genes for electron transport, subcellular trafficking, morphology, biogenesis, protein processing, signaling, and DNA replication. Interest here is on mitochondrial genes and mutations that might show a pattern of accumulative mutations, such as proposed for aging. Because MPTP discussed above targets the first enzyme complex (complex 1) in the electron transport chain, researchers have focused on one of the subunits of complex 1 encoded by a mitochondrial DNA gene (see Chapter 11). The rationale is that this gene is encoded by mtDNA and might serve as a reporter for energy stress in parkinsonian neurons. Mutations have been found in ND5, a mitochondrial gene encoding a complex 1 subunit (Parker and Parks, 2005). Numerous other genes causing mitochondrial dysfunction linked to Parkinson's have now been identified (Blok et al., 2009; Cassarino and Bennett Jr., 1999; Dawson and Dawson, 2003; Gautier, Kitada, and Shen, 2008; Haque et al., 2008; Luoma et al., 2007; Mandemakers, Morais, and DeStrooper, 2007; Obeso et al., 2010; Valente et al., 2004; Wang et al., 2006).

One of the most interesting areas of research on Parkinson's involves defects in the dynamics of parkinsonian mitochondria (VanLaar and Berman, 2009). The unique energy demands of these neurons require coordinated distribution and maintenance of mitochondria. Key dynamic properties of parkinsonian mitochondria include fission (growth), fusion, trafficking, biogenesis, and degradation. These properties of mitochondria are universal in all human cells but may be particularly important in Parkinson's. Dysfunction in mitochondrial dynamics has been increasingly linked to neurodegenerative diseases but particularly strong evidence is accumulating for roles in the molecular pathology of Parkinson's. Unique biochemical properties that might predispose parkinsonian neurons toward sensitivity to alterations in mitochondrial dynamics have already been discussed. Evidence from studies of Parkinson's-inducing chemicals supports the concept that mitochondrial fission, fusion, and transport may be involved in pathogenesis. In addition, increasing evidence suggests that two proteins linked to familial forms of the disease, Parkin and PINK1, interact in a common pathway to modulate mitochondrial fission/fusion. Parkin may also help target damaged mitochondria for degradation (mitophagy). Taken together these data suggest that mitochondrial dynamics may play an important role in Parkinson's (Banerjee et al., 2009; Beal, 2003; Büeler, 2009; Dodson and Guo, 2007; Thomas and Beal, 2007; Trimmer et al., 2009).

14.7 DUAL ENERGY-PACEMAKER MODEL OF AGE-DEPENDENT PARKINSON'S DISEASE

Recently, Clemens Scherzer, along with an international team of researchers have found that a root cause of Parkinson's disease may lie in 10 gene sets intimately related to energy production (Zheng et al., 2010). These genes include many that encode mitochondrial functions and others that encode bioenergetic systems besides those of mitochondria. These gene sets share the common property of modulating energy production rates in neurons. Lowering the energy threshold of neurons below a certain level is believed to be a death signal for parkinsonian neurons, which always seem to operate near full energy-producing capacity. With aging, these neurons, whose death triggers Parkinson's disease, seem to become even more vulnerable or sensitive to energy-triggered neuron death.

A master regulator called the PGC-1 alpha gene cuts a wide swath in neuron bioenergetics by modulating many energy-related genes. Abnormal expression of these genes is believed to occur during the early states of Parkinson's disease, long before symptoms appear. Thus, the PGC-1 alpha gene, its expression, and the biochemical activities of its gene product become new therapeutic targets for early treatment of Parkinson's disease. Mitochondria are especially important because PGC-1 alpha modulates hundreds of mitochondrial genes including many of those needed to maintain and repair mitochondria as the central power plants of neurons. In one sense these data open Pandora's box regarding the number of genes, chemicals, and conditions that can act as a trigger or enabler of neurodegenerative diseases, especially Parkinson's disease. These findings reinforce the widely held view that parkinsonian neurons are the "yellow canary" of the brain being predisposed to energy, oxidative or other forms of stress. This extreme sensitivity to stress is seen in the many hundreds of triggers of Parkinson's disease identified to date. This study highlights and opens a potential therapeutic strategy for slowing or stopping neuron damage before it becomes irreversible (i.e., at an early stage well before significant numbers of parkinsonian neurons are destroyed).

The current availability of U.S. Food and Drug Administration (FDA)–approved medications based on activation of PGC-1 alpha for the potential treatment of widespread diseases including diabetes may shorten the research time frame for development of a new parkinsonian drug (Muramatsu and Araki, 2002). A one-two punch against Parkinson's disease involving not only the currently available treatment of symptoms but also a preventive mechanism would be a giant step forward in the management of this disease. Zheng et al. (2010) tested their PGC-1a gene concept in rats showing that increasing expression of that protein in rat neurons grown in culture was enough to reduce toxic effects of MPTP and rotenone.

Researchers seeking the "holy grail" of aging will also find this result of great interest because these data are consistent with the revised mitochondrial theory of aging (Chapter 11). According to the mtDNA-mutator theory of aging, generalized energy decline occurring in a linear fashion is the result of point mutations in key energy-related genes encoded by mtDNA. Multiple mutations in mtDNA over time are envisioned to steadily lower energy levels toward a critical threshold. MtDNA mutations are considered to be occurring in all organs and cells of the body, more or less synchronizing aging in all cells. However, as proposed in this book we believe that neurons should be treated as a special case, subject not only to the mtDNA pacemaker of aging but also to a second membrane pacemaker mediated again by energy but involving the energy conservation side of bioenergetics (see Chapter 12). The dual pacemaker model highlights the risks of DHA as a weak link in energy conservation with aging. Parkinsonian neurons being predisposed to energy stress can be defined as a specialized subset of neurons whose program or programs for aging can readily be fast-forwarded by genetic, chemical, or physical means. With so many triggers being illuminated, is there a mechanism to unify various theories or concepts of Parkinson's disease? Data from Zheng and colleagues (2010) show that generalized energy status, especially at the level of mitochondrial energy production, is a unifying principle for Parkinson's disease with major implications for dementia in general. Above all, there seems to be a remarkable similarity between the aging of parkinsonian neurons and that of normal cells in that an energy deficit governs the aging of both.

In summary, research on Parkinson's disease has spearheaded the understanding of the importance of energy and oxidative stresses in neurodegeneration. Age-dependent Parkinson's is the normal pattern seen with onset of symptoms appearing after about 55 to 65 years of age. Thus, the major trigger or risk factor for Parkinson's disease is aging. Recent studies of the molecular basis of aging highlight the important role played by mutations in mtDNA and their likely causal relationship via energy stress to generalized death of cells occurring during aging. We propose that a second pacemaker of aging mediated by DHA present in neuron membranes is interlocked with the mtDNA pacemaker. Dual energy pacemakers are proposed to work together to set the life span of neurons and cells in general. Note that the mtDNA pacemaker is now considered to be linear in mechanism. Both pacemakers feature energy as the standard for setting these clocks, but the membrane-based pacemaker (i.e., DHA membranes in the case of neurons) is envisioned to be unique in its responsiveness to environmental signals. Finally, we propose that dual energy-based pacemakers govern premature aging in parkinsonian neurons. These combined pacemakers are envisioned to play the lead role in predisposing parkinsonian neurons to apoptosis, perhaps finalized by the recruitment of hyperactivated microglia undergoing an oxidative burst of ROS. A corollary of this idea is that any neuron in the brain finding itself on the cusp of energy-oxidative stress becomes an easy target for premature death caused by any further energy or oxidative stress. In late-stage Parkinson's disease the neuron death zone often spreads beyond the parkinsonian region of the brain and creates a larger field of devastation. Thus, advanced-stage Parkinson's disease begins to resemble Alzheimer's disease (AD). The dual pacemaker model as applied to Alzheimer's disease is discussed in Chapter 17.

REFERENCES

Banerjee, R., A. A. Starkov, M. F. Beal et al. 2009. Mitochondrial dysfunction in the limelight of Parkinson's disease pathogenesis. *Biochim. Biophys. Acta* 1792:651–663.

Beal, M. F. 2003. Mitochondria, oxidative damage, and inflammation in Parkinson's disease. *Ann. NY Acad. Sci.* 991:120–131.

Blok, M. J., B. J. van den Bosch, E. Jongen et al. 2009. The unfolding clinical spectrum of POLG mutations. *J. Med. Genet.* 46:776–785.

Büeler, H. 2009. Impaired mitochondrial dynamics and function in the pathogenesis of Parkinson's disease. *Exp. Neurol.* 218:235–246.

Cassarino, D. S., and J. P. Bennett Jr. 1999. An evaluation of the role of mitochondria in neurodegenerative diseases: Mitochondrial mutations and oxidative pathology, protective nuclear responses, and cell death in neurodegeneration. *Brain Res. Rev.* 29:1–25.

Chan, P., L. E. DeLanney, I. Irwin et al. 1991. Rapid ATP loss caused by 1-methyl-4-phenyl-1,2,3,6-tetrahydropyridine in mouse brain. *J. Neurochem.* 57:348–351.

Dauer, W., and S. Przedborski. 2003. Parkinson's disease: Mechanisms and models. *Neuron* 39:889–909.

Davey, G. P., and J. B. Clark. 1996. Threshold effects and control of oxidative phosphorylation in nonsynaptic rat brain mitochondria. *J. Neurochem.* 66:1617–1624.

Davey, G. P., S. Peuchen, and J. B. Clark. 1998. Energy thresholds in brain mitochondria. Potential involvement in neurodegeneration. *J. Biol. Chem.* 273:12753–12757.

Dawson, T. M., and V. L. Dawson. 2003. Molecular pathways of neurodegeneration in Parkinson's disease. *Science* 302:819–822.

Dodson, M. W., and M. Guo. 2007. Pink1, Parkin, DJ-1 and mitochondrial dysfunction in Parkinson's disease. *Curr. Opin. Neurobiol.* 17:331–337.

Fabre, E., J. Monserrat, A. Herrero et al. 1999. Effect of MPTP on brain mitochondrial H_2O_2 and ATP production and on dopamine and DOPAC in the striatum. *J. Physiol. Biochem.* 55:325–331.

Gautier, C. A., T. Kitada, and J. Shen. 2008. Loss of PINK1 causes mitochondrial functional defects and increased sensitivity to oxidative stress. *Proc. Natl. Acad. Sci. USA* 105:11364–11369.

Goodwill, K. E., C. Sabatier, C. Marks et al. 1997. Crystal structure of tyrosine hydroxylase at 2.3 A and its implications for inherited neurodegenerative diseases. *Nat. Struct. Biol.* 4:578–585.

Haque, M. E., K. J. Thomas, C. D'Souza et al. 2008. Cytoplasmic Pink1 activity protects neurons from dopaminergic neurotoxin MPTP. *Proc. Natl. Acad. Sci. USA* 105:1716–1721.

Hasegawa, E., D. Kang, and K. Sakamotoetal. 1997. A dual effect of 1-methyl-4-phenylpyridinium(MPP+)-analogs on the respiratory chain of bovine heart mitochondria. *Arch. Biochem. Biophys.* 337:69–74.

Hill, S., K. Hirano, V. W. Shmanai et al. 2011. Isotope-reinforced polyunsaturated fatty acids protect yeast cells from oxidative stress. *Free Radic. Biol. Med.* 50:130–138.

Jackson-Lewis, V., M. Jakowec, R. E. Burke et al. 1995. Time course and morphology of dopaminergic neuronal death caused by the neurotoxin 1-methyl-4-phenyl-1,2,3,6-tetrahydropyridine. *Neurodegeneration* 4:257–269.

Langston, J. W., P. Ballard, J. W. Tetrud et al. 1983. Chronic Parkinsonism in humans due to a product of meperidine-analog synthesis. *Science* 219:979–980.

Lotharius, J., and K. L. O'Malley. 2000. The parkinsonism-inducing drug 1-methyl-4-phenylpyridinium triggers intracellular dopamine oxidation. A novel mechanism of toxicity. *J. Biol. Chem.* 275:38581–38588.

Luoma, P. T., J. Eerola, S. Ahola et al. 2007. Mitochondrial DNA polymerase gamma variants in idiopathic sporadic Parkinson disease. *Neurology* 69:1152–1159.

Mandemakers, W., V. A. Morais, and B. DeStrooper. 2007. A cell biological perspective on mitochondrial dysfunction in Parkinson disease and other neurodegenerative diseases. *J. Cell Sci.* 120:1707–1716.

McGeer, E. G., K. Yasojima, and P. L. McGeer. 2001. Inflammation in the pathogenesis of Parkinson's disease. *BC Med. J.* 43:138–141.

Muramatsu, Y., and T. Araki. 2002. Glial cells as a target for the development of new therapies for treating Parkinson's disease. *Drug News Perspect.* 15:586–590.

Nicklas, W. J., S. K. Youngster, M. V. Kindt et al. 1987. MPTP, MPP+ and mitochondrial function. *Life Sci.* 40:721–729.

Obeso, J. A., M. C. Rodriguez-Oroz, C. G. Goetz et al. 2010. Missing pieces in the Parkinson's disease puzzle. *Nat. Med.* 16:653–661.

Parker, W. D. Jr., and J. K. Parks. 2005. Mitochondrial ND5 mutations in idiopathic Parkinson's disease. *Biochem. Biophys. Res. Commun.* 326:667–669.

Pavon, J. A., and P. F. Fitzpatrick. 2009. Demonstration of a peroxide shunt in the tetrahydropterin-dependent aromatic amino acid monooxygenases. *J. Am. Chem. Soc.* 131:4582–4583.

Przedborski, S., V. Kostic, V. Jackson-Lewis et al. 1992. Transgenic mice with increased Cu/Zn-superoxide dismutase activity are resistant to *N*-methyl-4-phenyl-1,2,3,6-tetrahydropyridine-induced neurotoxicity. *J. Neurosci.* 12:1658–1667.

Shchepinov, M. S., V. P. Chou, E. Pollock et al. 2011. Isotopic reinforcement of essential polyunsaturated fatty acids diminishes nigrostriatal degeneration in a mouse model of Parkinson's disease. *Toxicol. Lett.* 207:97–103.

Thomas, B., and M. F. Beal. 2007. Parkinson's disease. *Hum. Mol. Genet.* 16(R2):R183–R194.

Trimmer, P. A., K. M. Schwartz, M. K. Borland et al. 2009. Reduced axonal transport in Parkinson's disease cybrid neurites is restored by light therapy. *Mol. Neurodegener.* 17;4:26.

Valente, E. M., P. M. Abou-Sleiman, V. Caputo et al. 2004. Hereditary early-onset Parkinson's disease caused by mutations in PINK1. *Science* 304:1158–1160.

VanLaar, V. S., and S. B. Berman. 2009. Mitochondrial dynamics in Parkinson's disease. *Exp. Neurol.* 218:247–256.

Vila, M., V. Jackson-Lewis, S. Vukosavic et al. 2001. Bax ablation prevents dopaminergic neurodegeneration in the 1-methyl-4-phenyl-1,2,3,6-tetrahydropyridine mouse model of Parkinson's disease. *Proc. Natl. Acad. Sci. USA* 98:2837–2842.

Wang, D., L. Qian, H. Xiong et al. 2006. Antioxidants protect PINK1-dependent dopaminergic neurons in Drosophila. *Proc. Natl. Acad. Sci. USA* 103:13520–13525.

Zheng, B., Z. Liao, J. J. Locascio et al. 2010. PGC-1α, a potential therapeutic target for early intervention in Parkinson's disease. *Sci. Transl. Med.* 2:52ra73 (doi: 10.1126/scitranslmed.3001059).

15 Prion Diseases

In 1997, Stanley Prusiner was awarded the Nobel Prize in medicine for his discoveries on the molecular biology of prion diseases (Prusiner, 2001). Prion diseases in humans can be caused by eating meat (especially nervous tissue) that is contaminated by a remarkable infectious agent called a *prion*. Unlike all other infectious agents known previously, prions lack their own genetic material. Prions depend on a unique conversion process in the brain during which invading prions serve as protein templates to convert harmless and possibly beneficial cellular proteins located on the surface of neurons into prions. The conversion process ultimately floods the brain with prions killing massive numbers of neurons and eventually destroying brain function.

It is now clear that advances in understanding prion replication are applicable to neurodegenerative diseases in general. The purpose of this chapter is to explore the roles of energy, specifically energy stress, on prion replication.

15.1 A SPONGY-BRAIN DISEASE IN THE HIGHLANDS OF NEW GUINEA

The human form of prion disease is called *kuru*. Kuru reached national public attention following an outbreak in the highlands of New Guinea where cannibalism helped spread the disease. In the late 1950s, an outbreak of kuru occurred in the South Fore of the Okapa subdistrict of Papua New Guinea, though at the time the agent was unknown. Some researchers propose that kuru was spread easily and rapidly reaching epidemic proportions in the Fore people due to cannibalistic funeral practices. Relatives of the dead ate the remains, including brains, to guide the life force of the deceased back to their familiar homes. Male chauvinism was evident in this society because while the men of the village grabbed the choice cuts, women and children were left with the brain and the rest of the body. At the peak of the epidemic, kuru was eight to nine times more prevalent in women and children compared to men, a turn of events that turned out to be a blessing in disguise. Because of strict law enforcement against cannibalism by the Australian government, acting as protectorate of New Guinea, cannibalism disappeared and the epidemic of kuru rapidly declined.

Kuru is one of many prion diseases infecting animals (Table 15.1). In humans, Creutzfeldt-Jakob disease (CJD) is the most common type of transmissible spongiform encephalopathy (Table 15.2). CJD is still rare, found in about one per million people each year. It most often affects people between the ages of 45 to 75 but is most common in the age group 60 to 65. However, a variant CJD (vCJD) has recently been recognized that is lethal at a much earlier age. In more than 85 percent of cases the duration of CJD is less than one year (average of about four months) after onset of

TABLE 15.1
Selected Prion Diseases

Affected Animals	Disease	Shorthand	Infectious Prion
Sheep, goats	Scrapie		OrPrPSc
Cattle	Bovine spongiform encephalopathy	PscBSE	Bov PrPSc
Mink	Transmissible mink encephalopathy	TME	M PrPSc
White-tailed deer, elk, mule deer, moose	Chronic wasting disease	CWD	MDe PrPSc
Cat	Feline spongiform encephalopathy	FSE	Fe PrPSc
Human	Creutzfeldt-Jakob disease	CJD	HmPrPSc
Human	Iatrogenic Creutzfeldt-Jakob disease	ICJD	HmPrPSc
Human	Variant Creutzfeldt-Jakob disease	vCJD	HmPrPSc
Human	Familial Creutzfeldt-Jakob disease	fCJD	HmPrPSc
Human	Sporadic Creutzfeldt-Jakob disease	sCJD	HmPrPSc
Human	Gerstmann-Straussler-Scheinker syndrome	GSS	HmPrPSc
Human	Fatal familial insomnia	sFI	
Human	Kuru		

TABLE 15.2
Characteristics of the Most Common Familial Human Prion Disease: Creutzfeldt-Jakob Disease (CJD)

Characteristic	Classic CJD	Variant CJD
Average age at death	68 years	28 years
Average length of illness once symptoms appear	4 to 5 months	13 to 14 months
Clinical signs and symptoms	Dementia, early neurological signs	Prominent psychiatric/behavioral symptoms; delayed neurologic signs
Presence of amyloid plaques in brain tissue	May be present	May be present

symptoms. Like the case of Alzheimer's discussed in Chapter 17, events occurring prior to the onset of symptoms are poorly understood. However, unlike Alzheimer's, believed to start at a single or few focal points in the brain, CJD and related prion diseases have many disease points, essentially creating the spongy appearance of infected brain tissue. In the case of vCJD prions leave the brain via the lymph and bloodstream to infect other cells and tissues including tonsils. Thus, multiple sites in the brain seed neuron death, which eventually spreads exponentially throughout the infected brain. The presence of docosahexanoic acid (DHA) in neurons might enable the replication process as discussed later.

Note that following the kuru epidemic in New Guinea in the 1950s, two British scientists received Nobel Prizes for discovering that kuru is caused by an infectious agent and can spread to apes. The most publicized prion disease in the United States is known by its trivial name *mad-cow disease*. Prion diseases are now recognized as a family of fatal diseases destroying brain function in humans as well as animals (Table 15.1). For example, *wasting disease* of wild elk and deer and *scrapie* in domestic sheep are classified as prion diseases. In a recent outbreak recorded in Britain, "prion detectives" found that scraps of prion-infected meat, especially neural tissue, had been blended into cattle feed infecting these animals, which passed the infectious agent through their meat to humans.

Prusiner and colleagues found that the infectious agent did not belong to any known class of pathogenic agents and contained no genetic material but rather was a brand new disease agent without a name—the prion. Prions are so robust that sterilization in a conventional steam-pressurized autoclave at 274°F for 18 minutes may not be enough to kill infectivity. Powerful chemical procedures such as caustic soda are required, and it is believed that prions can persist for years in soil. Prions are peptides that are tough and are able to pass across the human gastrointestinal membrane and eventually cross the blood–brain barrier. Prions entering the brain cavity seem to linger on or replicate only slowly, in some cases for decades, before triggering disease symptoms. The replication process starts on the outside of neurons and might be started by a single prion.

Discoveries involving prion replication (Aguzzi, Baumann, and Bremer, 2008; Barria et al., 2009; Chiesa and Harris, 2009; Colby et al., 2009; Collinge and Clarke, 2007; LePichon et al., 2009) have opened up much new territory in the field of neurodegeneration and have provided important and universal information. Perhaps the most scientifically fascinating and revealing is the recruitment and conversion process. When harmless prion precursor protein is processed from a beneficial to a toxic state, each newly minted peptide is infectious, creating a snowballing effect, which overpowers and kills neurons (Elfrink et al., 2008; Hegde et al., 1999; Morillas et al., 1999).

15.2 PRION REPLICATION IS AN AMAZING PROCESS AND DEPENDS ON DYSFUNCTIONAL PROTEIN PROCESSING

Prion replication depends largely on biochemical processes carried out by the neuron, mechanisms that are pirated by prions to ensure their multiplication. There is reason to believe that a healthy neuron might be resistant to prion replication and that an unhealthy neuron is a more likely candidate to support the first round of prion replication. One scenario discussed in more detail later is that initial infective prions seek out neurons or axons with preexisting dysfunctional protein processing machinery. Background enzymes and other proteins are the workhorses in neurons. These molecular machines continuously wear out and must be degraded and resynthesized. Many mechanisms damage proteins including oxidation, alkylation, mutation, and misfolding. The apparatus that deliberately degrades aberrant proteins is called the *proteasome*, an abundant ATP-dependent protease that composes about

1 percent of total cellular protein. Proteasomes are widely dispersed in the cytoplasm and the nuclei of neurons. Each proteasome is formed of multiple subunits arranged as a hollow cylinder. Some of the subunits are distinct proteases whose active sites face the inner chamber of the cylinder. Proteins molecularly marked by quality control mechanisms as defective are threaded into the proteasome core where they are cleaved into smaller peptide segments. The threading reaction, which is energized by ATP, simultaneously unfolds the target protein, exposing the peptide bonds along the chain to proteases lining the proteasome core. The peptides produced by this process are further degraded, cleared from the brain, or accumulated as lipofuscin waste.

The main point emphasized here is that the mechanisms for monitoring and maintaining protein quality control along with systems for disposal of defective proteins burn up a significant fraction of the energy budget of neurons. Data from bacterial mutants harboring defects in the processing of defective proteins show that futile cycling of energy caused by dysfunctional protein processing can overwhelm the cell's energy supply, causing energy stress.

Whereas energy stress discussed below is a possible cause of dysfunctional protein processing in neurons, detailed mechanisms are poorly understood. But the products of defective protein degradation are readily apparent and include toxic peptides. Aggregates or tangles composed of peptide fragments are characteristic of neurodegenerative diseases, the most famous being amyloid plaque diagnostic of Alzheimer's disease.

Sophisticated mechanisms have evolved in neurons and other cells to ensure that the complex and unique folding patterns essential for the mode of action of specific proteins or enzymes are preserved. Some proteins are self-folding without help. Others require molecular chaperones to guide proper assembly. Note that ATP energizes a chaperone to ensure proper folding. Dysfunctionally folded proteins are threaded through proteasomes and digested, another ATP-dependent reaction as discussed above. Unfortunately, for long-term neural health all of these energy-dependent processes must continue to work smoothly to prevent the buildup of toxic protein aggregates from occurring.

It is now thought that the gradual decline of the neurons' protein quality control can cause neurodegenerative diseases by permitting normal and even essential proteins to form insoluble aggregates (Figure 15.1). In the case of Alzheimer's disease, the protein aggregates are released from dead neurons as amyloid plaque accumulating in the extracellular matrix. In other cases protein tangles accumulate inside neurons. Protein aggregates grow, survive, and trigger neuron damage because they have become highly resistant to proteolysis both outside and inside the neuron. Often toxic protein aggregates form fibrils built from a series of polypeptide chains that are layered one on top of another as a stack of β-sheets (Figure 15.1). The β-sheet structure blocks the access of proteases and thus builds up due to its resistance to proteolysis. Figure 15.1 is a simplified view of how a normally folded protein (ball) in the presence of a protein already converted to the β-sheet structure (flat sheet) can act as a protein template to seed the formation of more and more β-sheets starting with healthy proteins. A prion represented by the first β-sheet structure is believed to convert its precursor protein into newly minted prions by this type of mechanism.

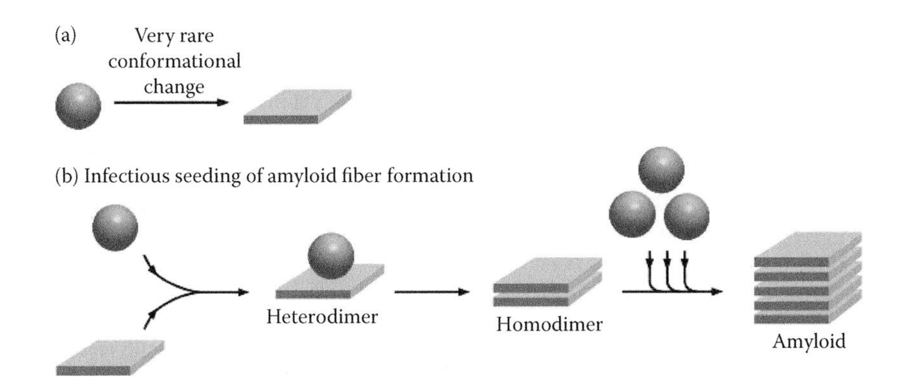

FIGURE 15.1 Synthesis of toxic protein aggregates including prions. A toxic peptide such as a prion is represented by the flat sheet, and its normal precursor protein is depicted as a sphere. A conformational change mediated by an infectious prion peptide converts a precursor protein to a toxic state. Note that a similar mechanism may result in the accumulation of toxic plaque in Alzheimer's disease with the exception that the initial infectious peptide needed to seed protein aggregate formation is generated internally in aging neurons. (From Alberts, B. et al., *Molecular Biology of the Cell*, Copyright 2008. Reproduced by permission of Taylor & Francis Group, LLC, a division of Informa plc.)

Prion diseases in general are caused by a misfolded aggregated form of prion protein (see Table 15.1 for nomenclature of specific prions). Prion precursor protein, PrPC, is located on the outer plasma membrane of neurons. There is increasing evidence that PrPC is beneficial, perhaps acting as a scavenger of copper ions protecting DHA membranes against lipid peroxidation (Hijazi et al., 2003; Hodak et al., 2009) (see Chapter 10). Copper is also an essential metal cofactor for building enzymes, which detoxify reactive oxygen species (ROS). Thus binding and transport of copper to the cytoplasm might be a normal beneficial function of PrPC.

Prion precursor protein (PrPC) and infectious prions (PrPSc) are internalized via clathrin-coated pits. With aging, levels of PrPC in the cytoplasm may build up to toxic levels. Entry of one or a few infectious prions into the cytoplasm of the cell is believed to facilitate the conversion process in which PrPC → PrPSc (see Figure 15.1). As discussed in the next section we propose that energy stress is the cause of dysfunctional protein processing and that energy-healthy cells can slow or postpone prion infection for many years. DHA membranes might play multiple roles including modulation of formation of clathrin-coated pits, generation of energy stress and oxidative stress, and anchoring of PrPC and PrPSc to the membrane surface. Note that lipid raft formation is considered to be important in the early stages of prion replication, and the dynamics of raft formation is modulated by DHA.

Once a prion in combination with its precursor PrPC enters the cytoplasm of neurons it becomes infectious because it recruits normally folded molecules of PrPC into prions. This template property of prion proteins creates a positive feedback loop that generates the abnormal form of PrPC. Like a chain reaction the pathological shape spreads rapidly from neuron to neuron, destroying brain function. It is extremely dangerous to eat the tissues of an animal which contain infectious prions, as seen in the recent highly publicized case of the spread of mad cow disease from cattle to humans in Great Britain. Fortunately, in the absence of prions as primer, the normal conformation of PrPC is very difficult to convert to its infectious form.

In a major breakthrough in the prion field, it has been found that yeast cells naturally produce prions, which surprisingly may play a beneficial role. Yeast has proven to be a powerful research tool and has helped clarify another unique property of prions. These protein molecules can form several different and specific types of aggregates from the same polypeptide chain. How a single polypeptide sequence of amino acids can adopt multiple aggregate forms is the subject of considerable research. Another question concerns the evolution of these "beneficial prions" believed to play positive roles in growth and development of yeast cells. Prion replication has now been engineered using recombinant DNA technology to occur inside recombinants of *Escherichia coli*.

15.3 DOES ENERGY PLAY A ROLE IN HOW PRIONS PICK THEIR FIRST NEURON TARGET?

Once crossing the blood–brain barrier and meandering on the outside of neurons for a variable amount of time, a single prion is able to multiply itself by the conversion process. The replication process starts on the membranes of neurons (Pinheiro, 2006; Taylor and Hooper, 2006; Westergard, Christensen, and Harris, 2007). Infectious prions are apparently free living for a time as they search among neurons for a victim for their replication. However, to multiply itself a PrPSc agent must feed on or recruit its parent protein PrPC, which is initially housed on the outer membrane surface of neurons. PrPC is a common protein on neuron membranes and is also located on membranes of other cells in the body (e.g., tonsil tissue, which also can be infected by prions). PrPC exhibits a membrane-loving tail called a *glycosyl phosphatidylinositol* (GPI) anchor, which is required for infectivity. The GPI anchor binds PrPC to the outer membrane surface specifically through association of the anchor held tight in cholesterol-rich lipid rafts. PrPC associates in rafts (Elfrink, Nagel-Steger, and Riesner, 2007) along with a variety of membrane proteins including signaling and adhesion molecules, stabilizing molecules and molecular chaperones, as well as scissoring enzymes or proteases. Thus, lipid rafts appear to provide a favorable membrane environment for conversion to begin. Because DHA likely governs lipid raft formation in neuron membranes, a link is established between DHA phospholipids and prion replication, a linkage reflecting the physical state of the membrane surface (see Chapter 5).

Researchers studying the sporadic form of Creutzfeldt-Jakob disease (sCJD) (Table 15.1) have uncovered an important missing link with Alzheimer's (Fernandez-Funez et al., 2009). This disease is called *sporadic* because it appears to occur without previous infection by prions from the environment but rather from prions originating spontaneously within the brain (prion precursor protein →→ infectious prion). These researchers used transgenic fruit flies carrying healthy prion precursor protein and showed that neurotoxic derivatives could be derived in the living fly.

They further showed that conversion is modulated by a second well-known protein that associates with and guides or chaperones prion precursor protein on the membrane surface. The generation of neurotoxic prion protein in living flies is greatly favored in old flies, a finding consistent with the view that aging cells expected to be energy stressed are more sensitive to prion infection.

The infectious cycle of prions is highly sophisticated and is believed to occur mainly in the cytoplasm or cytosol of neurons (Ma, Wollmann, and Lindquist, 2002). This involves extensive trafficking of both prion precursor protein and its infectious form PrPSc through lipid vesicles and their sites of synthesis and transformation in the neuron. As discussed in Chapter 5, DHA plays important roles in the dynamic of cycling of lipid vesicles. It is also well known that dysfunctional protein processing plays cardinal roles in prion replication and other forms of neurodegeneration, including Alzheimer's disease. According to one model the natural cycling of prion precursor protein and its load of copper from the outer membrane surface to the cytosol represents an essential cycle involving lipid vesicles. The onset of dysfunctional protein processing is envisioned to plug up the vesicle cycle and cause accumulation of toxic levels of PrPC in the cytoplasm, essentially turning healthy prion precursor protein to its infectious form. In the presence of invading infectious prions this natural tendency to convert PrPC → PrPSc is greatly accelerated. This model of prion replication raises numerous questions. What causes dysfunctional protein processing in the first place? This point is discussed next.

Whereas a great deal is known about prion recruitment, there is relatively little information about how a single prion targets its first neuron. The brain terrain is vast, and due to their robust stability there may be plenty of time for a prion to explore before finding a suitable site for replication. How a prion targets its victim might not be a random process, and it is possible that prions seek a specific physiological or genetic class of neurons. We propose that prion replication depends not only on a suitable supply of its precursor protein to feed on or replicate itself but also depends on a neuron predisposed to dysfunctional protein processing. This dependency on preexisting dysfunctional protein processing opens Pandora's box, because normal protein processing carried out by healthy neurons might prevent prion replication from taking hold. How rarely and how long it takes a prion to hone in on its first victim is an open question, but it may take decades (as discussed later). We propose that an energy-deficient neuron is the starting point for both prion disease as well as Alzheimer's, the latter is discussed in Chapter 17.

An anchorless form of prion precursor protein unable to attach to the membrane has been engineered in cells with the result that prions can no longer regenerate themselves, making these cells resistant to prion infection (Chesebro et al., 2005; McNally, Ward, and Priola, 2009). Transgenic mice expressing anchorless prion protein are also resistant to infection. Thus, prion replication might require not only a preexisting state of dysfunctional protein processing or energy stress but also a prion precursor protein that is first attached to the outer membrane surface. The mechanism of attachment might be influenced by DHA phospholipids known to break up lipid rafts. Interestingly, the anchorless PrP lacking the GPI anchor and blocking prion replication was found to be dysfunctionally processed yielding an abnormal peptide fragment deposited as amyloid plaque, similar to Alzheimer's. These data highlight the dependency of prion replication on conditions of the "host" cell including cellular energy homeostasis and energy stress, the physical state of the membrane influenced by DHA, and the dysfunctional protein processing.

In another study cells were protected from prion toxicity by the presence of excess chaperone proteins that interact with and fold amyloid peptides into shapes not recognized or converted by prions. Note that the process of proper folding of proteins by their chaperones requires a considerable amount of energy. Thus, healthy neurons with plenty of energy might retard replication, forcing prions to wait a long time before the energy supply of a target neuron weakens and allows the infection to begin.

Prions might be even smarter in selecting a cellular victim than discussed above. For example, prion precursor proteins are located on the membrane surface of different parts of neurons. Do prions target precursor proteins located at specific locations or domains of neuron membranes? In Chapter 14 we discuss how the enormous size and length of neurons might result in energy stress being localized in different regions of the cell. We suggest that cellular energy status, preexisting dysfunctional protein processing, membrane localization, and DHA homeostasis of membranes all deserve consideration in the context of early events in prion diseases.

15.4 PRION-INFECTED NEURONS MIGHT BE SUBJECTED TO MEMBRANE ENERGY UNCOUPLING CAUSED BY THE PRION

During the past decade or so remarkable progress has been made in understanding the molecular biology of prion replication. However, knowledge of how prions kill neurons has lagged behind. Data from studies of transgenic mice expressing prion protein molecules with deleted segments or with amino acid substitutions have provided insight into possible mechanisms of prion-induced neuron death. Data from transfected cells expressing the most toxic deletion mutants show that these peptides generate nonselective cation permeable channels or pores in the cell's membrane (Solomon, Huettner, and Harris, 2010). Interestingly, overexpression of wild-type protein silences membrane defects. Several different point mutations that cause familial prion diseases in humans also induce cation-permeable channels in the membrane. Solomon, Huettner, and Harris (2010) point out that cation pores defined in their experiments are distinct from experimental data generated by using synthetic peptides derived from the prion sequence targeting artificial lipid membranes. That is, cation pores in membranes of transfected cells are induced by membrane-anchored prions. These data are consistent with the concept that the anchored form of mutated prion protein synthesized *in vivo* creates cation channels in the membrane that are somehow closed by wild-type prion protein. One scenario to explain these data is that wild-type prion protein recruits mutant prion protein to lipid rafts for endocytosis to the cytoplasm, whereas mutant forms remain in liquid regions of the membrane where they tend to generate cation pores and cause futile energy cycling. According to a second scenario highlighting a role for DHA, deleted and mutant forms of PrPC create cation pores indirectly. The mutation can be envisioned to block the antioxidation function of wild-type prion proteins acting as copper scavengers important in preventing membrane peroxidation. The resulting membrane damage in DHA-enriched axon membranes is expected to generate lipid whiskers causing futile energy cycling.

15.5 NUMEROUS PINPOINT BRAIN LESIONS CAUSED BY PRIONS MAY SPREAD AS THE RESULT OF MICROGLIAL-MEDIATED NEURON DAMAGE

Brain death in prion diseases is apparently caused by numerous prion-infected foci initially involving relatively small lesions that spread, giving the classic spongiform look characteristic of these diseases. Much of the discussion so far has centered on the prion replication process in which one or a few prions multiply and spread out to seed many pinpoints of infection. Decades may elapse between the time an infectious prion enters the brain and the final stages where neuron death spikes sharply upward. It is not known what causes the apparent spiraling or exponential neuron death associated with prion diseases. One possibility is that prions initially seed pinpoint lesions that expand through participation of the inflammatory stage of activated microglia. A central role played by hyperactive microglia is gaining attention as a unifying mechanism applicable to all neurodegenerative diseases. DHA is proposed to contribute to microglial neuron death as discussed in more detail in Chapters 16 and 17.

15.6 A PROTECTIVE GENE AGAINST KURU FOUND IN SURVIVORS OF NEW GUINEA EPIDEMIC

In the mid-1990s a cure against prion replication in mice was developed in which prion replication was blocked (Büeler et al., 1992; Mallucci et al., 2003). During the next 16 years, attempts to prevent prion infection in humans based on this discovery were not successful. A recent flurry of scientific advances has once again fanned the flames of hope for a cure of prion diseases in humans. Following the outbreak of kuru in the 1950s, medical experts showed that the mean incubation period of the disease is 14 years and cases were reported with latent periods of 40 years. Recently, researchers have discovered that persons, mostly women, who survived the kuru epidemic were carriers of a prion resistance factor (Mead et al., 2009). The source of the immunity has been traced to the inheritance of a genetic variant of prion protein (Mead et al., 2009). The change of a single codon at the 129 position of the prion protein gene (PRNP) was found in persons exposed to kuru who survived the epidemic. This PRNP variant named G127V was found exclusively in people who lived in the region in which kuru was prevalent and was present in half of the otherwise susceptible women from the region of the highest exposure. This protective or tolerance gene variant is not found in patients with kuru or in unexposed population groups in the rest of the world. Families carrying the protective gene show a significantly lower incidence of kuru compared to families without the gene matched to geographically similar locals.

Thus, a natural genetic mutation targeted to the gene governing prion precursor protein results in protection against deadly kuru infections. However, in the case of variant Creutzfeldt-Jakob disease (see Table 15.2) the opposite is true because a different mutation in the same gene intensifies the disease. These mutations modulate prion diseases at the conversion process, which highlights the importance of dysfunctional protein processing perhaps mediated by energy stress on the infectious cycle of prions. These data have sparked great interest and raise hope for a cure for prion diseases in humans.

In summary, studies of the molecular pathology of prion diseases have led to a flurry of new discoveries. Many of these advances have occurred in the area of prion replication. Dysfunctional protein processing is a critical step of prion replication and might be linked to aging, energy stress, and membrane unsaturation now recognized as being risk factors for dementia. Mutations in mice and a natural mutation in humans that block prion replication stop the infectious cycle in mice and humans, respectively. The finding of a natural pool of resistance genes in survivors of the kuru epidemic in New Guinea opens a new window of opportunity for eventually curing prion diseases in humans. It is increasingly clear that there is major scientific cross-fertilization between research on prion diseases and neurodegeneration in general.

REFERENCES

Aguzzi, A., F. Baumann, and J. Bremer. 2008. The prion's elusive reason for being. *Annu. Rev. Neurosci.* 31:439–477.

Barria, M. A., A. Mukherjee, D. Gonzalez-Romero et al. 2009. *De novo* generation of infectious prions *in vitro* produces a new disease phenotype. *PLoS Pathog.* 5:e1000421.

Büeler, H., M. Fischer, Y. Lang et al. 1992. Normal development and behaviour of mice lacking the neuronal cell-surface PrP protein. *Nature* 356:577–582.

Chesebro, B., M. Trifilo, R. Race et al. 2005. Anchorless prion protein results in infectious amyloid disease without clinical scrapie. *Science* 308:1435–1439.

Chiesa, R., and D. A. Harris. 2009. Fishing for prion protein function. *PLoS Biol.* 7:e75.

Colby, D. W., K. Giles, G. Legname et al. 2009. Design and construction of diverse mammalian prion strains. *Proc. Natl. Acad. Sci. USA* 106:20417–20422.

Collinge, J., and A. R. Clarke. 2007. A general model of prion strains and their pathogenicity. *Science* 318:930–936.

Elfrink, K., J. Ollesch, J. Stöhr et al. 2008. Structural changes of membrane-anchored native PrP(C). *Proc. Natl. Acad. Sci. USA* 105:10815–10819.

Elfrink, K., L. Nagel-Steger, and D. Riesner. 2007. Interaction of the cellular prion protein with raft-like lipid membranes. *Biol. Chem.* 388:79–89.

Fernandez-Funez, P., S. Casas-Tinto, Y. Zhang et al. 2009. *In vivo* generation of neurotoxic prion protein: Role for hsp70 in accumulation of misfolded isoforms. *PLoS Genet.* 5:e1000507.

Hegde, R. S., P. Tremblay, D. Groth et al. 1999. Transmissible and genetic prion diseases share a common pathway of neurodegeneration. *Nature* 402:822–826.

Hijazi, N., Y. Shaked, and H. Rosenmann et al. 2003. Copper binding to PrPC may inhibit prion disease propagation. *Brain Res.* 993:192–200.

Hodak, M., R. Chisnell, and W. Luetal. 2009. Functional implications of multistage copper binding to the prion protein. *Proc. Natl. Acad. Sci. USA* 106:11576–11581.

LePichon, C. E., M. T. Valley, M. Polymenidou et al. 2009. Olfactory behavior and physiology are disrupted in prion protein knockout mice. *Nat. Neurosci.* 12:60–69.

Ma, J., R. Wollmann, and S. Lindquist. 2002. Neurotoxicity and neurodegeneration when PrP accumulates in the cytosol. *Science* 298:1781–1785.

Mallucci, G., A. Dickinson, J. Linehan et al. 2003. Depleting neuronal PrP in prion infection prevents disease and reverses spongiosis. *Science* 302:871–874.

McNally, K. L., A. E. Ward, and S. A. Priola. 2009. Cells expressing anchorless prion protein are resistant to scrapie infection. *J. Virol.* 83:4469–4475.

Mead, S., J. Whitfield, M. Poulter et al. 2009. A novel protective prion protein variant that colocalizes with kuru exposure. *N. Engl. J. Med.* 361:2056–2065.

Morillas, M., W. Swietnicki, P. Gambetti et al. 1999. Membrane environment alters the conformational structure of the recombinant human prion protein. *J. Biol. Chem.* 274:36859–36865.

Pinheiro, T. J. 2006. The role of rafts in the fibrillization and aggregation of prions. *Chem. Phys. Lipids* 141:66–71.

Prusiner, S. B. 2001. Shattuck lecture—Neurodegenerative diseases and prions. *N. Engl. J. Med.* 344:1516–1526.

Solomon, I. H., J. E. Huettner, and D. A. Harris. 2010. Neurotoxic mutants of the prion protein induce spontaneous ionic currents in cultured cells. *J. Biol. Chem.* 285:26719–26726.

Taylor, D. R., and N. M. Hooper. 2006. The prion protein and lipid rafts. *Mol. Membr. Biol.* 23:89–99.

Westergard, L., H. M. Christensen, and D. A. Harris. 2007. The cellular prion protein (PrP(C)): Its physiological function and role in disease. *Biochim. Biophys. Acta* 1772:629–644.

16 Brain Trauma–Induced Dementia

Traumatic brain injury is a leading cause of death and disability in the United States with an estimated incidence of 1.4 million persons each year (Centers for Disease Control and Prevention, 2006). Public awareness of the importance of brain trauma has increased greatly because of reports of increases in early dementia among athletes in trauma sports and the high incidence of brain injuries of combat troops.

Recently, a number of stars in the National Football League (NFL) have agreed to autopsies of their brains to gain a better understanding of the link between brain trauma and early dementia in athletes. Because of the celebrity status afforded NFL athletes, their case has attracted the attention of the public. NFL owners represented by the league president and the union head representing players were recently called before a U.S. congressional committee investigating an apparent rash of early dementia in retired players such as John Mackey, who was inducted into the Football Hall of Fame. In their summary statement to Congress, owners initially expressed "doubt" that sufficient scientific data were available to definitively link brain trauma to early dementia in retired NFL players. Early dementia in this case is defined as occurring in players about 15 to 20 years after retirement or in the range of 50 years of age. In fairness, the league has recently loosened its purse strings, and players diagnosed with early dementia now receive an added pension to cover some of their additional medical costs.

16.1 HIGH RATES OF BRAIN TRAUMA IN SPORTS AND COMBAT

Retired professional football players may have a higher rate than normal of Alzheimer's disease or other memory problems. This preliminary study provides more fuel for concerns about the long-term risk associated with contact sports including football, rugby, soccer, and boxing. Some 1063 ex-players in the NFL were asked if they have ever been diagnosed with dementia, Alzheimer's disease (AD), or other memory-related diseases. About 2 percent of former players aged 30 to 49 answered affirmatively. This is almost 20 times greater compared to the same age group in the general public (see Table 16.1). For retirees over age 50 the reported rate of 6 percent is five times higher than normal. This study performed at the University of Michigan and paid for by the NFL was first reported by the *New York Times*. Since the initial report became public, something of a firestorm has erupted raising concerns across many sports. High rates of brain trauma experienced by soldiers during combat have added fuel to this fire. This raised awareness has led to increased research in this area (Aoyama et al., 2008; Bayir et al., 2007; Giza and Hovda, 2001; Potts,

TABLE 16.1
Reported Rates of Dementia among Retired NFL Players

	U.S. Men Ages 30 to 49 Years	U.S. Men Ages 50+ Years	NFL Retirees Ages 30 to 49 Years	NFL Retirees Ages 50+ Years
Reported rates of dementia, Alzheimer's disease, or other memory-related diseases	0.1%	1.2%	1.9%	6.1%

Adwanikar, and Noble-Haeusslein, 2009; Shaw, 2002; Vagnozzi et al., 2005; Weber, 2007). The fuse or trigger for trauma-induced dementia is still poorly understood but likely shares properties with Alzheimer's disease and other forms of dementia as well as cancer and aging. Similarities in symptoms between sports versus blast-induced brain trauma are extensive and are reviewed by Lew et al. (2007).

Soldiers in combat share the same peril as athletes in contact sports. Army field studies show that greater than 10 percent of troops in Iraq and Afghanistan have suffered at least one concussion or brain injury, caused mainly by roadside bombs. About 5 to 15 percent of blast victims develop lasting problems with concentration, short-term memory, fatigue, and chronic headaches. There is a chance that these symptoms will later terminate in dementia (Trudel, MacKay-Brandt, and Temple, 2008). A summary of data made public by the Pentagon shows several similarities between multiple blows to the head in sports and brain injury from an explosion as follows:

- Damage to wiring in the brain from an explosion appears more severe and widespread compared to a single concussion in sports.
- Brain cell inflammation caused by blast waves lasts longer than inflammation caused by a blow to the head.
- In moderate or severe cases of blast-induced brain trauma, oxygen flow to the brain is more restricted due to apparent spasms in blood vessels compared to a blow to the head in sports (translated as meaning that many more neurons are subjected to energy stress).

The reasons behind the prominence of traumatic brain injury in the wars in Iraq and Afghanistan as opposed to previous conflicts are unclear. According to Moore et al. (2008), the physics of brain trauma may have changed as the result of increased survivability of blasts due to widespread use and improvements in body armor. This comprehensive review also traces the historical context of blast injury, which led to their theory that states that neurons of the brain are far more sensitive to blast trauma than neurons in the rest of the body. See Chapter 7 for a description of biochemical events occurring as axons are excessively stretched.

16.2 AXON STRETCH MODEL OF TRAUMA-TRIGGERED ALZHEIMER'S

In this and following sections molecular models explaining the linkage between brain trauma and Alzheimer's or Alzheimer's-like neurodegeneration are highlighted. The axon stretch injury model is discussed first. For background on basic aspects of axon stretching see Chapter 7. Figure 16.1 (top) shows a diagram of an apparatus that D. H. Smith at the University of Pennsylvania School of Medicine uses to gradually stretch axons. Axons that have been stretch-grown are shown in the inset (bottom). The main point here is that this research on the fundamental nature of stretch growth of neurons has implications in at least two important areas of applied neurology—regrowth of damaged or severed neurons (e.g., spinal injury) and treatment of brain damage from trauma. The latter area is now coming of age, and a model developed in Smith's laboratory is discussed next (Johnson, Stewart, and Smith, 2010). The cascade of events from brain trauma leading to the formation of amyloid plaque or amyloid-ß (AB) is described in Figure 16.2. This figure highlights what occurs shortly after traumatic brain injury from sports or combat. Injury is caused directly by mechanical forces that can damage the axon structure or indirectly by initiating harmful stress cascades or both. This model highlights mechanical events that clog axonal transport causing formation of *axonal bubbles* at the terminus of severed and damaged axons. Critical biochemical events occurring once

FIGURE 16.1 Apparatus used to mechanically stretch axons. Brain trauma results in damage to axon structure and biochemistry, events that are being reproduced in the laboratory of Smith and other neurobiologists. (Photo courtesy of D. H. Smith, University of Pennsylvania Medical School; Reprinted from *Prog. Neurobiol.*, 89, Smith, D. H., Stretch Growth of Integrated Axon Tracts: Extremes and Exploitations, 231–239. Copyright 2009, with permission from Elsevier.)

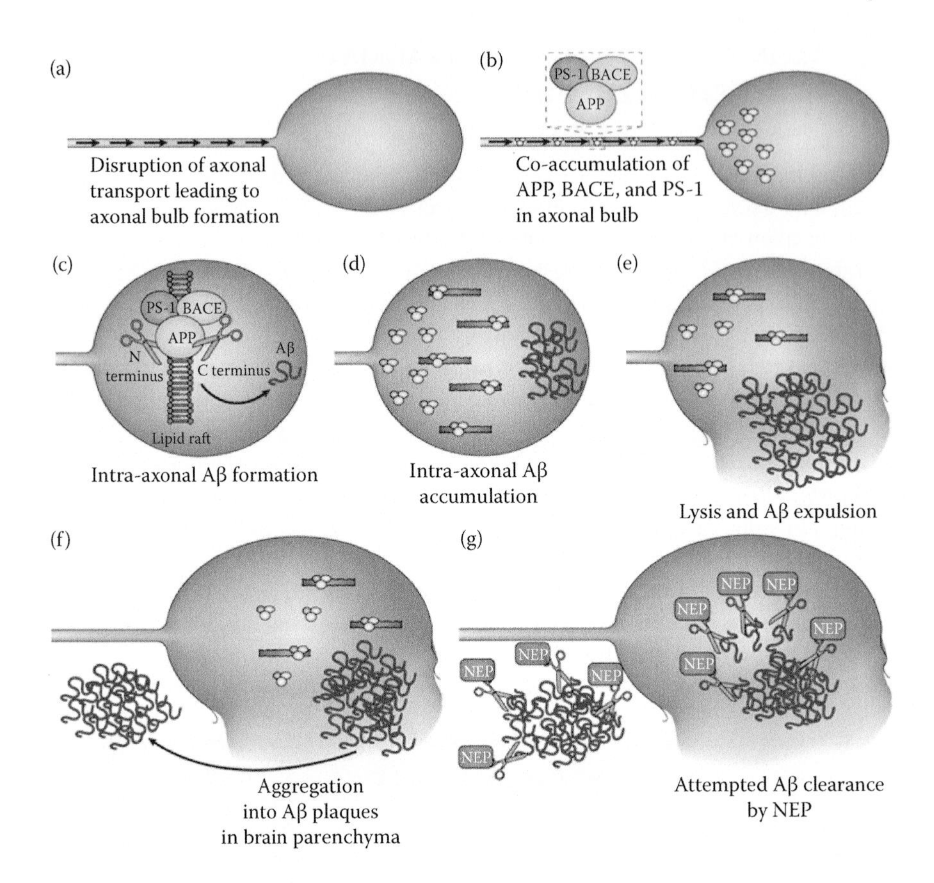

FIGURE 16.2 How mechanical damage triggers axons to produce amyloid plaque characteristic of Alzheimer's disease. Axonal transport is blocked or hindered in injured axons resulting in swelling at their disconnected terminals and accumulation of multiple misplaced proteins. Axonal bulbs contain the enzymes necessary for cleavage of amyloid precursor protein (APP) to amyloid-ß, including presenilin (PS-1), and ß-site, APP-cleaving enzyme (BACE). The enzyme that clears Aß, neprilysin (NEP), also accumulates in axonal bulbs, but the ability of this enzyme to clear amyloid-β might be overwhelmed by excessive production of this dysfunctionally produced peptide following trauma. Thus, localized brain damage caused by brain trauma is proposed to tip the balance between buildup and clearance of toxic peptides. (Reprinted by permission from Macmillan Publishers Ltd: *Nature Rev. Neurosci.*, Johnson, V. E., W. Stewart, and D. H. Smith, Traumatic Brain Injury and Amyloid-β Pathology: A Link to Alzheimer's Disease? *Nat. Rev. Neurosci.*, 11: 361–370, Copyright 2010.) **(See color insert.)**

bubbles form include accumulation of the substrate for formation of amyloid plaque (APP) as well as proteases that cleave APP → amyloid-ß. Amyloid-ß is depicted as tangled red strands accumulating inside the axon bulb. Lysis of the bulb releases this toxic peptide into the extracellular space where it can aggregate as amyloid plaque, the signature molecule of Alzheimer's disease. It is important to note that this cascade does not immediately trigger Alzheimer's, which may never occur or may develop some years later. Instead, the take-home lesson of this research is that brain trauma causes a rapid and momentary imbalance in production versus clearing of amyloid-ß. For example, during moderate trauma such as occurs during a hard hit in football, the clearing of amyloid-ß may relatively quickly return to normal or to a level that does not immediately create a focal point of the disease. However, any amyloid-ß remaining might constitute a smoldering threat that could "seed" more widespread neuron damage at a later time.

There are several remarkable features about the data that support this model. The first is the amazingly rapid onset of amyloid-ß accumulation following brain trauma. The brown spots seen in Figure 16.3a appeared within 10 days in the brain of a young patient with severe and fatal brain trauma (Johnson, Stewart, and Smith, 2010).

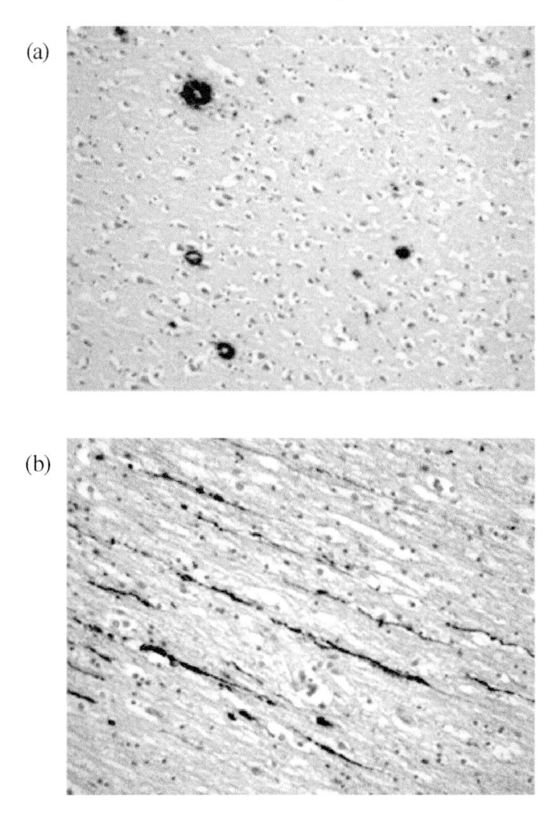

FIGURE 16.3 Focal points of brain damage can occur within hours of trauma to the brain. (a) Brown spots of amyloid-ß (Aß) peptide found in brain tissue following a single lethal incident of traumatic brain injury. The 18-year-old male's survival time after this injury was just 10 hours. Plaques were identified using an antibody specific for Aß. (b) A laboratory demonstration shows precursor protein for amyloid-ß (APP) accumulating along the length of stretch-damaged axons. APP (brown) was detected using an immunohistochemical technique. (Reprinted by permission from Macmillan Publishers Ltd: *Nature Rev. Neurosci.*, Johnson, V. E., W. Stewart, and D. H. Smith, Traumatic Brain Injury and Amyloid-β Pathology: A Link to Alzheimer's Disease? *Nat. Rev. Neurosci.*, 11: 361–370, Copyright 2010.)

Specific antibodies against amyloid-ß were used to detect these areas of excess production of this toxic peptide. The bottom photograph (Figure 16.3b) is a laboratory demonstration that shows the localization of APP, which accumulates along the length of damaged axonal tracts. APP was detected using immunohistochemical techniques. Accumulation of amyloid-ß plaque characteristic of age-dependent Alzheimer's disease may take 60 to 70 years to develop. Obviously, dysfunctional protein processing leading to amyloid-ß plaque is greatly fast-forwarded by brain trauma, a conclusion supported by studies of animal models (Johnson, Stewart, and Smith, 2010). A second difference between this model for brain trauma versus that of normal Alzheimer's is that age-dependent accumulation of toxic peptides is not reversible. Perhaps the most remarkable aspect of this model is that it pinpoints the axon as the site of dysfunctional protein processing leading to toxic peptide formation. This model highlights the linkage between brain trauma and formation of amyloid-ß. Smith addresses the question of whether an old brain injury can much later lead to dementia (Smith, 2010). The long latent period between brain trauma and dementia leads us to consider a second model for trauma-induced Alzheimer's.

16.3 INFLAMMATION MODEL

Inflammation, DHA membranes, and energy stress might be linked. Inflammation deserves more attention as the initial driver or trigger of dementia. The following chain of events is offered as a working model to guide discussion:

- Brain trauma causes localized inflammation (usually heals in healthy young athletes).
- Oxidative damage to DHA membranes of neurons generates energy-uncoupling water-wires.
- Repeated sharp but localized drops of energy caused by inflammation raise the probability of a "seed" for Alzheimer's.
- With aging more damaged regions arise and serve as "rotten apples in a barrel" leading to early symptoms of Alzheimer's.

For a general background, recall that inflammation is a life-saving process in the body attracting disease-fighting phagocytes that harness their potent oxidative chemicals to kill invading microbial pathogens and viruses. Unfortunately, too much or persistent inflammation can be harmful causing diseases such as rheumatoid arthritis, colitis, and even cancer and dementia. Inflammation may result in increased levels of reactive oxygen species (ROS) and raise the likelihood of damage to DHA membranes perhaps through formation of energy-robbing water-wires.

Chronic inflammation is well known as a risk factor in cancer, and research on this linkage has provided much of the insight on the importance of inflammation in age-dependent diseases. For example, in a study of 3000 older Australians, those with an exaggerated white cell count, a sign of inflammation, were more likely to die of cancer (Shankar et al., 2006). White blood cell (WBC) count or the measure of white blood cells in the blood is a reliable and widely used marker that reflects inflammation throughout the body. People with chronic infections have higher WBC counts. The Australian researchers asked the question of whether a WBC count can predict cancer. The main conclusions from this work are that individuals in the highest quarter of WBC count had an increased risk of cancer and that this risk decreased when anti-inflammatory drugs such as aspirin were taken routinely.

Other research shows an increased risk of certain cancers associated with chronic inflammation. For example, inflammatory bowel diseases including Crohn's disease and ulcerative colitis are linked to cancers of the intestinal tract. Like early dementia there is no general principle regarding the molecular pathology of these diseases except possibly the involvement of chronic inflammation (Farooqui, Horrocks, and Farooqui, 2007). Chronic inflammation impacts all stages of cancer development: initiation, progression, and metastasis. Inflammation from brain trauma induces the release of hormones that alert and recruit inflammatory cells into brain tissues causing release of ROS. Inflammatory macrophages recruited to damaged tissue are regarded as the primary source of ROS (Hanisch and Kettenmann, 2007). These toxic chemicals might be channeled through damage to DHA membranes to create localized but permanent damage to traumatized neurons—acting as a focal point for early dementia.

A trauma-induced brain lesion of a football player in the prime of health may heal quickly enough for next week's game. However, this may be a game of Russian roulette played with long-term brain health at stake. That is, each successive hit or concussion might increase the probability of a wounded neuron eventually triggering a cascade leading to early dementia. Once again evidence is accumulating that the Alzheimer's disease pacemakers of athletes suffering repeated trauma in sports are set about 15 to 20 years earlier. The fuse or trigger for trauma-induced dementia is still poorly understood, but inflammation is a likely candidate and might set into motion an inflammation-DHA-energy cascade of events leading to dementia.

A linkage between inflammation, DHA, and Alzheimer's disease is discussed in Chapter 17. As summarized in Figure 16.4, microglia, which sense and guard against excessive oxidative damage to neuron membrane structure, can become hyperactivated and become executioners of neurons. As shown in Figure 16.4, microglial cells in the surveillance (sentry) state—traditionally termed *resting state*—constantly scan for signals that would indicate a potential threat to central nervous system (CNS) homeostasis. The appearance of such "activating" signals (in infection, trauma, or cell impairment) or the loss of constitutive "calming" signals triggers a transition to an alerted state (see Figure 16.4 figure legend for details).

Oxidized DHA phospholipids are believed to modulate the activity states of microglia by signaling through the CD36 signaling pathway (Chapters 8 and 17). Excessive inflammation by brain trauma may do more irreversible localized damage to the brain than previously thought, with each trauma-induced wound healing but raising the probability with aging that a focal point for AD can be seeded from this damaged region.

An inflammation model is consistent with recent data on lipid peroxidation following traumatic brain injury in rats (Bayir et al., 2007). Molecular species analysis demonstrated the presence of highly oxidizable species of DHA phospholipids in

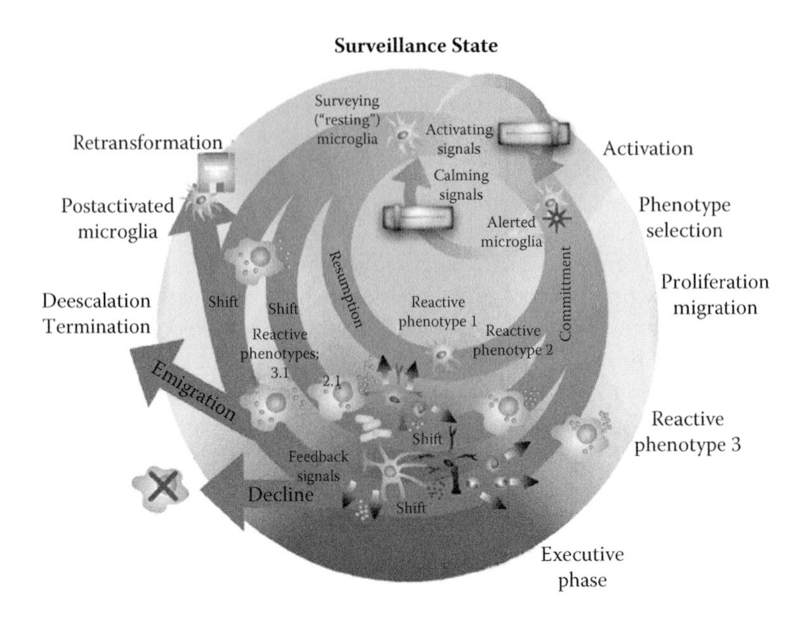

FIGURE 16.4 Microglial cells in the surveillance (sentry) state— traditionally termed *resting state*—constantly scan for signals that would indicate a potential threat to central nervous system (CNS) homeostasis. The appearance of such "activating" signals (in infection, trauma, or cell impairment) or the loss of constitutive "calming" signals triggers a transition to an alerted state. Signals and their context are interpreted and converted to an initial response of "activation." Cells commit to distinct reactive phenotypes as they enter their executive phase (e.g., release of cytokines and chemokines, phagocytic activity). Three examples are depicted (phenotypes 1, 2, 3), but the diversity could be larger. Throughout the subsequent period, the reactive behavior of microglia may change (reactive phenotypes 2.1, 3.1), largely controlled by a fading (or elimination) of the initial activating signals as well as influences from resident CNS and invading immune cells (illustrated as feedback signals). Reactive phenotypes may thus shift, eventually leading to a more repair-orientated profile. While some cells (indicated by an "X" over the cell) may emigrate to the blood system or die, others may revert to a "resting" (surveying) state. Some cells may not retransform to a completely naive status and may remain as postactivated microglia. These cells could keep subtle changes (e.g., in transcriptional activity) that affect their sensitivity to constitutive (calming) signals or alter responses to subsequent stimulation. Postactivated microglia could thus have acquired some experience (indicated as memory in the figure by a floppy disk icon). (Reprinted by permission from Macmillan Publishers Ltd: *Nature Neurosci.*, Hanisch, U.-K., and H. Kettenmann, Microglia: Active Sensor and Versatile Effector Cells in the Normal and Pathologic Brain, 10: 1387–1394, Copyright 2009.)

mitochondrial and synaptosomal membranes. These data show that as early as three hours after injury DHA-cardiolipin is the only molecular species to display high levels of peroxidation followed 24 hours later by other phospholipids. Thus, peroxidation of DHA-CL preceded the appearance of other biomarkers of apoptosis including caspase-3 activation. These data are consistent with oxidation of DHA-CL being one of the earliest detectable events in a cascade that eventually may lead to dementia.

16.4 SUICIDE PEPTIDE MODEL

The mechanical-biochemical cascade discussed above does not trigger an acute chain reaction mechanism leading to Alzheimer's and instead still depends on the process of aging. The events occurring in late stages of Alzheimer's are poorly understood. In Chapter 14 we discussed how a highly toxic neurotoxin (MPTP) was found to trigger an acute cascade leading in days to symptoms of Parkinson's disease. Recently, Nikolaev et al. (2009) discovered a potent suicide peptide produced during dysfunctional protein processing of axons, which binds to naturally occurring suicide receptors on healthy axons and quickly triggers dramatic axon destruction and lysis (Figure 16.5a). This experiment was carried out using axons cultured in the

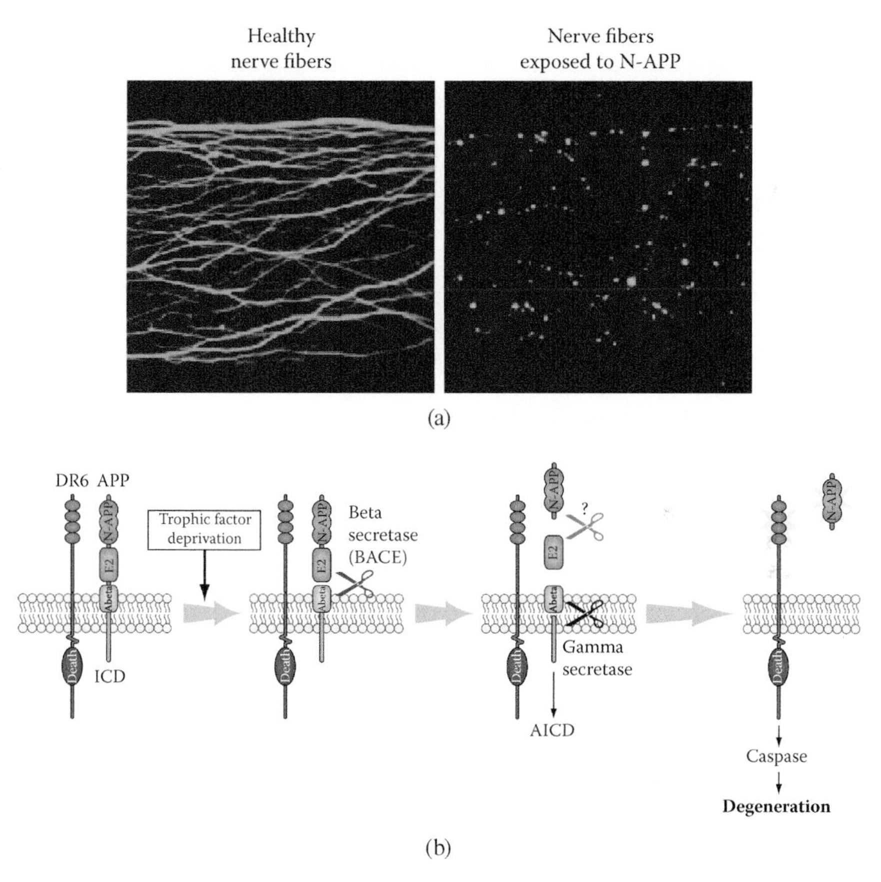

FIGURE 16.5 A potent suicide peptide N-APP generated by dysfunctional protein processing rapidly destroys axons. (a) Healthy neurons grown in the laboratory (left panel) are rapidly degraded by the toxic peptide N-APP (right panel). (b) Mode of action of a suicide peptide. DR6 is a suicide receptor that is activated when N-APP binds to its active site. Note that DR6 suicide receptors on neighboring axons can be activated by N-APP. Severe brain trauma might activate this destructive process. (Reprinted by permission from Macmillan Publishers Ltd: *Nature*, Nikolaev, A., T. McLaughlin, D. D. M. O'Leary et al., APP Binds DR6 to Trigger Axon Pruning and Neuron Death via Distinct caspases, 457: 981–989, Copyright 2009.)

laboratory. The left panel is the control and the right panel shows how the peptide fragment N-APP acts as a potent axon apoptotic agent. A model of N-APP as a potent suicide molecule against axons causing their destruction and lysis is shown in Figure 16.5b. Note that N-APP is generated by dysfunctional protein processing of APP, the precursor protein parent of amyloid-ß. DR6 is a suicide receptor protein embedded in an inactive form on the nearby axon membrane surface. N-APP stands for the peptide fragment produced during dysfunctional protein processing from the *N*-terminal end of APP. According to this model N-APP has strong binding affinity to the active site of DR6, activating a potent apoptotic signal through the caspase cascade leading to axon lysis.

Theoretically, the release of significant levels of N-APP from a defective axon into the extracellular environment would be expected to destroy surrounding axons, perhaps creating a chain reaction of destruction. One scenario is that generation of N-APP and its destructive power against neighboring axons is a relatively rare event in a young, healthy brain even after trauma, but is somehow favored by aging.

16.5 ENERGY STRESS CAUSED BY DHA MIGHT LINK THE THREE MODELS

Brain injuries from moderate trauma generally heal with symptoms of Alzheimer's disease only appearing after a long latent period. The three models discussed above do not adequately address the effects of aging, and this is where a DHA theory for neurodegeneration ties to brain trauma. According to the DHA pacemaker model, the presence of DHA chains in axon membranes causes energy stress in neurons especially important during aging. We have discussed how the DHA cascade putatively enabling neurodegeneration is often modulated by environmental signals, with mechanical stress from trauma being added to the list of triggers for Alzheimer's. Recall that it has been known for years that repeated brain trauma can trigger the death of parkinsonian neurons, though symptoms usually appear some years after brain injury. It is not possible at this stage to develop a molecular picture explaining the 15- to 20-year latent period between brain injury and extensive neurodegeneration. However, we suggest that the dual energy pacemaker model developed in this book is a reasonable guide or model helping to fill in the stages between initial brain trauma and dementia.

In summary, symptoms of dementia caused by the excessive death of neurons typically strike athletes repeatedly traumatized in contact sports and soldiers injured in combat a decade or two following the brain trauma. This long-term cascade governing neurodegeneration following brain trauma can be divided into several stages as follows:

- In athletes initial trauma events last a week or two marked by an accumulation and rapid clearing of amyloid plaque.
- Trauma caused by blast injury follows a similar course but is more pronounced.
- Healing following repeated brain trauma in athletes or blast trauma might take months.

- Long latency period between brain trauma and dementia might take years.
- Neuron death might accelerate after the first symptoms of dementia appear.

The mechanical stress model discussed above helps illuminate early events, whereas the suicide peptide model provides an example of how massive neuron death might eventually be sparked. We suggest that the DHA pacemaker model of neurodegeneration and aging helps explain the long latent period before symptoms appear. During this period, mutations accumulating in mtDNA steadily lower energy levels, likely increasing dysfunctional protein processing and slowing repair of membrane damage. The high levels of DHA in membranes of neurons are predicted to potentiate the long-term danger of brain trauma by increasing energy-oxidative stresses, perhaps via chronic inflammation. A linkage between DHA-microglia generating both energy and oxidative stresses is possible. The accelerating pace of research on brain trauma will likely help elucidate key principles of neuron vitality and degeneration important for all forms of dementia.

REFERENCES

Aoyama, N., S. M. Lee, N. Moro et al. 2008. Duration of ATP reduction affects extent of CA1 cell death in rat models of fluid percussion injury combined with secondary ischemia. *Brain Res.* 1230:310–319.

Bayir, H., V. A. Tyurin, Y. Y. Tyurina et al. 2007. Selective early cardiolipin peroxidation after traumatic brain injury: An oxidative lipidomics analysis. *Ann. Neurol.* 62:154–169.

Centers for Disease Control and Prevention (CDC). 2006. Incidence rates of hospitalization related to traumatic brain injury—12 states, 2002. *MMWR Morb. Mortal. Wkly. Rep.* 55:201–204.

Farooqui, A. A., L. A. Horrocks, and T. Farooqui. 2007. Modulation of inflammation in brain: A matter of fat. *J. Neurochem.* 101:577–599.

Giza, C. C., and D. A. Hovda. 2001. The neurometabolic cascade of concussion. *J. Athl. Train.* 36:228–235.

Hanisch, U.-K., and H. Kettenmann. 2007. Microglia: Active sensor and versatile effector cells in the normal and pathologic brain. *Nature Neurosci.* 10:1387–1394.

Johnson, V. E., W. Stewart, and D. H. Smith. 2010. Traumatic brain injury and amyloid-β pathology: A link to Alzheimer's disease? *Nat. Rev. Neurosci.* 11:361–370.

Lew, H. L., D. K. Thomander, K. T. Chew et al. 2007. Review of sports-related concussion: Potential for application in military settings. *J. Rehabil. Res. Dev.* 44:963–974.

Moore, D. F., R. Radovitzky, L. Shupenko et al. 2008. Blast physics and central nervous system injury. *Future Neurol.* 3:243–250.

Nikolaev, A., T. McLaughlin, D. D. O'Leary et al. 2009. APP binds DR6 to trigger axon pruning and neuron death via distinct caspases. *Nature* 457:981–989.

Potts, M. B., H. Adwanikar, and L. J. Noble-Haeusslein. 2009. Models of traumatic cerebellar injury. *Cerebellum* 8:211–221.

Shankar, A., J. J. Wang, E. Rochtchina et al. 2006. Association between circulating white blood cell count and cancer mortality: A population-based cohort study. *Arch. Intern. Med.* 166:188–194.

Shaw, N. A. 2002. The neurophysiology of concussion. *Prog. Neurobiol.* 67:281–344.

Smith, D. H. 2010. Ask the Brains: Can an old head injury suddenly cause detrimental effects much later in life? *Sci. Am. Mind.* April 12.

Trudel, T. M., A. MacKay-Brandt, and R. O. Temple. 2008. Traumatic brain injury and dementia: A systematic review. *Brain Injury Professional* 5:12–13.
Vagnozzi, R., S. Signoretti, B. Tavazzi et al. 2005. Hypothesis of the postconcussive vulnerable brain: Experimental evidence of its metabolic occurrence. *Neurosurgery* 57:164–171.
Weber, J. T. 2007. Experimental models of repetitive brain injuries. *Prog. Brain Res.* 161:253–261.

17 Alzheimer's Disease

Among the four neurodegenerative diseases discussed in this section, Alzheimer's disease (AD) is the most important in terms of numbers of patients and is the most studied, yet it is the least understood. One reason for this enigma is that there is no known cause for AD. Prion diseases are clearly caused by prions. Distinct chemicals can cause Parkinson's disease. Multiple concussions or one big one are now believed to cause brain trauma–induced dementia. The main risk factor for Alzheimer's is aging, and the nature of the cascade of events causing aging is still poorly understood. However, the end result of AD is well known and is a spreading death of billions of neurons across critical regions of the brain. There is increasing data that reactive oxygen species (ROS) play important roles in AD, Parkinson's disease (PD), and other neurodegenerative diseases (Lull and Block, 2010.)

As discussed in the introductory chapter, Alzheimer's can be portrayed as a calendar of events marked by increasingly devastating symptoms. By the time symptoms appear the brain has sustained such serious damage that it is difficult to imagine how to restore billions of dead neurons. The current wisdom is that with the long-term goal of managing or eventually curing AD, intervention must take place long before symptoms appear. For these reasons we choose to focus on possible early stages of AD with the recognition that initiation events are poorly understood.

17.1 AN INDIVIDUAL ENERGY-STRESSED OR DAMAGED NEURON MIGHT GENERATE A FOCAL POINT FOR ALZHEIMER'S

There are considerable data in the literature that show how a single or a few damaged cells in an essential human organ can start a focal point of disease. In many cases the tissues used in these studies are composed of cells capable of growth and division, allowing the identification of clonal populations or sectors of cells originating from an altered or mutated cell line. This effect has been demonstrated for several kinds of human tissue as follows:

- Heart
- Limb muscle
- Diaphragm
- Colonic crypt stem cells
- Hippocampus/choroid plexus
- Hippocampus neurons in Alzheimer's
- Substantia nigra neurons in aging and Parkinson's

Finding the first cells forming a focal point for Alzheimer's is a daunting task. Scientists studying focal points of disease in other tissues have used a powerful trick to get around this problem. That is, analysis of mitochondrial DNA (mtDNA) mutations is carried out after damaged cells have had a chance to multiply, providing enough biomass to permit biochemical testing or isolation of sufficient mtDNA for analysis. Using histochemical techniques, a focal respiratory deficiency was spotted in heart cells with aging (Müller-Höcker, 1989). Age-related focal respiratory dysfunction in muscle specimens has also been documented (Müller-Höcker, 1990). Age-associated focal respiratory dysfunction has also been demonstrated in colonic crypts (Taylor et al., 2003). In studies on the role of mtDNA in aging muscles, deletions were found to be reporters of both aging and respiratory dysfunction (Fayet et al., 2002). These researchers also found that a number of muscle cells from old human subjects contain high levels of one or a few mtDNA point mutations. Scientists have adapted similar techniques to studies of focal lesions in neurodegenerative diseases. For example, analysis of mtDNA isolated from single substantia nigra neurons from Parkinson's patients (age 76 to 77) showed accumulations of about 50 percent mtDNA deletions (Bender et al., 2006). Thus, high levels of mtDNA mutations are linked to respiratory chain dysfunction in dementia. Different cells carried unique mtDNA deletions that were observed to be clonally expanded. Clones of mitochondrial enzyme-deficient neurons have been found to be greater in quantity in Alzheimer's patients compared to age-matched controls (Cottrell et al., 2002). Clonal expansion of prematurely damaged neurons as a cause of Alzheimer's is made difficult because neurons seldom divide. Thus, alternative mechanisms are required to explain how a single defective neuron might spread damage to neighboring neurons. Dufour et al. (2008) reported that respiratory chain–deficient neurons in mice have adverse effects on normal adjacent neurons and induce what these authors call "*trans*-neuronal degeneration." These data seem to be the first in support of what might be considered a neuron version of the "rotten apple in a barrel" concept of how a focal point of Alzheimer's might spread, inducing normal neighboring neurons into a pathological state (Figure 17.1). These data are consistent with the idea that a single defective neuron might in essence poison its neighbors. Scientists recently used finely tuned lasers to cut out and analyze the relatively localized cluster of parkinsonian neurons in the brains of deceased patients (Zheng et al., 2010). Numerous new genes involved in energy stress were identified in this manner (see Chapter 14).

How could a focal point arise? As background, recall that neuron mitochondria constantly age. These senescing mitochondria are transported back to the neural cell body region to grow and be refurbished. Two separate and highly sophisticated molecular conveyor belts are used for long-distance transport of aging and refurbished mitochondria. Note that mitochondria are selected for rebuilding when their energy status is low (Hollenbeck and Saxton, 2005; Kann and Kovács, 2007; Miller and Sheetz, 2004). Only newly divided mitochondria with a healthy energy status are returned from the neuron cell body region back to the synapse. Thus, long-distance axonal transport in combination with mitochondrial growth provide a steady supply of healthy mitochondria to different and often distant regions of axons.

However, defects or plugging of fast axonal transport of mitochondria might lead to energy stress in axons. In one study a mutation in a mitochondrial enzyme involved in

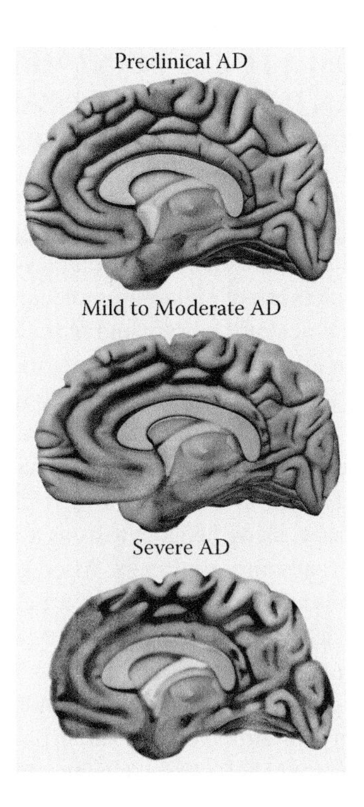

FIGURE 17.1 "Rotten apple in a barrel" concept of Alzheimer's disease. Three stages of brain damage leading to symptoms are shown—early, intermediate, and late. Massive neuron destruction and brain shrinkage occur during late states of Alzheimer's disease. (Image courtesy of the National Institute on Aging/National Institutes of Health.) **(See color insert.)**

protection against damage by ROS (SOD1) perturbs fast axonal transport and reduces axonal mitochondrial content (De Vos et al., 2007). This mutation in superoxide dismutase causes Lou Gehrig's disease. Later studies show that mutant SOD1 causes toxicity and mitochondrial dynamics abnormalities (Magrané et al., 2009; Morfini et al., 2009). Although Lou Gehrig's disease occurs in the peripheral nervous system, lessons learned on the importance of mitochondrial dynamics likely apply to the brain as well. Thus, recent discoveries have highlighted that neurons are particularly dependent on the dynamic properties of mitochondria (Chen and Chan, 2009) and provide a mechanism to explain how the first "rotten apple" neuron might arise.

A second scenario we are considering to explain the origin of the first "rotten apple" neuron involves dysfunctional targeting of DHA allowing the formation of DHA-CL. Incorporation of DHA into human mitochondria is considered to be a sign of a pathological state. Because relatively heavy DHA trafficking occurs over a lifetime in a human brain, any breach in mechanisms disturbing selective targeting of DHA away from neuron mitochondria could sow the seeds for an unhealthy or damaged neuron. What we are saying here is that preventing the incorporation of DHA into mitochondria of neurons is considered as a critical antiaging mechanism (see Chapter 11). In contrast, allowing DHA incorporation into DHA-cardiolipin of neurons is considered as a pro-aging mechanism.

17.2 SPREADING FROM A FOCAL POINT MIGHT BE CAUSED BY AN INFLAMMATORY CASCADE LINKING DHA AND ALZHEIMER'S DISEASE

Microglia are an amazing class of macrophages that roam throughout the brain in search of pathogens, changes of homeostasis, or structural defects in neurons. Because the brain lacks white blood cells, microglia play this role but are more generalized in responding to a plethora of signaling molecules besides invading pathogens. Microglia are the policemen of the brain and have numerous receptors on their ever-changing processes that allow them to constantly monitor chemical changes in the environment and probe the surfaces of neurons for defects. Microglia not only sense their environment but also quickly adapt to counteract imbalances or structural defects with the goal of restoring homeostasis. Microglia are more than policemen, being undertakers as well. That is, microglia and their glial partners the astrocytes are undertakers when necessary, using phagocytosis as a means to dispose of broken neuron body parts or even senescing glial cells. Microglia can also be executioners, sending powerful apoptotic signals to destroy a neuron deemed too far gone to save (Figure 17.2). For a description of the versatile powers of microglia see the articles by Hanisch and Kettenmann (2007) and Lull and Block (2010). As shown in Figure 17.2, activated microglia produce toxic compounds including ROS (e.g., H_2O_2 and superoxide radical) that can damage neurons, especially their DHA-enriched membranes. The potent neurotoxin MPTP (see Chapter 14) is shown directly damaging neurons. Soluble neuron-injury factors produced by damaged neurons are sensed by and cause activation of microglia, including upregulation of the powerful phagocytic oxidase system (PHOX), which can produce damaging levels of ROS.

In Chapter 8 truncated oxidation products of DHA phospholipids were shown to flip to the membrane surface where this DHA derivative acts as a signaling molecule for triggering phagocytosis by microglia. Microglia hone in on the oxidized membrane surface of neurons via the scavenger receptor CD36, which locks onto the lipid whisker and triggers phagocytosis (see references in Chapter 8). In this manner senescing axons, membrane fragments, and spent synaptosomes are cleaned up. Microglia also engulf amyloid-beta plaque and have receptors for this tangle of toxic peptides. So far the work of microglia in sensing, engulfing, and degrading defective neuron membranes falls into the routine category. However, as signal strength through the CD36 scavenger receptor intensifies and is perhaps reinforced by parallel signaling pathways, microglia adapt by becoming increasingly active, entering a gray zone between being beneficial and harmful. Harmful in this case is defined as being too aggressive or hyperactive to the point of initiating a state of sterile inflammation. Sterile inflammation contrasts to pathogen-induced inflammation, which is induced to kill the harmful invader. Unfortunately, in the case of sterile inflammation created by hyperactive microglia, there is no pathogen present, but the inflammatory environment created around neurons can destroy synapses and axons and eventually trigger complete neuron apoptosis. Now, it is important to keep in mind that microglia as well as neurons are subject to the stresses of aging, which can shift their abilities to adapt to change.

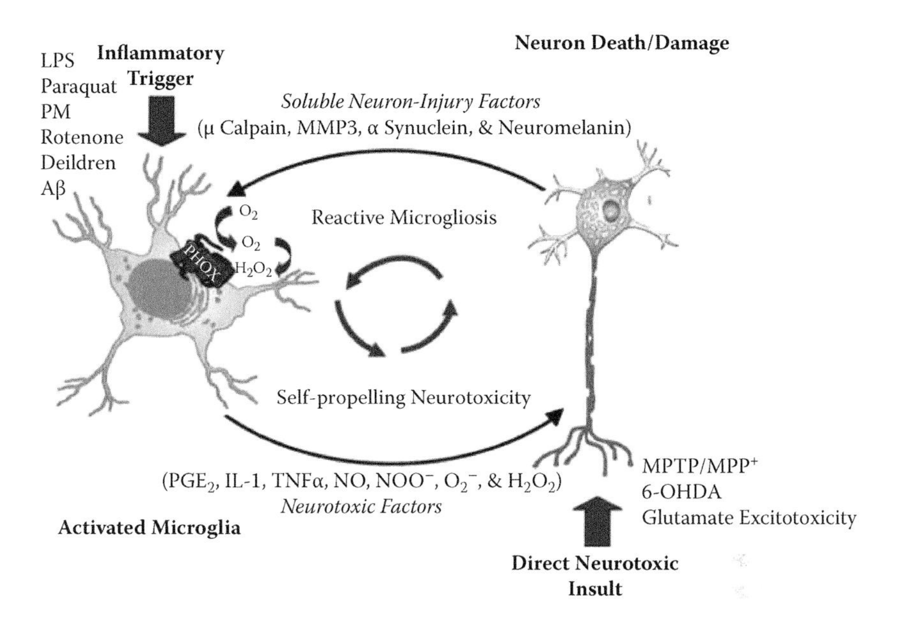

FIGURE 17.2 Reactive microgliosis drives chronic neuron damage. Microglia contribute numerous essential functions for maintaining long-term health of neurons but when activated can damage or even kill neurons. This diagram shows the sophisticated network of signaling pathways leading to activation of microglia. Both stimulation of microglia with pro-inflammatory triggers (e.g., LPS) and direct neuron damage (e.g., glutamate excitability) result in microglial activation causing the release of neurotoxic factors, such as IL-1β, NO, TNFα, $ONOO^-$, $O_2\bullet^-$, and H_2O_2. Subsequently, following damage with either a pro-inflammatory trigger or a direct neurotoxin, the neuron releases microglial activators (soluble neuron-injury signals) such as μ calpain, MMP3, α-synuclein, and neuromelanin which activate microglial cells and propagate the cycle. This self-perpetuating cycle of neurotoxicity is known as *reactive microgliosis*. Abbreviations: hydrogen peroxide (H_2O_2); interleukin 1 beta (IL-1β); lipopolysaccharide (LPS); matrix metalloproteinase 3 (MMP3); 1-methyl-4-phenyl-1,2,3,6-tetrahydropyridine (MPTP); 1-methyl-4-phenylpyridinium ion (MPP^+); nitric oxide (NO); peroxynitrite ($ONOO^-$); prostaglandin E2 (PGE_2); superoxide ($O_2\bullet^-$); tumor necrosis factor alpha (TNFα); phagocytic oxidase (PHOX). (With kind permission from Springer Science+Business Media: *Neurotherapeutics,* Microglial Activation and Chronic Neurodegeneration, 7: 354–365, Copyright 2010, Lull, M. E., and M. L. Block, Figure 1.)

Recently, Stewart et al. (2010) defined in molecular terms how the CD36 receptor promotes sterile inflammation. It is established that in Alzheimer's disease deposition of amyloid-beta (Maezawa et al., 2011) and oxidized lipids (Kunjathoor et al., 2004) triggers a protracted sterile inflammation response. Although chronic stimulation of microglia was believed to underlie the pathology of AD, the molecular mechanisms of activation were unclear. In this seminal study Stewart et al. (2010) show that the scavenger receptor CD36, which receives signals from amyloid-beta and importantly for our discussion, oxidized DHA phospholipids (see Chapter 8), stimulates sterile inflammation. This and other data from this laboratory (see Stuart et al., 2007) define a molecular cascade of events beginning with DHA membranes and leading to sterile inflammation. The DHA principle states that DHA can eventually harm neurons, and a cascade has now been identified to link DHA and Alzheimer's disease.

The loss of DHA in the hippocampus and other critical regions of the brain at an early stage of Alzheimer's disease can now be explained on the basis of the CD36 (scavenger receptor) signaling pathway. According to this model, oxidized-truncated DHA phospholipids, behaving as lipid whiskers, overstimulate microglia via the CD36 cascade. Once activated beyond their normally benign state microglia can become overly aggressive, switching roles from beneficial to harmful. A state of sterile inflammation is created that can endanger and kill neurons. Axons and synapses are at the highest risk of inflammation-mediated membrane peroxidation. Ironically, DHA phospholipids are not only especially vulnerable to the oxidizing environment created during inflammation but can also sustain and increase the intensity of inflammation through the formation of ROS. Thus, inflammation becomes a vicious cycle that can ignite an oxidative chain reaction (Wyss-Coray, 2006). According to this model, DHA is both a cause and an effect of Alzheimer's disease.

17.3 BLOCKING NEURON DEATH IN A MOUSE MODEL OF ALZHEIMER'S DISEASE BY KNOCKOUT OF A KEY SIGNALING RECEPTOR ON MICROGLIA

Microglia are the cells responsible for the immune surveillance in the brain, and they initiate protective inflammatory reactions in response to tissue damage and infections. Recent data suggest that overzealous microglia may actually make a significant contribution to the loss of neurons associated with Alzheimer's disease (Fuhrmann et al., 2010; McGeer and McGeer, 2002). Herms and his colleagues at the Center of Neuropathology and Prion Research, Ludwig-Maximilians-University, Munich, Germany, were able to look directly into the brains of genetically modified mice that develop many of the symptoms characteristic of AD in humans. The mice have also been engineered to synthesize florescent forms of proteins that are specific for neurons and microglia. The imaging technique allowed the researchers to monitor the fate of identifiable neurons and microglia over periods of weeks and months. This approach made it possible for the first time to visualize the loss of nerve cells in the brains of living mice. Nerve cell loss was found to be preceded by the activation of microglia.

It is thought that senescing nerve cells near plaques secrete a chemical messenger that induces the microglia to concentrate around focal points of neurodegeneration. The best candidate for the messenger responsible was predicted to be the chemokine fractalkine, which docks onto a receptor protein on the surface of the microglial cells. When this receptor was genetically eliminated, nerve cell loss was prevented (Fuhrmann et al., 2010). These results demonstrate that microglia are not only involved in the removal of amyloid aggregates typical of Alzheimer's disease, but they also contribute actively to the catastrophic loss of nerve cells. These data show that stressed nerve cells in mice secrete a chemical messenger that attracts microglia. Unfortunately, when large numbers of microglia gather around neurons the ensuing inflammatory reaction can become too strong, resulting in the elimination of neurons. This implies that chemical signaling between nerve cells and microglia plays an important role in mediating neuron loss during the course of neurodegeneration. It now appears that a battery of chemical signals is exchanged between neurons and microglia. DHA-mediated lipid whiskers discussed in Chapter 8 fall into the category of signaling molecules between senescing neuron membranes and receptors on microglia (Silverstein, 2009).

17.4 IS THE TOXIC PEPTIDE TAU THE LONG-AWAITED INFECTIVE AGENT CAUSING THE SPREAD OF ALZHEIMER'S DISEASE?

Once a focal point for AD is triggered in the entorhinal cortex of the brain, the disease spreads slowly outward to encompass areas that involve remembering and cognitive processing. Neurons in the initial focal point of AD are known to generate and store the toxic peptide *tau*. Tau production is caused by dysfunctional protein processing and accumulates inside unhealthy neurons. Recent data suggest that like a prion, tau behaves as an infective agent. It is transmitted along anatomically connected networks from tau-producing neurons to contaminate healthy neurons (Liu et al., 2012). This study involved a mouse model of AD genetically engineered to express human tau in the same region of the brain in which AD starts in humans. Knowing the path for the spread of Alzheimer's and the identification of tau as a potential infective agent represents an important advance, opening up new targets for preventing AD.

17.5 DHA OXIDATION PRODUCTS IN CEREBROSPINAL FLUID HELP VALIDATE THE DHA PRINCIPLE

Oxidative damage to DHA-enriched membranes of neurons has been validated quantitatively using F4-neuroprostanes as biomarkers (Dalle-Donne et al., 2006; Montine et al., 2002, 2005, 2007; Poon et al., 2004). In essence this sensitive analytical procedure detects the presence of oxidized by-products of DHA in cerebrospinal fluid (CSF). The presence of F4-neuroprostanes in CSF shows that DHA phospholipids of specialized membranes of neurons including axon membranes and synaptic vesicles are being oxidatively attacked. Billions of neurons constantly repair their oxidatively damaged membranes by removing damaged DHA chains using repair phospholipases (see Chapter 10).

Development of chemical procedures to detect and quantitate F-4 neuroprostanes in CSF is a major breakthrough with important implications for understanding the molecular pathology of Alzheimer's disease and other neurodegenerative diseases. Consistent and reproducible data show that F4-neuroprostane levels in CSF increase in AD patients. Of particular interest is the possibility that neuroprostane levels begin to rise ahead of symptoms of AD (Praticò et al., 2002, 2004). This is consistent with the view that peroxidation of DHA is a driver of AD. Clinical applications of this technology for several other diseases besides AD are gaining a foothold. As discussed in Chapter 9, these methods also offer important tools for understanding numerous fundamental aspects of neuron health and degeneration including the important roles played by antioxidants in protecting DHA-enriched membranes of neurons.

From a basic perspective the detection of increased levels of DHA oxidation products in CSF of age-dependent Alzheimer's patients is consistent with and seems to validate the DHA principle as applied to neurons. In humans DHA-derived F4-neuroprostanes report what is happening regarding oxidative stress specifically in the brain and neurons where most of the DHA in the human body is stored as phospholipids. This is not the case in mice where DHA is widely distributed in organs besides the brain. In contrast, arachidonic acid is widespread among organs of both humans and mice, and its oxidation products provide a snapshot of oxidative stress at the whole-body level. The presence of F2-isoprostanes detected in plasma and urine validates the concept that arachidonic acid is not only a target of oxidative damage but is also an oxidative threat. In mice, DHA is a whole-body marker for oxidative stress and is considered an oxidative threat in most cells of this animal. Gathering comparative data for levels of F2-isoprostanes versus F4-neuroprostanes in mice might indicate whether DHA or arachidonic acid (ARA) is the greater oxidative threat. This question has been answered in chemical studies where the rates of production of F4-neuroprostanes from DHA are faster than the synthesis of F2-isoprostanes, as expected from lipid peroxidation theory.

17.6 DEMYELINATION OCCURRING DURING AGING MIGHT EXPOSE DHA-ENRICHED MEMBRANES OF AXONS OF WHITE MATTER TO LIPID PEROXIDATION

As already introduced in Chapter 10, demyelination proceeds by a genetic program activated roughly in the fourth decade of life and proceeding along a timescale measured in the decades preceding death. The central question addressed in this section concerns the myelin model of dementia championed by Bartzokis (2004) and supported by recent data using triple-transgenic Alzheimer's mice (Desai et al., 2009, 2010). Data from transgenic mice show that region-specific patterns of demyelization precede the appearance of toxic peptides, amyloid plaque, and tau tangles, the latter a second major class of putative toxic peptides found intracellularly in neurons of dementia patients. Thus, the mouse model of Alzheimer's and the myelin model developed for human dementia seem to be trending in the same direction. Bartzokis, Lu, and Mintz (2004) show that the pattern of age-related myelin breakdown in humans measured by magnetic resonance imaging (MRI) is in agreement with data from neuropsychology and neuropathology. These data suggest that the process of myelin breakdown begins in adulthood and accelerates as aging progresses. These researchers hypothesize that demyelination underlies both age-related cognitive declines and the most powerful risk factor of dementia (causing disorders such as Alzheimer's), a person's age. These data have given rise to the myelin-centered model that makes it possible to measure the trajectory of myelin breakdown related to dementia. This model also provides molecular neurologists with a framework for asking questions and developing research strategies toward exploring still earlier events along the chain of events leading to neurodegeneration.

Peters and colleagues (Peters, 2002; Peters, Moss, and Sethares, 2000; Peters and Sethares, 2002) have pioneered studies of demyelination linked to aging using the Rhesus monkey as a model. Seminal findings include evidence that normal aging results in widespread damage to myelin sheaths that insulate axons. This loss of myelin correlates significantly with a decline in cognitive status. In the cerebral cortex there is no significant loss of nerve fibers with age, in contrast to white matter where there is evidence that large numbers of nerve fibers may be lost. This loss is presumed to compromise connectivity between various parts of the monkey brain and also is believed to contribute to cognitive decline. Peters and coworkers distinguish certain myelination events during normal aging in monkeys, which might be applicable to understanding how human neurons degenerate. For example, in the cerebral cortex where there is a breakdown of myelin and little loss of nerve fibers, the principal phagocytic mechanism appears to involve astrocytes. In contrast, microglial cells seem to be activated in white matter perhaps triggered by more extreme losses of axon connections along with their myelin sheaths. According to a scenario developed in Chapter 13, microglial cells forced to handle high levels of damaged axons might themselves be subject to oxidative damage—energy stress due to DHA.

Recent studies of myelination patterns in triple-transgenic Alzheimer's mice support a myelin-model of dementia (Desai et al., 2009, 2010). These data show that triple-transgenic mice exhibit region-specific abnormalities in brain myelination prior to the appearance of amyloid plaque and tau tangles.

Just as the "cause versus effect" status of toxic peptides has been the subject of considerable debate in the past, a similar argument might be applied to the myelin model. The question of what causes demyelination can be asked. One possibility is energy stress caused by normal aging of mitochondria in concert with energy uncoupling at the level of age-damaged DHA membranes. One of the most interesting aspects of the myelin model is that the time frame for the loss of myelin measured in the human brain coincides with that of dementia. This provides a glimpse of and stimulates thinking about how such long-term pacemakers or biological clocks governing aging and dementia evolved in the first place. From a practical standpoint recent papers from both Bartzokis' laboratory and Bowers' group (Desai et al., 2009, 2010) highlight potential novel therapeutic targets based on their research.

17.7 NEURON MEMBRANES SEEM TO REQUIRE EXTRAORDINARY PROTECTION: THE FOXO STORY

Recent data suggest that neurons are especially sensitive to stresses linked to aging and require extraordinary protective mechanisms for long-term health (Bishop, Lu, and Yankner, 2010; Soerensen et al., 2010; Walter et al., 2011). We hypothesize that lifelong protection of neurons against oxidative stress and energy stress caused by aging are due to the need for extraordinary membranes essential for neuron function. We further propose that aging neurons become increasingly vulnerable to an oxidative chain reaction caused in part by their unusual membranes. How can neuron membranes be given added protection against the stresses of aging?

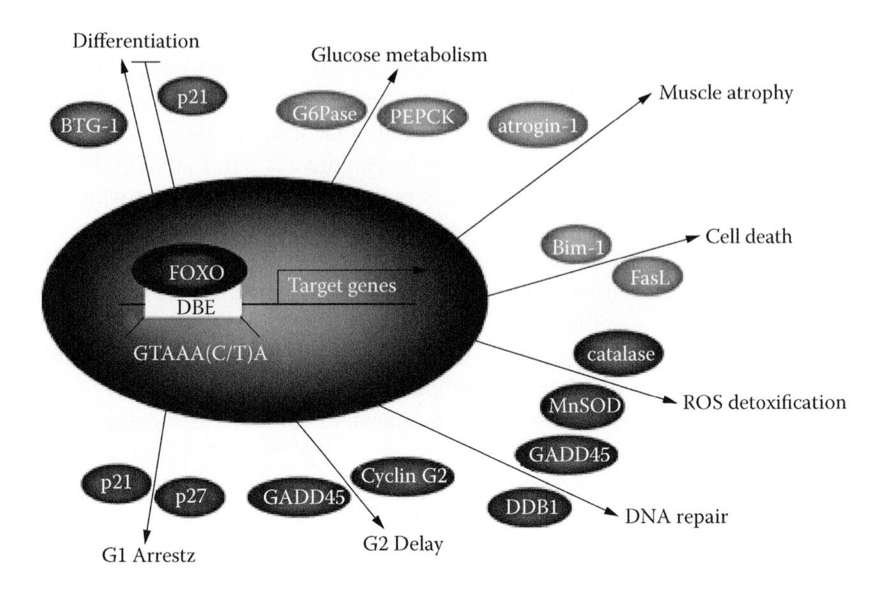

FIGURE 17.3 FOXO target genes include several important in neuron development, health, and brain span. Genes upregulated by FOXO transcription factors, and likely involved in prolonging neuron health and increasing longevity, include reactive oxygen species (ROS) detoxification, DNA repair, cell death, and glucose metabolism. BTG-1, B-cell translocation gene 1; p21, cyclin-dependent kinase inhibitor 1A; p27, cyclin-dependent kinase inhibitor 1B; MnSOD, manganese superoxide dismutase; G6Pase, glucose-6-phosphatase; PEPCK, phosphoenolpyruvate carboxykinase; FasL, Fas ligand; GADD45, growth arrest and DNA damage-inducible protein 45; DDB1, damage-specific DNA-binding protein 1; DBE, DAF-16 family member-binding element. (Reprinted by permission from Macmillan Publishers Ltd: *Oncogene*, 24: 7410–7425, Greer, E. L., and A. Brunet, FOXO Transcription Factors at the Interface between Longevity and Tumor Suppression, Copyright 2005.)

The answer may be emerging from data on how cells in general are protected during the long human life span. This research involves the understanding of longevity genes in humans, a field receiving increasing attention. The remainder of this section deals with the case history of a potential longevity mechanism called the FOXO system.

The molecular biology of FOXO transcription factors is understood in considerable detail (Accili and Arden, 2004; Burgering and Kops, 2002; Greer and Brunet, 2005; Hoekman et al., 2006; Lehtinen et al., 2006; Miyamoto et al., 2007; Paik et al., 2009; Renault et al., 2009; Salih and Brunet, 2008; Tothova et al., 2007). Figure 17.3 shows the diverse set of genes modulated by FOXO factors and includes genes essential for neuron bioenergetics, development, cell cycle, death, and protection. FOXO factors upregulate genes essential for the long-term health of all human cells including neurons.

FOXO transcription factors are modulated by elaborate regulatory cascades that sense and respond to environmental signals including signals coming from both outside and inside the neuron. Figure 17.4 summarizes how FOXO factors change their structures in response to changes in the world around them (Greer and Brunet, 2005). In essence, the most sophisticated kinds of regulatory circuitry or biochemistry in

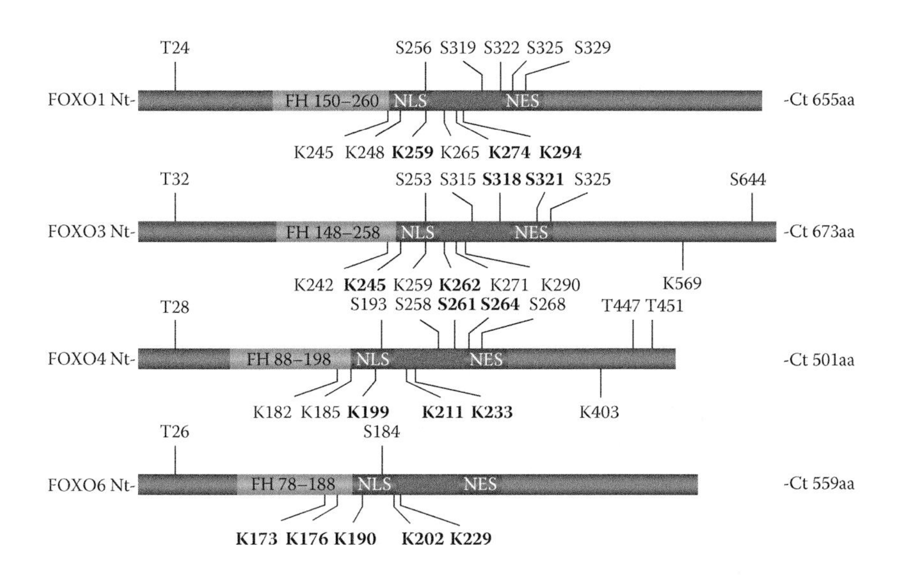

FIGURE 17.4 FOXO transcription factors bind to DNA and sense and respond to stress signals. Environmental signals are transduced, resulting in changes in patterns of phosphorylation and acetylation of specific amino acid residues along the protein chain of these transcription factors. These changing patterns alter nuclear localization, FOXO degradation, DNA-binding ability, transcriptional activity of stress tolerance genes, and protein-protein interactions. Akt sites (black); SGK, serum and glucocorticoid inducible kinase (black); IKKb, IkB kinase b (orange); JNK, Jun N-terminal kinase (green); DYRK, dual-specificity tyrosine (Y) phosphorylation-regulated kinase (red); CK1, casein kinase 1 (purple); acetylation sites (blue); FH, Forkhead domain; NLS, nuclear localization signal; NES, nuclear export sequence. Stress-induced phosphorylation sites of FOXO3 (not shown) occur at Ser90, Ser284, Ser294, Ser300, Ser413, Ser425, Thr427, and Ser574. (Reprinted by permission from Macmillan Publishers Ltd: *Oncogene*, 24: 7410–7425, Greer, E. L., and A. Brunet, FOXO Transcription Factors at the Interface between Longevity and Tumor Suppression, Copyright 2005.) **(See color insert.)**

nature are seen in this family of regulatory proteins. FOXO factors are regulated by environmental signals as diverse as hormones, glucose levels, or oxidative stress. It is little wonder that genetic variation in these regulatory proteins results in a mind-boggling number of phenotypes. FOXO proteins reside and are active in the nucleus where they bind to DNA as monomers via the forkhead domain or box, which is a 110-amino acid region located roughly in the central position of the molecule. When present in the nucleus and attached to DNA, FOXO regulators behave as potent transcriptional activators, upregulating synthesis of specific messenger RNAs, later translated into stress-protective proteins. The activation domain of FOXO factors is located in the C-terminal portion of the protein. FOXO proteins can switch from being transcriptional activators to repressors of gene activity depending on the specific DNA promotor they bind to along with existing environmental conditions. Thus, the presence of FOXO proteins in all human cell types along with their exquisite ability to sense and respond to changes in their surroundings make them ideal candidates for pushing back against neurodegeneration.

Following the discovery of FOXO3A as a potential longevity gene (Willcox et al., 2008), several laboratories worldwide are actively screening DNA samples of centenarians in search of genes enhancing life span and brain span. Some selected advances on FOXO transcription factors beginning in 1993 (Kenyon et al., 1993) are as follows:

- 1993—*Caenorhabditis elegans* mutant with double the life span of wild type reported
- 2002—FOXO found to play critical roles in cell cycle and death control
- 2004—FOXO role in cellular metabolism, differentiation established
- 2005—Molecular biology of FOXO transcription factors elucidated and linkage to longevity proposed
- 2005—FOXO role in longevity of *C. elegans* reviewed with implications for aging in other animals
- 2005—Unique patterns of expression of multiple forms of FOXO determined in mouse brain
- 2006—Linkage of FOXO to oxidative stress and life span described
- 2007—FOXO found to mediate stem cell resistance to oxidative stress
- 2008—Unified concept of FOXO transcription factors in maintaining cellular homeostasis during aging proposed
- 2008—FOXO3A genotype found to be strongly associated with human longevity
- 2009—FOXO shown to regulate neural stem cell homeostasis
- 2009 through 2011—FOXO link to longevity in centenarians confirmed

A review of this field has recently been published (Soerensen et al., 2010).

17.8 EFFECT OF INSULIN ON SYMPTOMS OF ALZHEIMER'S DISEASE

FOXO transcription factors are part of the famous insulin cascade that governs glucose-dependent, energy homeostasis (Kaletsky and Murphy, 2010). Insulin levels in the brain tend to decrease or become less effective during the progression of Alzheimer's (reviewed by Craft et al., 2011). This fact led these researchers to test the effects of a twice-daily dose of insulin in the form of a nasal spray as a treatment for AD (Craft et al., 2011). In this small pilot study insulin was squirted deep into the nose where it readily travels to the brain. This simple treatment seems to stabilize or improve cognition and function in a selected group of patients in the early phase of Alzheimer's. These data come at a time when there are no effective ways to prevent or delay the progress of Alzheimer's. Experts consider these data to be provocative, yet at the pilot stage, and requiring confirmation using a larger group of subjects. However, these data are consistent with a growing body of information that highlights the importance of energy homeostasis on the pace of neurodegeneration. Some key points regarding the important roles played by insulin in brain function are as follows (Craft et al., 2011):

- Insulin enhances memory in healthy persons.
- Insulin levels and insulin activity in the central nervous system decline with aging and are reduced in AD.
- Persons with type 2 diabetes have an increased risk of AD.
- Insulin preserves cerebral metabolic rate of glucose utilization in regions of the brain affected by AD.

Thus, there is a solid body of data that documents the important roles of insulin in brain health and decline. FOXO transcription factors act downstream from the conventional insulin receptor pathway. What seems to be missing is an overarching framework or unified view of the powerful insulin-FOXO cascade and its function during neurodegeneration. Perhaps the answer is in front of our noses, being energy itself. As emphasized in this book we believe that energy conservation, especially at the membrane level, is far more important for overall energy homeostasis than currently appreciated.

Will insulin therapy help cure AD? There is a glimmer of hope that insulin treatment might reverse symptoms of AD in its earliest stages. We predict that insulin therapy will have little effect on patients in advanced stages of AD. The reasoning is that the massive death of neurons occurring by the time symptoms appear is driven by out-of-control, oxidative chain reactions. However, this concept should not distract attention from the provocative data that energy enhancement mediated by insulin may benefit patients in the early stages of AD.

17.9 WORKING MODEL OF ALZHEIMER'S AS A MEMBRANE DISEASE

In our first book (*Omega-3 Fatty Acids and the DHA Principle*) it was concluded that polyunsaturated fatty acids offer important benefits balanced against considerable risks. We proposed that oxidatively damaged membranes might be a cause of neurodegenerative diseases including Alzheimer's (see a recent review by Murphy et al., 2011, on this subject). The purpose of this section is to develop a molecular model to explain how oxidatively damaged neuron membranes may link to neurodegeneration.

As background, arachidonic acid (ARA, 20:4) is found in fish oil and is classified as an omega-6 fatty acid with four double bonds. ARA is present in significant amounts in axon phospholipids of neurons and is distributed widely in plasma membranes of most cells. The phospholipid form of ARA is cleaved by phospholipase to yield the free acid. A low amount of the free acid is present in brain tissue. Arachidonic acid is best known as a precursor for a series of eicosanoid hormones. Eicosanoids are needed only in trace amounts and are highly unstable. In the brain eicosanoids are produced and function locally. As hormones, eicosanoids play a variety of roles including acting as triggers of inflammation. The linkage between eicosanoids and inflammation was established during studies of the mode of action of common aspirin (acetylsalicylic acid) acting as a painkiller. The pain cascade is blocked after the acetyl moiety of aspirin is transferred to and blocks the active site of the eicosanoid-producing enzyme cyclooxygenase. Thus, aspirin stops pain and inflammation by blocking the first step in the enzymatic synthesis of eicosanoids, which shows that eicosanoids act in a cascade to trigger pain and inflammation.

For long-term brain health, too much eicosanoid production may be harmful, essentially creating a state of chronic inflammation, which is a form of oxidative stress. This is where the small, highly regulated apolipoprotein D (ApoD) protein molecule comes into the picture. The structure of ApoD shows a deep binding pocket that is highly specific for certain lipids—the steroid hormone progesterone and arachidonic acid (Eichinger et al., 2007). A single ARA molecule fits tightly in the pocket such that ApoD has a powerful ability to sequester and remove ARA from the brain. Free ARA is present in low amounts in brain tissue though its levels may rise or fall with aging, disease, or after an oxidative challenge. It is interesting to speculate that a regulatory system allowing dramatic fluctuation of expression of ApoD evolved in the brain as a mechanism to sequester ARA and modulate its levels below a critical threshold causing neurodegeneration. The main point is that too much ARA likely creates a state of persistent inflammation-oxidative stress that is not compatible with a long brain span. We hypothesize that ApoD, like aspirin, blocks the inflammatory cascade caused by ARA, thus decreasing oxidative stress and membrane peroxidation in the brain, a pro-longevity mechanism.

The following data (Eichinger et al., 2007; Elliott, Weickert, and Garner, 2010; Ganfornina et al., 2008; Muffat, Walker, and Benzer, 2008) are consistent with this model:

- ApoD is ubiquitous in distribution and function and highly regulated among animals ranging from flies to humans (e.g., 500-fold increase in ApoD levels during smash-induced neuron damage in rats).
- Loss-of-function mutants of ApoD in *Drosophila* decrease longevity, whereas overexpression increases longevity.
- Human ApoD overexpression in *Drosophila* increases longevity.
- Membrane peroxidation levels are predicted by levels of ApoD with membrane damage being low when ApoD levels are elevated, and vice versa.
- Biochemical mode of action of ApoD is consistent with tight sequestering of ARA to block conversion of ARA → eicosanoids.

The linkage between inflammation and DHA has already been discussed above in the context of oxidation products of DHA phospholipids acting as signals to attract microglia to sites of membrane damage (Figure 17.2). There is also the possibility that eicosanoids signal the activation of microglia, increasing the probability of a disease focal point spreading as the result of overactive microglia. DHA and its oxidation products may play multiple roles in the ARA-inflammation cascade, acting not only to signal and activate microglia but also being the most vulnerable target of oxidative damage. Once again we propose that oxidatively damaged membranes are a direct source of energy stress caused by futile cycling of cations, especially important in bioenergetics and energy homeostasis in neurons.

The last topic deals with evidence linking ApoD with Alzheimer's disease (reviewed by Elliott, Weickert, and Garner, 2010; see their extensive reference list). In AD, increased levels of ApoD are found in cerebrospinal fluid and in diseased regions of the brain including the hippocampus, frontal cortex, and temporal cortex. ApoD immunostaining of cortical neurons suggests a correlation with the presence of neurofibrillary tangle density (a signature of AD pathology). ApoD levels are increased in the parkinsonian region of the brain during PD and in the CSF of patients with meningoencephalitis, stroke, dementia, and motor neuron disease. In an animal model of the human cholesterol storage disorder Niemann-Pick disease, ApoD levels were found to increase about 30-fold. Recent data support a generalized role of ApoD in preventing membrane peroxidation.

Our model focuses on protecting neurons, specifically neuron membranes as a mechanism enabling long brain span. An alternative model has recently been proposed in which the protective role of ApoD occurs at the level of astrocytes (Bajo-Grañeras et al., 2011). Clearly, protecting astroglial cells against oxidative stress is important, as is protecting neurons; protecting astrocytes against oxidative stress may also protect neurons. Another explanation is that ApoD plays multiple roles in the brain protecting both astrocytes and neurons, respectively.

In summary, an international effort is underway and gaining momentum, with the goal of understanding the molecular pathology of Alzheimer's disease. While the cause remains elusive, molecular studies are providing insight and are on the path to identifying causes. Advances outside the field of neurodegeneration, especially discoveries on the mechanism of aging, define a close relationship between aging and neurodegeneration. In this book we have emphasized the relationship among aging, energy, oxidative stress, and neurodegeneration and propose that the specialized DHA membranes of neurons predispose these vital cells to energy-oxidative stresses.

Alzheimer's disease is portrayed here as a membrane-based disease that is triggered in an age-dependent manner starting with only one or a few neurons that prematurely reach a critical threshold of energy stress. Recent advances defining the revised mtDNA pacemaker of aging provide a fresh approach toward understanding the earliest events of Alzheimer's. A two-stage but interlocking process determining energy status is envisioned. The mtDNA pacemaker is believed to steadily lower energy output with aging in neuron mitochondria. Simultaneously, the harmful effects of DHA in the brain and nervous system are thought to cause steady levels of oxidative stress → energy stress during normal aging. However, as seen with the case of parkinsonian chemicals including paraquat, a manageable level of oxidative stress can be dramatically amplified as a lethal oxidative chain reaction, creating a rapidly spreading zone of neuron death (see Chapter 14). This view is consistent with early intervention as the best strategy for slowing or managing AD.

Understanding of the molecular biology of the unique microglial system of brain immunity has opened up a new window and exposed new targets for managing AD. The selective targeting of DHA in neurons, discussed in Chapter 11, deserves more attention as a major mechanism for long-term protection of neurons against neurodegeneration. Some recent data on a variant form of a FOXO transcription factor (Willcox et al., 2008) that might protect neurons against the initiation of Alzheimer's disease provide fresh hope for understanding this devastating neurodegenerative disease.

Finally, a working model of AD linking ARA, inflammation, and DHA membranes is developed. This model features the biochemical role of the apolipoprotein D in sequestering ARA and blocking eicosanoid hormone production for the purpose of decreasing levels of membrane peroxidation and enabling neuron longevity.

REFERENCES

Accili, D., and K. C. Arden. 2004. FoxOs at the crossroads of cellular metabolism, differentiation, and transformation. *Cell* 117:421–426.

Bajo-Grañeras, R., M. D. Ganfornina, E. Martín-Tejedor et al. 2011. Apolipoprotein D mediates autocrine protection of astrocytes and controls their reactivity level, contributing to the functional maintenance of paraquat-challenged dopaminergic systems. *Glia* 59:1551–1566.

Bartzokis, G. 2004. Age-related myelin breakdown: A developmental model of cognitive decline and Alzheimer's disease. *Neurobiol. Aging* 25:5–18.

Bartzokis, G., P. H. Lu, and J. Mintz. 2004. Quantifying age-related myelin breakdown with MRI: Novel therapeutic targets for preventing cognitive decline and Alzheimer's disease. *J. Alzheimers Dis.* 6:S53–S59.

Bender, A., K. J. Krishnan, C. M. Morris et al. 2006. High levels of mitochondrial DNA deletions in substantia nigra neurons in aging and Parkinson disease. *Nat. Genet.* 38:515–517.

Bishop, N. A., T. Lu, and B. A. Yankner. 2010. Neural mechanisms of ageing and cognitive decline. *Nature* 464:529–535.

Burgering, B. M., and G. J. Kops. 2002. Cell cycle and death control: Long live Forkheads. *Trends Biochem. Sci.* 27:352–360.

Chen, H., and D. C. Chan. 2009. Mitochondrial dynamics—Fusion, fission, movement, and mitophagy—In neurodegenerative diseases. *Hum. Mol. Genet.* 18(R2):R169–R176.

Cottrell, D. A., G. M. Borthwick, M. A. Johnson et al. 2002. The role of cytochrome c oxidase deficient hippocampal neurones in Alzheimer's disease. *Neuropathol. Appl. Neurobiol.* 28:390–396.

Craft, S., L. D. Baker, T. J. Montine et al. 2012. Intranasal insulin therapy for Alzheimer disease and amnestic mild cognitive impairment: A pilot clinical trial. *Arch. Neurol.* 69(1):29–38 (doi:10.1001/archneurol.2011.233).

Dalle-Donne, I., R. Rossi, R. Colombo et al. 2006. Biomarkers of oxidative damage in human disease. *Clin. Chem.* 52:601–623.

Desai, M. K., M. S. Mastrangelo, D. A. Ryan et al. 2010. Early oligodendrocyte/myelin pathology in Alzheimer's disease mice constitutes a novel therapeutic target. *Am. J. Pathol.* 177:1422–1435.

Desai, M. K., K. L. Sudol, M. C. Janelsins et al. 2009. Triple-transgenic Alzheimer's disease mice exhibit region-specific abnormalities in brain myelination patterns prior to appearance of amyloid and tau pathology. *Glia* 57:54–65.

De Vos, K. J., A. L. Chapman, M. E. Tennant et al. 2007. Familial amyotrophic lateral sclerosis-linked SOD1 mutants perturb fast axonal transport to reduce axonal mitochondria content. *Hum. Mol. Genet.* 16:2720–2728.

Dufour, E., M. Terzioglu, F. H. Sterky et al. 2008. Age-associated mosaic respiratory chain deficiency causes trans-neuronal degeneration. *Hum. Mol. Genet.* 17:1418–1426.

Eichinger, A., A. Nasreen, H. J. Kim et al. 2007. Structural insight into the dual ligand specificity and mode of high density lipoprotein association of apolipoprotein D. *J. Biol. Chem.* 282:31068–31075.

Elliott, D. A., C. S. Weickert, and B. Garner. 2010. Apolipoproteins in the brain: Implications for neurological and psychiatric disorders. *Clin. Lipidol.* 51:555–573.

Fayet, G., M. Jansson, D. Sternberg et al. 2002. Ageing muscle: Clonal expansions of mitochondrial DNA point mutations and deletions cause focal impairment of mitochondrial function. *Neuromuscul. Disord.* 12:484–493.

Fuhrmann, M., T. Bittner, C. K. Jung et al. 2010. Microglial Cx3cr1 knockout prevents neuron loss in a mouse model of Alzheimer's disease. *Nat. Neurosci.* 13:411–413.

Ganfornina, M. D., S. DoCarmo, J. M. Lora et al. 2008. Apolipoprotein D is involved in the mechanisms regulating protection from oxidative stress. *Aging Cell* 7:506–515.

Greer, E. L., and A. Brunet. 2005. FOXO transcription factors at the interface between longevity and tumor suppression. *Oncogene* 24:7410–7425.

Hanisch, U.-K., and H. Kettenmann. 2007. Microglia: Active sensor and versatile effector cells in the normal and pathologic brain. *Nature Neurosci.* 10:1387–1394.

Hoekman, M. F. M., F. M. J. Jacobs, M. P. Smidt et al. 2006. Spatial and temporal expression of FoxO transcription factors in the developing and adult murine brain. *Gene Express. Patt.* 6:134–140.

Hollenbeck, P. J., and W. M. Saxton. 2005. The axonal transport of mitochondria. *J. Cell. Sci.* 118:5411–5419.

Ilieva, H., M. Polymenidou, and D. W. Cleveland. 2009. Non-cell autonomous toxicity in neurodegenerative disorders: ALS and beyond. *J. Cell Biol.* 187:761–772.

Kaletsky, R., and C. T. Murphy. 2010. The role of insulin/IGF-like signaling in *C. elegans* longevity and aging. *Dis. Model. Mech.* 3:415–419.

Kann, O., and R. Kovács. 2007. Mitochondria and neuronal activity. *Am. J. Physiol., Cell Physiol.* 292:C641–C657.

Kenyon, C., J. Chang, E. Gensch et al. 1993. A *C. elegans* mutant that lives twice as long as wild type. *Nature* 366:461–464.

Kunjathoor, V. V., A. A. Tseng, L. A. Medeiros et al. 2004. beta-Amyloid promotes accumulation of lipid peroxides by inhibiting CD36-mediated clearance of oxidized lipoproteins. *J. Neuroinflammation* 1:23.

Lehtinen, M. K., Z. Yuan, P. R. Boag et al. 2006. A conserved MST-FOXO signaling pathway mediates oxidative-stress responses and extends life span. *Cell* 125:987–1001.

Liu, L., V. Drouet, J. W. Wu et al. 2012. Trans-synaptic spread of tau pathology *in vivo. PLoS ONE* 7:e31302 (doi:10.1371/journal.pone.0031302).

Lull, M. E., and M. L. Block. 2010. Microglial activation and chronic neurodegeneration. *Neurotherapeutics* 7:354–365.

Maezawa, I., P. I. Zimin, H. Wulff et al. 2011. Amyloid-beta protein oligomer at low nanomolar concentrations activates microglia and induces microglial neurotoxicity. *J. Biol. Chem.* 286:3693–3706.

Magrané, J., I. Hervias, M. S. Henning et al. 2009. Mutant SOD1 in neuronal mitochondria causes toxicity and mitochondrial dynamics abnormalities. *Hum. Mol. Genet.* 18:4552–4564.

McGeer, P. L., and E. G. McGeer. 2002. Local neuroinflammation and the progression of Alzheimer's disease. *J. Neurovirol.* 8:529–538.

Miller, K. E., and M. P. Sheetz. 2004. Axonal mitochondrial transport and potential are correlated. *J. Cell. Sci.* 117:2791–2804.

Miyamoto, K., K. Y. Araki, K. Naka et al. 2007. Foxo3a is essential for maintenance of the hematopoietic stem cell pool. *Cell Stem Cell* 1:101–112.

Montine, T. J., J. F. Quinn, J. Kaye et al. 2007. F(2)-isoprostanes as biomarkers of late-onset Alzheimer's disease. *J. Mol. Neurosci.* 33:114–119.

Montine, T. J., J. F. Quinn, D. Milatovic et al. 2002. Peripheral F2-isoprostanes and F4-neuroprostanes are not increased in Alzheimer's disease. *Ann. Neurol.* 52:175–179.

Montine, T. J., J. F. Quinn, K. S. Montine et al. 2005. Quantitative *in vivo* biomarkers of oxidative damage and their application to the diagnosis and management of Alzheimer's disease. *J. Alzheimers Dis.* 8:359–367.

Morfini, G. A., M. Burns, L. I. Binder et al. 2009. Axonal transport defects in neurodegenerative diseases. *J. Neurosci.* 29:12776–12786.

Muffat, J., D. W. Walker, and S. Benzer. 2008. Human ApoD, an apolipoprotein up-regulated in neurodegenerative diseases, extends lifespan and increases stress resistance in Drosophila. *Proc. Natl. Acad. Sci. USA* 105:7088–7093.

Müller-Höcker, J. 1989. Cytochrome-c-oxidase deficient cardiomyocytes in the human heart—An age-related phenomenon. A histochemical ultracytochemical study. *Am. J. Pathol.* 134:1167–1173.

Müller-Höcker, J. 1990. Cytochrome c oxidase deficient fibres in the limb muscle and diaphragm of man without muscular disease: An age-related alteration. *J. Neurol. Sci.* 100:14–21.

Murphy, M. P., A. Holmgren, N. G. Larsson et al. 2011. Unraveling the biological roles of reactive oxygen species. *Cell Metab.* 13:361–366.

Paik, J. H., Z. Ding, R. Narurkar et al. 2009. FoxOs cooperatively regulate diverse pathways governing neural stem cell homeostasis. *Cell Stem Cell* 5:540–553.

Peters, A. 2002. The effects of normal aging on myelin and nerve fibers: A review. *J. Neurocytol.* 31:581–593.

Peters, A., M. B. Moss, and C. Sethares. 2000. Effects of aging on myelinated nerve fibers in monkey primary visual cortex. *J. Comp. Neurol.* 419:364–376.

Peters, A., and C. Sethares. 2002. Aging and the myelinated fibers in prefrontal cortex and corpus callosum of the monkey. *J. Comp. Neurol.* 442:277–291.

Poon, H. F., V. Calabrese, G. Scapagnini et al. 2004. Free radicals and brain aging. *Clin. Geriatr. Med.* 20:329–359.

Praticò, D., C. M. Clark, F. Liun et al. 2002. Increase of brain oxidative stress in mild cognitive impairment: A possible predictor of Alzheimer disease. *Arch. Neurol.* 59:972–976.

Praticò, D., J. Rokach, J. Lawson et al. 2004. F2-isoprostanes as indices of lipid peroxidation in inflammatory diseases. *Chem. Phys. Lipids* 128:165–171.

Renault, V. M., V. A. Rafalski, A. A. Morgan et al. 2009. FoxO3 regulates neural stem cell homeostasis. *Cell Stem Cell* 5:527–539.

Salih, D. A., and A. Brunet. 2008. FoxO transcription factors in the maintenance of cellular homeostasis during aging. *Curr. Opin. Cell Biol.* 20:126–136.

Silverstein, R. L. 2009. Type 2 scavenger receptor CD36 in platelet activation: The role of hyperlipemia and oxidative stress. *Clin. Lipidol.* 4:767.

Soerensen, M., S. Dato, K. Christensen et al. 2010. Replication of an association of variation in the FOXO3A gene with human longevity using both case-control and longitudinal data. *Aging Cell* 9:1010–1017.

Stewart, C. R., L. M. Stuart, K. Wilkinson et al. 2010. CD36 ligands promote sterile inflammation through assembly of a Toll-like receptor 4 and 6 heterodimer. *Nat. Immunol.* 11:155–161.

Stuart, L. M., S. A. Bell, C. R. Stewart et al. 2007. CD36 signals to the actin cytoskeleton and regulates microglial migration via a p130Cas complex. *J. Biol. Chem.* 282:27392–27401.

Taylor, R. W., M. J. Barron, G. M. Borthwick et al. 2003. Mitochondrial DNA mutations in human colonic crypt stem cells. *J. Clin. Invest.* 112:1351–1360.

Tothova, Z., R. Kollipara, B. J. Huntly et al. 2007. FOXOs are critical mediators of hematopoietic stem cell resistance to physiologic oxidative stress. *Cell* 128:325–339.

Walter, S., G. Atzmon, E. W. Demerath et al. 2011. A genome-wide association study of aging. *Neurobiol. Aging* 32:2109.e15-28.

Willcox, B. J., T. A. Donlon, Q. He et al. 2008. FOXO3A genotype is strongly associated with human longevity. *Proc. Natl. Acad. Sci. USA* 105:13987–13992.

Wyss-Coray, T. 2006. Inflammation in Alzheimer disease: Driving force, bystander or beneficial response? *Nat. Med.* 12:1005–1015.

Zheng, B., Z. Liao, J. J. Locascio et al. 2010. PGC-1α, a potential therapeutic target for early intervention in Parkinson's disease. *Sci. Transl. Med.* 2:52ra73 (doi: 10.1126/scitranslmed.3001059).

Index

A

D

E

P

Q

R

U

V

W

X

Y

Z